# NEW NATURES

# NEW NATURES

JOINING
ENVIRONMENTAL HISTORY
WITH
SCIENCE AND TECHNOLOGY STUDIES

EDITED BY
DOLLY JØRGENSEN,
FINN ARNE JØRGENSEN,
AND SARA B. PRITCHARD

UNIVERSITY OF PITTSBURGH PRESS

Published by the University of Pittsburgh Press, Pittsburgh, Pa., 15260
Copyright © 2013, University of Pittsburgh Press
All rights reserved
Manufactured in the United States of America
Printed on acid-free paper
10 9 8 7 6 5 4 3 2 1

Library of Congress Cataloging-in-Publication Data

New natures: Joining environmental history with science and technology studies / edited by
Dolly Jørgensen, Finn Arne Jørgensen, and Sara B. Pritchard.
    p. cm.
ISBN 978-0-8229-6242-7 (pbk.)
1. Human ecology—History. 2. Nature—Effect of human beings on. 3. Environmental sciences—Study and teaching. 4. Science—Study and teaching. 5. Technology—Study and
teaching. 6. Interdisciplinary approach in education. I. Jørgensen, Dolly. II. Jørgensen, Finn
Arne. III. Pritchard, Sara B.
GF13.N48 2013
304.2—dc23                                          2013007084

# CONTENTS

# PREFACE

IN OCTOBER 2008, Dolly Jørgensen and Finn Arne Jørgensen, then affiliated with the Norwegian University of Science and Technology, proposed the idea for a small conference on the contributions of science and technology studies (STS) to environmental history at the annual meeting of the Society for the History of Technology in Lisbon, Portugal. We submitted numerous grant applications on both sides of the Atlantic during 2009 and early 2010. Fortunately, generous sponsors made it possible to move forward with the workshop, which we titled "Bringing STS into Environmental History."

We were overwhelmed by the response to the call for papers. We received over seventy-five abstracts from scholars across all ranks working in various subspecialties, geographies, and time periods. They proposed diverse empirical topics and equally varied analytical tools. Both the remarkable response and the many fascinating proposals suggested that the intellectual *problématique* was indeed timely. We ended up discussing sixteen papers at the workshop held in Trondheim, Norway, in August 2010. The chapters appearing in this collection are drawn from those papers.

We would like to extend our sincere thanks to the institutions and programs that supported the conference: the Research Council of Norway; the Norwegian University of Science and Technology's Faculty of Humanities and Department of Interdisciplinary Studies of Culture; Cornell University's Institute for European Studies, Institute for the Social Sciences, David R. Atkinson Center for a Sustainable Future, and Ethical, Legal, and Social Issues Initiative; and the Network in Canadian History and Environment. This intellectual project and the multifaceted ways in which it has contributed to our own research, teaching, and professional collaborations would have been impossible without their generous financial support. We are also grateful to the Norwegian University of Science and Technology for serving as our local institutional sponsor.

We thank all of the authors for sharing their insights during the workshop and for their hard work in revising their essays to further this intellectual project. We would also like to extend our thanks to Clapperton Mavhunga, Per Østby, and Benjamin Wang, all of whom participated in the conference and helped sharpen the chapters appearing in this edited volume, as well as two anonymous reviewers who provided thoughtful comments on individual chapters and the work as a whole. Finally, Cynthia Miller at the University of Pittsburgh Press expressed early interest in this project. We greatly appreciate her support of the resulting edited volume.

# NEW NATURES

# CHAPTER 1

# JOINING ENVIRONMENTAL HISTORY WITH SCIENCE AND TECHNOLOGY STUDIES

## PROMISES, CHALLENGES, AND CONTRIBUTIONS

### SARA B. PRITCHARD

THIS EDITED VOLUME is the product of recent dialogue within and between the fields of environmental history and science and technology studies (STS). It is also the outcome of a workshop that examined one piece of this larger intellectual puzzle: how perspectives gleaned from STS might facilitate and ultimately extend the contributions of environmental history. Indeed, disciplinary hybridity has marked the professional identities and trajectories of the three editors of this collection (not to mention many of the authors whose chapters are included here). We self-identify as environmental historians who were also trained and publish in the history of technology and science and technology studies. At the beginning of this project, we all held positions in STS departments.

Although there has been growing interest in how environmental history and science studies have engaged with and can contribute to one another, and stimulating scholarship has begun to develop at their nexus, we were interested in fostering more explicit theoretical dialogue between the fields. In particular, we wanted to think deeply about the ways in which our skills, developed from our experience in these fields, could enhance our work as environmental historians. For example, how might fundamental STS tenets

such as knowledge production as a social process, the politics of professionalization, and negotiations over expertise help us gain a richer understanding of how "the environment" is constructed, perceived, contested,
and (re)shaped by historical actors? How might unpacking the processes of
knowledge making and technological development illuminate human interactions with nonhuman nature and therefore enrich our analyses of those
relationships? Most broadly, how might conceptual STS tools such as black
boxes, boundary-work, and technological systems offer insights that enable,
but also deepen and sometimes even transform, our understanding of past
human-natural interactions? The title of this book, *New Natures*, seeks to
suggest how new natures emerge from studies that join environmental history with science and technology studies.

From the very outset, then, this project has been premised on an asym-
metrical relationship between environmental history and STS. Indeed, it is
appropriate that we here use the concept of symmetry to frame the dynamic
between the two fields, since it is basic to STS.[1] Yet in framing the disciplines
and their relationship in this way, it is essential that we make two crucial
caveats.

First, we acknowledge that we are certainly not the first scholars to engage STS in the writing of environmental history. To the contrary, as I have
already suggested, this volume builds on conversations within and between
the fields that emerged during the 1990s and 2000s. During those years,
a number of conferences, publications, and institutional changes not only
reflected but also fostered growing interest at the intersection of STS and
environmental history, both relatively new fields.[2] Conference panels at the
American Society for Environmental History, the History of Science Society (HSS), and the Society for the History of Technology (SHOT) explored
these concerns empirically and analytically. In 1997, the theme of the Max
Planck Institute for the History of Science's Summer Academy was "Nature's Histories," which focused specifically on the history of science and
environmental history. Influential monographs such as Gregg Mitman's *The
State of Nature*, Robert Kohler's *Lords of the Fly*, Conevery Bolton Valenčius's
*The Health of the Country*, Michelle Murphy's *Sick Building Syndrome and the
Problem of Uncertainty*, and Linda Nash's *Inescapable Ecologies*, among others, made crucial interventions in these discussions.[3] Meanwhile, hybridity
and actor-network theory had become increasingly prominent within environmental history.[4] Research at the intersection of science, technology, and
the environment eventually became institutionalized within subspecialties
affiliated with relevant professional organizations. In 2000, James C. Williams and I cofounded Envirotech, a special-interest group within SHOT; the
Earth and Environment Forum, a parallel group within HSS, was officially
established the following year.[5] Meanwhile, several PhD programs further

institutionalized the convergence of the fields, while environmental history began to have a stronger presence in some STS departments.[6] As this brief overview suggests, significant dialogue, scholarship, and professionalization efforts have emerged over the past two decades. This volume therefore both reflects the conversation thus far and seeks to develop additional contributions.

Second, readers familiar with even a few of the authors, publications, and professional communities mentioned above will already know that intellectual traffic between the fields has not been one-way—far from it. Scholars trained in STS and particularly those specializing in the history of science, technology, and medicine have enriched environmental history; but environmental historians have also offered critical insights to those working in science studies.[7]

Thus, although the scholarship in this volume, like that of authors not included here, is predicated upon the productive, synergistic effect of integrating environmental history and science and technology studies, ultimately we decided to retain our original goal: a focused, sustained discussion of how concepts, methods, and approaches taken from STS might develop the aims, narratives, and insights of environmental history. In other words, this volume foregrounds the contributions of STS to environmental history, even as we, editors and authors alike, assume in our larger work that they inform and enhance one another. To borrow another concept from STS, individually and collectively, we work from the assumption that the disciplines have shaped—and should shape—one another.[8] However, the chapters in this collection isolate and develop one part of that reciprocal relationship.

## ENVIRONMENTAL HISTORY'S CONTRIBUTIONS

As an introduction to the heart of our discussion, it is worth highlighting some of environmental history's established insights writ large, with an eye to examples of their relevance to STS. Put another way, this section briefly shifts foci: it summarizes the background, given the foreground stated earlier, to the rest of this collection.

One foundational contribution of environmental history is that human perceptions of and interactions with nonhuman nature are valuable objects of historical inquiry. One might say that environmental historians helped put nature into the past (history), as well as into studies of the past (historiography).[9] Furthermore, environmental historians have offered detailed understandings of the ways in which influential historical phenomena such as capitalism, consumerism, and industrialization are modes of production or cultural values predicated not solely on social relations but also on assumptions about the environment and particular relationships between humans and the natural world. By doing so, environmental historians have shown how human-natural interactions are fundamental to what are often seen as purely "social" processes.[10] In the process, they have exposed intended and unintended consequences for both humans and nonhumans.

Environmental history has also invited us, scholars and citizens alike, to consider how natural entities and processes are not only legitimate objects of historical study but also important factors that shape historical phenomena, an idea often abbreviated as "nature's agency": even if, as several scholars have ably demonstrated, human agency is a problematic notion, the "nonhuman" is actually entangled with the "social," and human knowledge always mediates our representation and understanding of the natural world.[11] Recent events such as the triple disaster at Fukushima drive home the point. Of course, the development of nuclear power in Japan and the government's regulation of TEPCO (the owner of the fated nuclear reactors) are crucial to understanding exactly how events began to unfold on March 11, 2011. Yet the massive earthquake and destructive tsunami are also key, and environmental historians push us to remember this vital point: not all historical contingencies emanate from humanity.[12] To recall this insight is not just an academic exercise. Stressing the presence and dynamics of the natural world in human history has the potential, for instance, to alter the selection, design, and use of technologies and to shape policy making.[13] One of environmental history's most valuable contributions, then, has been to call attention to the role of the material world—from genes and organisms to disease and hydrology—in shaping the past. As such, our accounts of historical processes need to change, and to some extent have already changed, accordingly.[14]

However, environmental historians have also convincingly shown how "natural" entities and processes like these have, in fact, usually been mediated by human activities. Native Americans shaped the evolution of corn and cotton in the so-called New World, while flies, mice, dogs, and viruses were carefully selected and bred to facilitate scientific and medical research.[15] These examples thus highlight another central premise of the field that is also one of its most important contributions: the *reciprocal* dynamics between human and nonhuman nature. Edmund Russell's recent analysis of the ongoing interplay between human choices and evolutionary processes in antibiotic resistance offers an excellent illustration of these dynamics at work. Indeed, as his example shows, that very reciprocity calls into question tidy categories and entities. It also challenges simplistic representations of causality. Overall, environmental historians have shown how the terms *human* and *nonhuman nature* are convenient yet also problematic abbreviations for much more complicated objects that encapsulate the complex, dynamic, ongoing interactions between people and the environment.[16]

Placing such complexity at the center of historical analyses, rather than relying on a reductive understanding of the past, is one of environmental history's real strengths. Using ecology as both science and metaphor has been one important tool for environmental historians to achieve this goal. Indeed, ecology has been especially influential within the field.[17] There are several probable explanations for this pattern, including the field's strong political and moral origins, particularly its ties to late-twentieth-century environmentalism and the relevance and utility of the ecological sciences in helping to delineate, understand, and explain environmental change—and humans' role in it—over time.[18] Moreover, because ecology is fundamentally about dynamics and interrelationships, rather than seeing things in isolation, as metaphor, it has offered a particularly useful way to describe such complex, ongoing dynamics of reciprocal shaping that many environmental historians seek to capture in their studies.[19] While ecology as science and metaphor aids environmental historians in their work, I also suggest several cautions regarding this practice later in this chapter.[20]

A final contribution I will emphasize here is the broad temporal and spatial scale of many environmental histories. Environmental historians often tackle big issues and write "big histories."[21] This is not to imply that other historians or STS scholars shy away from significant, "meta" processes such as imperialism, slavery, capitalism, industrialization, or the emergence of the atomic age.[22] Nonetheless, many STS studies use specific, bounded sites or communities of knowledge production to delineate their analyses. Some of the field's earliest contributions, for instance, emerged from rich, fine-grained studies of individual laboratories and specific scientific controversies.[23] William Cronon's influential books *Changes in the Land* and *Nature's Metropolis* are emblematic (and, needless to say, exemplars) of the ways in which many environmental historians have taken up wide historical and geographical scales in their analyses. Other scholars across the field's forty-year history in the United States reinforce the point. Alfred Crosby's influential *Ecological Imperialism* spans several centuries and several European imperial powers. More recently, David Blackbourn has shown how managing water was central to the making of modern Germany, and Russell's *Evolutionary History* uses evolutionary theory to help explain human history over *la longue durée*, in the broadest sense of the term. Clearly, these are not microhistories.[24]

Together, these insights help both scholars and citizens understand the roots of contemporary environmental dilemmas, in the process deepening yet simultaneously transforming our understanding of the past. They have also enriched historical and contemporary studies of knowledge making and technological development in several important ways. Environmental historians have begun to show how human-natural interactions both shape and are shaped by knowledge and technical change. Environmental history's interest in "nature's agency" has also called attention to the ways in which nonhumans such as biological organisms literally matter in the practice of science, technology, and medicine.[25] In addition, the field's commitment to nature as a material object has pushed scholars to refine social constructivist approaches, inviting them to consider the materialization of ideas and assumptions, as well as the ways in which the material world constrains what can be known.[26] These contributions, along with hybridity and complexity, are some of the vital contributions of environmental history.[27] With this overview of the collection's background, let us return to its focal point.

## THEORY AND ENVIRONMENTAL HISTORY

In organizing this edited volume, we sought to engage *explicitly* with theoretical approaches and concepts developed in STS. Each author in this collection uses a particular analytical tool or school of thought to frame the study and ultimately deepen the analysis. Thus, while the empirical research and historical analyses in these chapters are contributions in their own right, the volume has been conceptualized and organized primarily in terms of how the authors engage with key STS theories, in an effort to elaborate the wider implications and contributions of these concepts to environmental history as a field.[28]

Of course, historians, including environmental historians, have developed their own theoretical approaches and conceptual tool kits. After all, history is not the past, but the *study* of that past.[29] As such, theorization is inherent to historical inquiry. Historical analyses are predicated, for instance, upon theories of historical agency and causality. Other historians, such as Joan Scott, have shown how scholars' categories of analysis such as gender alter our understanding of the past. Such categories do not simply add to historical analyses; they can fundamentally transform them.[30] Overall, however, many historians adopt more subtle, implicit theoretical frameworks in their studies and tend to be cautious about generalizing from historical specificity. In many ways, history is a narrative-driven discipline that values a rich elaboration of context and contingency over theoretical arguments that seek primarily to extrapolate generalizations from historical phenomena.[31]

In contrast, the role of theory in science studies, both as a mode of analysis and as an objective of scholarly production, is generally more explicit. Indeed, STS scholars have formulated a number of useful concepts and sophisticated theories to describe the relationship between knowledge and society, and they have sought to use such analytic tools to understand the contested social and historical processes of knowledge making in particular contexts. The influential role of sociology within science studies, particularly during the discipline's early years, may offer one reason for the field's theoretical orientation.[32]

We suggest that using theory more extensively and deliberately can enhance analyses in environmental history in at least four ways beyond the specific insights of a given concept. First, analytic tools can encapsulate and thus *crystallize* the central lessons that emerge from the rich details of empirical studies. For instance, in *Irrigated Eden,* Mark Fiege traces the transformation of agricultural landscapes in late-nineteenth- and early-twentieth-century Idaho. He shows how irrigation networks altered the land. Yet he also demonstrates how these "artificial" systems were imposed on earlier creeks and streams that cut across the landscape. In this sense, irrigation networks associated with early industrial agriculture shaped—but were also shaped by—existing hydrologic processes. Fiege's influential notion of "hybrid landscape" captures this ambiguous, complicated dynamic between nature and culture (and therefore between nature and technology) in few words. Fiege's concept thus emerges from his particular study, yet it offers a way to synthesize and distill those wider insights in an astute shorthand. Various science studies scholars have sought to make parallel arguments through their concepts of hybridity, nature-culture, "nature-cultures," and envirotech.[33]

Second, building off this point, concepts such as hybrid landscape or nature-cultures provide a specific *language* for describing significant yet complicated historical phenomena. In other words, they can help make complex processes, not to mention historians' nuanced interpretations of those processes, more legible and therefore comprehensible.[34]

Third, using theoretical approaches and conceptual tools to frame a given historical case may facilitate *comparative* analysis. To return to Fiege's example, one might contrast Idaho's "irrigated Eden" with irrigation schemes in colonial Sri Lanka, post-1945 France, or nations in the global South "aided" by "technical assistance" during the Cold War, or compare different kinds of hybrid landscapes—from forests and rivers to cities—in an attempt to consider their similarities and their differences.[35]

Finally, adopting specific analytic frameworks makes one's theoretical assumptions *explicit*, rather than implicit. As a leading historian of modern France once declared, all historians have theories; the question is whether they are deliberate and explicit or unreflective and implicit. Making one's theoretical assumptions evident to readers enables them to begin assessing a study's premises and contributions, as well its limits.[36]

However, in conceptualizing the collection as a whole and the individual chapters in this way, we do not seek to reify theory—in environmental history or any other field, for that matter. To the contrary, it would be rather ironic to fetishize theory, especially unreflexively, in a volume that advocates, among other things, paying attention to hierarchies of knowledge, in part because those hierarchies have historically had significant implications for both humans and nonhumans.[37] Rather, as Fiege's book illustrates, the particularities of an empirical study can foster the reconsideration of existing conceptual frameworks and even spur the development of new analytic tools. The formulation of envirotechnical analysis at the intersection of environmental history and the history of technology offers another recent example of the ways in which empirical and historical studies have driven the formulation of theory, rather than the other way around.[38] In other words, conceptual frameworks and empirical material are always in ongoing dialogue with one another. As Paul Edwards has shown in the case of climate change models, this is certainly true for scientists, but it is true for other scholars as well.[39] Furthermore, it is worth paying attention to the context in which new theoretical approaches and conceptual tools are developed. After all, we, as analysts, are situated as well, and our concepts undoubtedly reflect our own cultural and historical contexts—for better and for worse.[40]

## STS'S THEORETICAL CONTRIBUTIONS

To foster this volume's intellectual coherence, we have organized the chapters around three central concerns in science and technology studies. These issues thread through many of the chapters—at times explicitly, at other times implicitly—although they are particularly prominent in the chapters included in their respective sections.

Part I tackles questions of epistemology by examining ways of knowing. Taking knowledge out of the "black box," STS scholars seek to understand and tease out the specific ways and contexts in which knowledge is produced. Below, I discuss constructivism to highlight the social shaping of knowledge production and thus knowledge itself. As such, contextualizing knowledge making stresses how knowledge systems always mediate representations and understandings of the environment.

Part II focuses on constructions of environmental expertise and signals not only the historical emergence and making of the modern sciences, including the ecological sciences, but also how "science" and particularly "expertise" become differentiated from "mere" knowledge. This section therefore examines categories, categorization, and hierarchies of knowledge, all central to the construction of expertise. Such processes matter, in part, because they define whose knowledge of the environment counts and therefore whose ends up forming the basis of environmental policies and practices.

Part III examines networks, mobilities, and boundaries. These themes allude to actor networks, assemblages, and boundary-work, which together highlight the heterogeneity of knowledge systems, as well as the ways that historical actors construct various borders through their work. Part III thus highlights the diverse dimensions of environmental knowledge making and the boundaries that both shape and are shaped by these heterogeneous processes. Here I focus on the concept of boundary-work, which helps illuminate both the processes and the stakes of boundary making. The creation, maintenance, and erosion of borders—all central to dynamic and mobile networks—have significant implications for both humans and nonhumans.

Finally, in the epilogue, Sverker Sörlin uses contemporary urban nature to consider the generative, "real-world" possibilities of fully embracing nature-culture, or the deep entanglement of people and the environment.[41]

## CONSTRUCTIVISM

Nature and ecology are central analytical tools within environmental history that play crucial roles in driving, organizing, and ultimately enriching analyses in the field. Paying attention to the natural world and incorporating knowledge from the ecological sciences have, for instance, helped generate new questions about the past, as well as fresh understandings of historical phenomena and causality.

At the same time, the environment and ecology are historical categories and objects to be examined and understood. In other words, they are not simply unproblematic *explanas*.[42] For instance, Valerie A. Olson demonstrates in her chapter in this volume how astronomers' recent research on "Near Earth Objects" in space led them to radically reconceptualize the boundaries and scale of "the environment." Instead of seeing Earth *as* environment, these scientists reframed the planet within a larger, *cosmic* environment that potentially posed dire threats to it. Olson's analysis offers a particularly powerful illustration of the ways in which concepts like "environment" are situated and historical. Olson also opens up new questions for environmental historians by suggesting how, based on this definition of the environment, environmental history could actually extend beyond the boundaries of planet Earth.[43]

Dolly Jørgensen's chapter examines a critical implication of the environment's constructed character: multiple understandings of the "same" nature. In her analysis of the recent "rigs-to-reefs" debate in California, Jørgensen challenges the idea that one side was "pro-environment" and the other "anti-environment." She instead shows how their different assumptions and practices led to quite different understandings of the ocean, which ultimately informed their positions in the controversy. Studies such as these emphasize how nature and knowledge are both analytical tools and historical objects in the field of environmental history.[44]

Constructivist frameworks call attention to this dual character and invite environmental historians to remain attuned to this critical point; they also offer powerful ways to investigate their historical particularities. At its core, adopting a constructivist approach to knowledge and technology means not treating them as "black boxes" and instead studying them as social and historical phenomena. Although there are several constructivist schools of thought within science studies, they share an interest in examining how complex social and historical processes shape knowledge making and ultimately knowledge itself by studying what research questions are asked, which methods are used, who is included in (and excluded from) a given knowledge community, and so on.[45] This approach applies to even the intriguing case discussed by Frank Uekotter in this collection, in which agricultural knowledge in twentieth-century Germany is intentionally absent. Thus, tools of constructivist analysis provide useful ways to explore nature and ecology as historical objects that merit their own analysis, even as scholars simultaneously use these concepts to help frame their studies.[46] In many ways, this means taking the fundamental strengths of history, including its attention to contexts and contingencies, and applying them to various forms of knowledge (including science, technology, engineering, and medicine), even if such knowledge is often represented as outside society, politics, or culture and therefore beyond scholarly inquiry.[47]

Constructivist approaches can also be extended to studies of environmental problems, whether in the past or the present. The methodology is particularly fruitful here because it opens up the pivotal question of how an "environmental problem" became just that: conceived by certain historical actors as a concern and constituted specifically as a natural problem, rather than, for instance, as a social, political, or technical issue. Such categorizations imply differential policy and other solutions.[48] Moreover, STS scholars have shown how the making of environmental problems entails considerable work, rather than being self-evident. As John Law puts it, "successful large-scale heterogeneous engineering is difficult. Elements in the network prove difficult to tame or difficult to hold in place."[49] Kevin C. Armitage's chapter provides an instructive example of this process at work. He uses frame analysis to tease out how government scientists and bureaucrats carefully mobilized resources, institutions, and eventually farmers around the issue of soil erosion during the New Deal. Finn Arne Jørgensen offers another interesting case in his story of how beverage container recycling moved toward systemization at a particular moment in time and required institutionalization through systems and scripts so that consumers acted properly. Constructivist approaches, then, help environmental historians tease out how and why concerns over soil erosion, garbage, DDT, or endocrine disruptors emerged thanks to particular groups at specific historical junctures, even if these objects existed long before they were perceived as environmental problems.[50]

Influenced by these insights, scholars have thus teased out the complex processes by which environmental problems come into existence, emphasizing, in fact, how they are *brought* into existence by a constellation of historical, cultural, material, and epistemological factors. Michelle Murphy traces, for example, how surveys by women office workers enabled them to identify patterns in illness and therefore mobilize around previously imperceptible contaminants. Peter Thorsheim examines how pollution was "invented" in industrializing Britain, while Scott Frickel's research shows how certain 1960s scientists recast chemical mutagens from a useful research tool into a potent threat: environmental mutagenesis.[51] In his chapter in this collection, Michael Egan demonstrates how Swedish scientists not only constituted mercury pollution as a pressing issue but framed it in ways designed to reach a wider, public audience, to increase its likelihood of being taken up by government regulators and policy makers.

Exploring how particular groups constructed environmental problems as such may therefore help environmental historians understand how and why they were perceived, received, mitigated, or, as Frank Uekotter shows for German agriculture, ultimately ignored.[52] As Christopher Jones has argued with respect to BP's Deepwater Horizon oil spill and Dolly Jørgensen shows in the rigs-to-reefs case here, defining and framing environmental problems in specific ways is not just a question of rhetoric or semantics. Rather, doing so simultaneously shapes and thereby constrains the solutions proposed and eventually selected.[53] For these reasons, the particular construction of environments and environmental problems matters—for humans and non-humans alike.[54]

Constructivist tools of analysis thus offer ways for environmental historians to study both nature and knowledge. However, the relationship between them is complicated; to paraphrase Kim Fortun and Douglas Weiner, knowledge systems always mediate human understandings and representations of the natural world. For example, as we see in Anya Zilberstein's chapter, colonial settlers in eastern North America sought to "improve" the landscape. This concern drove many of their studies of the region. Confidence in human abilities to transform and improve the environment thus shaped the questions naturalists asked and the kind of research they conducted. These studies were not, then, neutral descriptions of the natural world. Rather, they were wholly entwined with colonial political economy and culture. Bruno Latour might use Zilberstein's example to question the traditional boundaries between "matters of fact" and "matters of concern," or "science" and "politics."[55] It is therefore impossible to entirely separate nature from *knowing* nature.[56]

Overall, constructivism enables environmental historians to use key analytic tools—nature and knowledge—while remaining attentive to their very historicity. In other words, constructivism offers a way to disentangle actors' views from analysts' own terms, thereby providing the distance necessary to facilitate mindfulness of both.

EXPERTISE

As we have seen, constructivism highlights and opens up the making of knowledge. In the second part of this volume, we turn to a related topic—expertise—focusing specifically on environmental expertise because it is most relevant to the field of environmental history. Experts and expertise, in both historical and contemporary settings, are generally associated with significant authority and its attendant power. Part of that power comes from the assumption that expertise is self-evident and beyond question. STS scholars instead study and contextualize expertise, investigating exactly how certain areas of knowledge became perceived as expert, how specialists in these fields acquired and maintained authority, and who benefited from (or was harmed by) these moves. STS analyses of expertise thus examine the categories, categorization, and hierarchies of knowledge in given contexts.[57] These processes matter, in part, because different forms and echelons of knowledge are generally associated with different levels of power. To examine the definition, production, and maintenance of experts and their associated expertise, then, is to explore the making of influential social relations and dynamics.[58]

Such STS insights regarding the production of expertise stress the politics of environmental expertise. Actors' views of ecological knowledge in general, and environmental expertise in particular, are not, then, merely abstract debates about "the best" knowledge of the environment.[59] Defining and negotiating what is perceived as cutting-edge knowledge may determine, for example, who gets to speak for nature in environmental controversies—what Latour calls the spokesperson.[60] It may also decide whose knowledge serves as the basis for both formal and informal environmental practices. Consequently, the construction of environmental experts and expertise and, in the process, who is not an expert and what is mere knowledge has significant consequences for not only the formulation and enforcement of environmental management strategies but also the environment itself, as well as for the people who have historically depended on those landscapes. In short, environmental experts play powerful roles in shaping what counts as the environment or specific natures such as ecosystems, species, or wetlands, as well as proper interactions with them.[61]

Several chapters in this collection explore these central questions in environmental history through detailed studies that expose the history and politics of environmental expertise. First, studying the contested definitions and negotiations over such expertise offers a richer understanding of how exactly specific environments are conceived, contested, and ultimately shaped by historical actors. Environmental experts and expertise are fundamental, rather than incidental, to this process. Second, Michael Egan's chapter raises an important related issue: the opposition of lay and expert knowledge and how experts seek to translate specialized knowledge to the public.[62] Finally, examining the history and politics of expertise opens up the constitution of human-natural relations and particularly how knowledge regimes with differential levels of power shape what is sanctioned and, conversely, what is criticized—if not criminalized. In her chapter, Eunice Blavascunas traces negotiations over the management of Poland's Białowieża Forest, a rare old-growth forest in central Europe. As she shows, local people perceived forest-ers, who had historically been members of their communities, as experts, while they remained skeptical about wildlife biologists' claims to that posi-tion.

Examples from other environmental historians' work show how close studies of expertise can help explain the framing and mitigation (or lack thereof) of socioenvironmental problems. For one, expert/nonexpert status can mediate which environmental problems are perceived and treated as such. Michelle Murphy has demonstrated, for example, how scientific and medical experts tended to dismiss women office workers' complaints about the modern office building and people experiencing multiple chemical sensitivity, while Nancy Langston's recent book has also shown how expert forms of knowledge such as toxicological models of risk can make certain en-vironmental problems such as endocrine disruptors invisible because they do not conform to these models.[63] In other words, although expert forms of knowledge have contributed to environmental regulations, reforms, and ultimately protection, at other times these very knowledge systems have ne-glected other issues, leading to longer periods of exposure and detrimental effects on both people and the environment.

Analyzing expertise thus often opens up contestation over nature: what it is, who knows it best, how it should be managed, and by whom. Being attentive to the social and historical contingencies of environmental exper-tise also suggests why some environmental problems are made visible and taken seriously, while others remain invisible or are dismissed entirely. As such, teasing out the workings and implications of expertise within partic-ular environmental histories provides a valuable lens onto influential power dynamics both shaping and constituted through environmental conflicts in the past and the present.

## BORDERS AND BOUNDARY-WORK

Borders and what sociologist of science Thomas Gieryn calls "boundary-work" are important themes in this volume, especially in the third and final section.[64] Both STS and environmental history share a strong interest in analyzing the making, remaking, and unmaking of boundaries. In many ways, it is a premise of both fields. Environmental historians have generally focused on the porous relationship between nature and culture, with some scholars focusing recently on the complicated dynamics between nature and technology.[65] Meanwhile, STS scholars have developed concepts such as "nature-culture," "sociotechnical," and "technopolitics" to describe the entanglement of other prominent modernist dualisms.[66] As an alternative to such binaries, several recent scholars in both fields have instead emphasized hybridity and multiplicity.[67] Overall, scholarly analysis of borders and boundaries has suggested that they are less self-evident, more unstable, and more multifaceted than historical actors often assert.

Gieryn's concept of boundary-work offers a useful way to critically examine the creation, maintenance, and erosion of borders, both physical and rhetorical, in environmental history. Gieryn originally proposed the concept to describe how emergent disciplines demarcate a specific terrain for their expertise and assert their authority over that area. These moves usually form a key step in discipline formation and professionalization.[68] For example, in *Fathoming the Ocean,* Helen Rozwadowski traces how "amateur" naturalists, including whalers and women, were eventually excluded from oceanography, even though their knowledge and collections had been vital to the field's development.[69] Most broadly, then, the concept of boundary-work calls attention to the ways in which historical actors not only differentiate between forms of knowledge but also forge hierarchies among them. In the process, they produce critical dichotomies such as science/not science and expertise/ knowledge, as well as delineate professional terrains for given disciplines. The concept of boundary-work thus highlights the ways in which historical actors strategically construct intellectual, disciplinary, and professional boundaries, thereby revealing fundamental connections between knowledge and power.[70] As such, various borders and boundary-making should be studied and analyzed, rather than taken at face value.

As several chapters in this collection show, the concept of boundary-work can help environmental historians tease out the processes by which actors define and negotiate borders—literal and metaphorical, both consciously and unconsciously—as well as understand the larger stakes of these moves. These insights particularly hold for analyzing environmental knowledge, but they can also be extended to several related contexts.

First, as we have seen, the constitution of environmental expertise—who knows and can know the natural world—provides an illustration of Gieryn's original argument. Complementing Blavascunas's study of contestation over Poland's Białowieża Forest is Stephen Bocking's chapter on Arctic ecologists, in which he demonstrates how these scientists strategically framed their research as both particular and universal. These scientists worked to maintain that their scientific investigations contributed to understandings of northern environments as unique places, while simultaneously creating knowledge that was germane far beyond those locales in order to bolster their relevance to wider intellectual and professional communities.

In addition, environmental knowledge and increasingly environmental "expertise" have played critical roles in environmental management strategies, including what are deemed appropriate interactions with the natural world. Boundary-work in environmental management can be both metaphorical and literal. When French hydraulic engineers naturalized multipurpose dams and described the Rhône River in technical terms, they conveniently justified the large-scale transformation of the river.[71] Yet some environmental management has also depended on quite literal boundary-work, such as the creation of national parks, the exclusion of peoples who had historically lived in those areas, and the criminalization of earlier practices such as hunting (now deemed poaching).[72]

Because boundary-work is process oriented (after all, it is called boundary-*work*), the concept helps environmental historians examine and therefore reveal the construction, contestation, and negotiation over borders. It thus foregrounds power and authority both reflected in and constituted through the regulation of "nature" and the establishment of norms in environmental management. This process is often exclusionary, as a number of environmental historians, political ecologists, and other scholars have convincingly shown.[73] Yet, as David Tomblin establishes in his study of recent ecocultural restoration efforts by the White Mountain Apache included in this volume, boundary-work also has the potential to be generative and empowering, sometimes even counter-hegemonic.

Together, these examples allude to the larger stakes of boundary-work: various borders and boundaries are not made (and unmade) for any old reason; they are profoundly political. As environmental historians know, nature—the category, object, and management thereof—is fundamentally political. Animals like the karakul sheep discussed by Tiago Saraiva have been molded by political concerns, including the overt political ideologies of fascism and imperialism in early-twentieth-century Europe. Meanwhile, Thomas D. Finger describes how grains were transported from New World to Old, breaking down historic natural boundaries, in the quest for economic profit (which is, of course, inherently political). As these cases suggest, the categories and borders of nature and culture are maintained (or erased) by historical actors for interested reasons.

Overall, Gieryn's notion of boundary-work offers environmental historians a productive way to explore the various processes by which historical actors preserve (or erode) borders, both literally and metaphorically. It also alludes to the broader professional, intellectual, and political stakes of these moves. Boundary-work may therefore help explain, in part, why wilderness and nature-culture are so politicized.

The chapters in this volume illustrate how specific STS concepts such as enactment, systems, and model organisms can enrich environmental historians' analyses, but they also show how joining environmental history with science and technology studies yields several wider insights. Collectively, these authors historicize and contextualize knowledge. This approach complicates the place of scientific and technical knowledge in environmental history, but the authors demonstrate the importance of analyzing knowledge and knowledge making, rather than treating them as being outside scholarly investigation. By considering the construction and politics of expertise, these chapters also highlight the vital role of experts and expertise in defining, shaping, and mitigating environmental conflicts, as well as shaping human-natural interactions more broadly in complex ways. In addition, boundary-work particularly reveals the processes and wider stakes of critical borders, including nature and culture in environmental history. Overall, by being explicit about using STS theories in environmental history, the chapters in this collection exemplify the fruitfulness of cross-disciplinary thinking, regardless of the direction.

Given the value of this engagement, we hope that conversations between environmental historians and STS scholars will continue, and we especially hope that some of the specific concepts and approaches discussed here (as well as others not featured in this volume) are considered and extended in future studies. We also suggest four paths for additional research and dialogue with both academic and wider communities that might emerge from and build on some of the issues raised in this collection.

First, this volume speaks to ongoing debates within history over the role and utility of theory in the discipline. Although the editors and authors here are situated within the field of environmental history, we hope that historians will continue to consider how theory, writ large, is an analytic tool that can help sharpen historical analyses and their wider implications and specifically how concepts and approaches from STS might enrich historical studies of knowledge and technology, including within so-called mainstream history.[74]

Second, although the focal point of this edited volume is the contributions of STS to environmental history, many of the chapters in this collection actually speak to the productive synergies between the fields and thus to the benefit of bringing questions, methods, and insights from science and technology studies *and* environmental history to a given study. We hope that more scholarship will develop *both* dimensions in a single analysis and thereby advance the potential for even more fruitful dialogue and reciprocity between the fields.

Third, given many pressing environmental issues in the contemporary world, environmental historians are well situated to engage with these debates, helping policy makers and citizens understand the deep roots and complex causes of these dilemmas. Indeed, some have already begun to do so. STS methods offer productive ways to analyze science, technology, engineering, and medicine, including their central roles in these environmental issues. For instance, in a recent editorial, historian Alan Brinkley defended the humanities and liberal arts education, arguing that "science and technology teach us what we can do. Humanistic thinking can help us understand what we should do."[75] Yet, as the chapters in this collection and STS insights more broadly demonstrate, values, politics, and power already thoroughly permeate science and technology. Thus, Brinkley's stereotypical divide between facts and ethics, knowledge and politics, is not only inadequate but problematic: it reproduces the common argument that science is apolitical, which is itself a political claim.[76] Many environmental debates, including those discussed in these chapters, powerfully demonstrate this false divide. STS scholarship therefore offers critical perspectives in terms of teasing out the complex histories and politics of both knowledge and nature, and environmental historians can harness these contributions to deepen their studies of the past while also contributing to contemporary debates.

Finally, as I have just begun to suggest, STS approaches are undoubtedly relevant in the "real world." Thus, STS concepts and methods offer both analytical and political work. Concepts such as agnotology and enactment help us, as scholars *and* as citizens, understand the social, cultural, economic, and political conflicts over environmental knowledge and management. They help explain the development of specific policies, resistance to changing current systems, and so on. Meanwhile, nature-culture helps us understand genetically modified organisms; boundary-work elucidates the efforts of environmentalists to maintain the idea of pristine nature, despite ample evidence to the contrary; and systems and momentum help explain the persistence and even maintenance of a carbon economy. Such theoretical tools enhance our ability to ascertain underlying assumptions and politics, enabling us not only to analyze current environmental debates but also to be more engaged in contemporary issues. These tools are not therefore just modes of *Kritik* (in the German sense of the term), but—as Sverker Sörlin suggests in the epilogue—they can also help us create generative possibilities. In this sense, the title of this edited volume, *New Natures*, alludes to the creative, productive opportunities for meaningful work in the scholarly world but also beyond.

PART I

WAYS OF KNOWING

# CHAPTER 2

# THE NATURAL HISTORY OF EARLY NORTHEASTERN AMERICA

## AN INEXACT SCIENCE

### ANYA ZILBERSTEIN

IN 1795 YALE College president Timothy Dwight decided to write a book about New England and surrounding areas. It was a project motivated by two recuperative goals. The first was personal: his job was too sedentary, and he needed more exercise. To counteract these defects, he devoted vacations to exploring the region on horseback, and his health was "preserved." The second goal was public: to rehabilitate the region's image. Dwight bristled at the "misrepresentations, which foreigners, either through error or design, had published of my native country." These outsiders—both Europeans and other Americans—drew a depressing "caricature" of the local environment as poor, barren, and unpromising. His book would serve as a corrective by depicting greater New England "as it truly appeared, or would have appeared, eighty or one hundred years before." By tracing its natural history and documenting current conditions—mineral deposits, mountain ranges, forests, pastures, croplands, wild and domesticated plants and animals—he would publicize an accurate account of the area's bounty and potential. But reaching this goal proved elusive.

Dwight found the landscape interrupted by numerous "dismal," "rough, lean, and solitary" tracts, all in "different degrees of improvement." The variety of scenes and developments challenged exact description. He observed that "the state of this country changes so fast, as to make a picture of it, drawn at a given period, an imperfect resemblance of what a traveler will find it to be after a moderate number of years have elapsed. The new settlements particularly, would in many instances scarcely be known, even from the most accurate description, after a very short lapse of time." The environment, he wrote, was as changeable as a cloud and just as challenging to represent: "the form, and colors, of the moment must be seized; or the picture will be erroneous." To adequately account for this changeability, Dwight had to adjust his original objective. Published posthumously in 1821, *Travels in New England and New York* offered a natural history of the region with "a good degree of exactness" because anything more definitive, he concluded, was "unattainable."[1] As a result, Dwight (like those he criticized) admittedly produced "misrepresentations" of the regional environment, sometimes through error and sometimes by design. Like other writers he realized that the exigencies of exploring and capturing the changes in early northeastern America's settler landscapes made writing their natural history an unavoidably imperfect science.

Instead of precise accounts of the region's nature and geography, writers provided—and often explicitly defended—best-effort, provisional descriptions of environments in flux. They offered modest explanations of rapid environmental alterations, generating a language of empirical imprecision to convey the changeable character of nature in colonial or settler states.[2] This chapter shows how and why inexact description was a characteristic feature of natural histories and geographical texts written about the Northeast and other colonial regions in North America during the long eighteenth century. Despite a growing culture of precision in eighteenth-century science, descriptions of change in early American environments remained mainly indefinite and provisional records of place. Precise methods and results were sometimes of limited use or simply not feasible for natural historians writing about northeastern America. In the borderlands of Native and new European settlement or in wilder terrain outside the controlled spaces of villages, towns, farms, or gardens, naturalists became particularly sensitive to the inexact methods or vague results, explanations, and predictions they used to produce and communicate their surveys and reports.[3] Historians of science have cogently analyzed the growing cultural, material, and epistemological significance of precision and accuracy in this period, concentrating on the process by which late Enlightenment scientists promoted a modern culture of measurement, quantification, and explanation that produced certain, durable, and universal facts. But they have mostly neglected the history of their counterparts—ambivalent hypotheses, approximate measurements, fuzzy data, and indeterminate conclusions. These concepts proved especially important for naturalists who attempted to describe colonial environmental transformations.[4]

By highlighting the limitations, contradictions, inconsistent geographies, and temporal shifts in a range of textual descriptions of early northeastern American landscapes, this chapter poses a question at the intersection of environmental history and science studies. How can rapid environmental changes be accurately described? I examine this question in the context of the production and circulation of descriptions of New England and Nova Scotia in the scientific community of the eighteenth-century Atlantic world, analyzing three prominent tendencies of imprecision in the structure and content of these texts. First, their authors often assumed that the environmental geography and the political geography of a province, empire, or nation were equivalent in extent. This assumption was implicit in the titles of numerous "natural and civil histories" published in the eighteenth century, such as Thomas Jefferys's *The Natural and Civil History of the French Dominions in North and South America* (1760) and Samuel Williams's *A Natural and Civil History of Vermont* (1794), which suggested that ecological habitats coincided with political borders. Second, while they seemed to be natural histories of an entire domain (French America, Vermont), writers were willing or able to describe only selective features of the region's biota. Such samples of limited areas were thus emblems, microcosms, or synecdoches—rather than comprehensive catalogs—of a province's environment. Finally, natural history narratives often abruptly shifted time frames; moved backward and forward among past, current, and projected landscape conditions; or merged these sequential states in a single description, blurring temporal distinctions between stasis and change in the natural world.

Taken together, these three themes of spatial and temporal ambiguity reveal that, just as natural philosophers increasingly valued exactitude, naturalists describing colonial environmental conditions came to appreciate the value of imprecision. The natural history of early America was therefore an unavoidably, and sometimes a deliberately, inexact science.

## NATURAL HISTORY IN EARLY AMERICA

Most early American naturalists were, like Dwight, educated elites who were local officials, clergymen, lawyers, merchants, and large landowners —and the women associated with them—who sought, elevated, or reinforced their genteel social status through scientific pursuits. Like their counterparts elsewhere in the Atlantic world, they practiced a learned but largely informal natural history—which before the mid-nineteenth century could include travel writers, geographers, and surveyors of regional climate, weather, botany, zoology, mineralogy, and topography. In the eighteenth century, no colleges or universities on either side of the Atlantic offered degrees in the sciences of natural history—agricultural improvement, botany, chemistry, entomology, geography, geology, ornithology, or zoology. Many North American physicians were educated in Edinburgh, where they would have heard botanical lectures on materia medica; by the late 1780s, the University of Pennsylvania, Columbia, and Harvard had established professorships or offered lecture series by subscription in natural history. But most scientific Americans were dabblers, gentlemen and ministers with a general education or strivers like Benjamin Franklin and Benjamin Thompson (Count Rumford), who were passionately engaged in experimentation and theorizing but entirely self-taught. Their intellectual credentials derived not from professional scientific training, but rather from broader political, institutional, or social authority in their local communities.[5]

They were linked to local elites and naturalists elsewhere by translocal, imperial, and international networks in the colonial Americas and the Atlantic world. Although these networks remained largely intact through the American and French revolutions, until recently, this persistence was not evident in the historiography of early American science. Historians tended to argue or assume that the Revolutionary War with Britain marked a rupture between colonial and national science. Most studies examined scientific culture in the Thirteen Colonies up to 1776 or in the early nation after 1783, neglecting the continuities across conventional periods of national history or the regional and transatlantic ties among naturalists in Britain, the new United States, and the remaining British colonies.[6] However, in the last quarter of the eighteenth century, scientific activity—field surveys, specimen collecting, correspondence, publications, and institution building—increased markedly across the Atlantic world in the British, French, and Spanish empires and for Americans in the new United States.[7] Despite the profound political, economic, and geographical reconfigurations of the revolutionary era, early American naturalists made sure that the intellectual and affective connections across national and imperial boundaries survived what Harvard professor of medicine Benjamin Waterhouse called the "unnatural" political break with Britain. Waterhouse had supported American political independence, but in 1793 he wrote to Royal Society president Sir Joseph Banks: "I really wish to see (as you express it) the claims of consanguinity renewed which subsisted before the war, especially among the men of Science, who if I mistake not, ultimately govern both countries."[8] American naturalists like Waterhouse remained crucially dependent on the material and symbolic support of metropolitan patrons in Edinburgh, London, Madrid, Paris, and Leiden who were, in turn, reliant on dispatches from foreign environments. Dwight's cautious approach to his subject reflected an epistemic humility that was characteristic of the patronage and exchange culture of transatlantic scientific networks.[9]

As these networks took shape during the seventeenth and eighteenth centuries, precision and accuracy emerged as key values in scientific culture. Monarchs, modernizing bureaucrats, and merchant elites enthusiastically sponsored the development of quantitative methods because enumerating populations, resources, goods, and expenses proved useful for managing them. In turn, mathematics, statistics, predictive models, and the development of better measuring instruments became more central to a range of administrative, commercial, and scientific enterprises. Exact figures seemed to augur more objective, consistent, and transferable facts, lending greater prestige and authority to scientific knowledge production. The emphasis on ever more precise methods and accurate results also reflexively intensified debates about what counted as such, especially because these debates were embedded in a broader Enlightenment culture of mastery over the natural world and its improvement. As the practical meanings of *precision* and *accuracy* were continually subjected to revision, these terms gained enormous rhetorical potency.

The valorization of precision and accuracy also pervaded early American society.[10] Samuel Williams, Hollis Professor of Mathematics and Natural Philosophy at Harvard College from 1780 to 1788, taught his students that the "peculiar advantage" of mathematics was its "exactness" and the "certainty of the *Data*" it produced. In order to defend his controversial theory about radical climate change in colonial America, Williams knew that it would be "of much importance to state and note this change of climate with accuracy and precision," by which he meant the difference in average seasonal temperatures in New England between 1630 and 1780.[11] Historian Jeremy Belknap worried that available calculations of Mount Washington—the highest peak in New England—had been determined without the "requisite precision" and called for further measurements by someone who could "exercise judgment with mathematical precision."[12] He also publicly criticized the inaccuracy of several late-eighteenth-century maps of New Hampshire, including one he considered misleadingly titled *An Accurate Map of the Four New England States*.[13] In these examples, disputes about the meaning of precision and accuracy could be addressed in terms of measurable quantities.

But in subjects that were less compatible with measurement, such as their circumstantial accounts of regional flora and fauna—Williams's *A Natural and Civil History of Vermont* and the third volume of Belknap's *The History of New Hampshire*—both men took a less rigorous approach. Belknap told his readers that his knowledge of natural history was "imperfect," and so it was not his "intention to write, systematically, the natural history of the country, or to describe with botanical accuracy, the indigenous vegetables which it contains."[14] His admission of the imperfection and incompleteness of his work might seem to reinforce the traditional view of colonial or early American science as essentially inconsequential: simplistic, exceedingly deferential to metropolitan expertise, or entirely derivative.[15]

Instead, such avowed imperfection was an instrumental compromise. Early American naturalists contributed to the transatlantic republic of scientific letters in part by providing empirical reports about the American environment and in part by explaining their dominant role in improving it. Belknap, Williams, Dwight, and other writers were fully aware of the tension between their dual roles—on the one hand as disinterested experts on local nature and on the other as self-interested land settlers, investors, and improvers. The language of empirical imprecision was a way for them to reconcile these competing roles. They aimed to prove that the natural characteristics of the Northeast were profoundly destabilized by colonial settlement, and they offered a contemporary description just as a provisional document of its impermanent attributes.

As Waterhouse had suggested to the president of the Royal Society, "men of science" did not practice natural history merely for its own sake. Waterhouse lectured to his students at Harvard College that "the ultimate ends" of science were utility and economy. "Commerce," he declared, was "the friend of science." Naturalists in northern North America committed themselves to data gathering and looked forward to their potential contributions to theory, but their aims were broader. The nature of New England, scientific descriptions of it, and technologies for transforming it each shaped and were shaped by prevailing political and economic concerns. Like learned elites throughout the Atlantic world, they considered natural history "increasingly a science of natural economy"; field surveys, economic botany, and agricultural improvement were closely related pursuits. They conflated or justified the study of natural history in the economic terms of agricultural improvement or the politics of territorial expansion.[16]

Since their primary interests in and views of nature were shaped by an ideal of economic development based on resource exploitation and an expanding frontier of settlement, surveying the nature of a newly subjugated, supposedly uninhabited, or improperly settled place was a preliminary step in developing it.[17] Because of this continuum between studying nature as it was and reengineering it, ecological, economic, and political geography subtly converged in written accounts of New England and are nearly indistinguishable in retrospect. In the following section, I disentangle them, showing that those writing about early America often assumed and exploited such convergences to produce inexact accounts of its environmental history and geography.

## STATES OF NATURE

Uncertainties about greater New England's natural history followed first of all from imperial fiat and the subsequent shifts in Native and European political geography in the region. From the late sixteenth through the early nineteenth centuries, New England continually expanded and contracted, at various times encompassing parts of present-day Quebec, New York, Connecticut, Rhode Island, Massachusetts, Vermont, New Hampshire, Maine, New Brunswick, Nova Scotia, and Prince Edward Island. In 1623–24, James I created a Council of New England, which negotiated an internal division of the region into twenty districts but left unresolved the question of its outer bounds. Later in the century, during James II's brief revocation of the Massachusetts Bay Colony's proprietary charter from 1686 to 1689, the Dominion of New England included all British lands from the Delaware River to New France.[18]

Because the region grew and shrank so dramatically over the course of the colonial and early national period, the charters, maps, and geographical references sometimes simply situated New England somewhere in the northern part of North America. The monolithic but geographically indeterminate names "the North," "northern colonies," or "northern states" were convenient shorthand for New England. The signers of the Mayflower Compact invoked it in 1620, when they asserted that Plymouth plantation would be "the first Colony in the northern Parts of Virginia." Boston physician and historian William Douglass also substituted "North Virginia" as a collective term for New England in his widely circulated geographical history published in the middle of the eighteenth century, which noted that maps and other references to both Nova Scotia and New England continued to use this terminology. During the American Revolution, a detailed map depicting Vermont's contested boundaries with "part of Canada" and surrounding states, including Connecticut and Pennsylvania, named this broad region "the Northern Department of North America." Such geographical lumping could also be done in negative terms, as when Arthur Young disparaged the commercially unpromising lands of "the Northern Colonies," adding the clarification "that is, north of tobacco." As these terms imply, through the period between the Seven Years' War and the independence of the United States, it might have made as much sense to imagine "New England" as any British colonial territory lying north or northeast of someplace else—Virginia, the Hudson River, southern plantations—rather than as any more specific place.[19]

The most problematic borderland between New England and someplace else, however, was still farther to the northeast, in Acadia–Nova Scotia. Until the Acadian expulsion of 1755, the British and French engaged in a series of half-hearted and inconclusive imperial contests over the peninsula. Unlike the far more lucrative offshore claims in the Gulf of St. Lawrence and the Gulf of Maine, land in what is now eastern Canada was never a top priority for European imperial powers. Even after the British took possession of Nova Scotia at the turn of the seventeenth century (formalized in the Treaty of Utrecht of 1713), they only intermittently attempted to bring order to their claims in the area. While Mi'kmaq, Acadians, and New Englanders competed for the limited arable land in Nova Scotia, territorial sovereignty continued unresolved.[20]

These ongoing disputes produced a persistent geographical uncertainty that was reflected in reports and surveys of the peninsula, especially since individual surveyors who were charged with recording official boundary lines were reluctant to act decisively in settling rival borderland claims. In 1720 the British commissioned French-Huguenot engineer Paul Mascarene to survey Acadia–Nova Scotia's boundaries. Mascarene described a territory that stretched "from the Limits of the Government of Massachusetts Bay in New England or Kennebeck River, about the 44th degree North latitude to Cape des Roziers on the South side of the entrance of the River of St. Lawrens in the 49th degree of the same latitude, and . . . from ye Easternmost part of the Island of Cape Breton to the South side of the river of St. Lawrens." But he cautioned that his account portrayed the colony only "according to the Notion the Brittains have of it." Since "the Boundaries having as yet not been agreed on between the Brittish and French Governments in these parts," he maintained that "no just ones can be settled in this Description."[21] The scope of eighteenth-century geographical or natural history surveys was therefore necessarily imprecise in situations of disputed sovereignty or ambiguous jurisdiction.

Well into the late eighteenth century, the northern borderlands of New England presented problems for geographers. Though Jedidiah Morse vowed that his gazetteer *American Geography* (1789) was "as accurate, compleat, and impartial as the present state of American Geography and History could furnish," he could not definitively describe the geography of New England. In the early national period, the region was unquestionably comprised of five states—Connecticut, Massachusetts, New Hampshire, Rhode Island, and Vermont—and the district of Maine, which until 1820 was a noncontiguous part of Massachusetts. But Morse was unsure how to address the disputed territorial borderlands between northeastern Maine and New Brunswick and the waters off the coast of Nova Scotia. While his map of the southern states was the "most accurate yet published respecting that country," he warned that the map of the northern states was "compiled principally by the Engraver, from the best Maps that could be procured; it was chiefly designed to give the reader an idea of the relative situation, and comparative extent of the several states and countries comprehended within its limits."[22]

If his political map of New England was only relatively accurate, Morse nevertheless provided discrete natural histories for each state, relying on secondary sources such as Williams's and Belknap's "natural and civil histories," which described aspects of the environment enclosed within putatively stable political borders.[23] By choosing such units for their natural histories, these American writers followed the conventions of early modern chorography. Chorography was the practice of surveying a place on a local or regional scale (as opposed to the global scale of geography or the universal scale of cosmography). Only those features that coincided with and were limited by some given outline were identified. This approach was undoubtedly pragmatic. For the purpose of isolating salient ecological features, however, it was rather arbitrary because weather, seeds, and animals tended to range beyond the fiat borders of the state. When political borders or property lines were very new or long disputed, as was the case over a long period in northern New England, it was an especially arbitrary—if ideologically advantageous—approach to natural history. Yet this contents-in-a-container approach was so typical that biogeography could seem to reinforce and naturalize political geography.[24]

Aware of such problems, in 1787 David Ramsay, a South Carolina naturalist, suggested to Morse that he try using climatic geography, rather than state and national borders, to organize information on natural history. He asked if it might be "better to describe the natural history of the globe by zones than in the common way of following the political division? e.g. to describe the animals minerals face of the country climate & etc. of the torrid zone & of the other zones would be in my opinion better than to describe the animals minerals & etc. of Egypt Abyssinia & etc."[25] Perhaps in response to this advice, in the second edition of *American Geography* Morse offered "a natural history of the globe" organized by torrid, temperate, and frigid zones and lines of latitude, including thirty climates of "our habitable world" and a list of cities or regions that fell into each division. However, even in this format, Morse insisted on a tidy climatic distinction between New England and Nova Scotia that simultaneously reinforced their political boundaries: the former was in the seventh climate with Rome, Constantinople, and the Caspian Sea; the latter, joined to the rest of British North America, was in the eighth climate with Paris and Vienna.[26] These groupings were facile and largely wrong for purposes of comparing the weather in coastal North America and inland Europe. And despite Morse's inclusion of a chart of climatic geography, this apparently alternative view had little impact on the overall organization of his book's subsequent editions. He retained his narrative

entries on the climate and natural history of individual states, colonies, nations, and empires, demarcations that assumed a perfect overlap between
natural and political geography.

Such reifying overlaps were also common in chorographies at national
and continental scales. French traveler Constantin-François Volney's widely
read gazetteer of the early United States took Morse's work as authoritative,
but Volney (whom Dwight criticized for "scientific vanity") observed that the
natural and political geographies of the states were already interchangeable.
To explain the relationships between cultural and natural regions within the
country, he divided it into four climates: cold, middle, hot, and western. The
"coldest climate" was coincident with the outer collective boundary of the
New England states, a "natural boundary" that could be "traced by the southern side of Rhode-Island and Connecticut on the ocean, and interiorly by the
chain of hills, that furnishes the waters of the Delaware and Susquehanna."
These interior limits, however, were only temporary, since Volney further
posited that "the most simple idea" of the nation as a whole could be encapsulated in terms of the major bodies of water that surround the North American continent. "In an age when the advantage of natural boundaries are so
well known," he argued, "we can scarcely question, that these will sooner or
later form the limits of the country, as they are so distinctly marked." The
complementarities between the United States and the watery edges of North
America seemed to him so strong that the nation would inevitably extend
to the entire land mass and, as a result, subvert altogether the northern and
western boundaries of New England.[27]

Some naturalists manipulated the natural and political geography of New
England and North America on still grander scales. In the 1770s, Thomas
Pennant (who had published zoologies of Britain and India) was working
on a "sketch" of North American animals until the American Revolutionary
War abruptly curtailed his work. Unwilling to quit the project, he repack-
aged his research material as the natural history of a transnational northern
region and changed the title. His *Arctic Zoology* integrated other naturalists'
work on the Arctic, Europe, and Asia with that of the "northern part of the
New World" as far south as the Carolinas. Most editions included two global
maps of this idiosyncratic Arctic. The first map, which centered on the Atlantic Ocean, pictured North America from the North Pole to Long Island.
In this view, New England was usurped into the natural history of a nearly
hemispheric zone.[28]

These examples demonstrate that eighteenth-century natural histories
and gazetteers pretended that ecological habitats coincided with political geography on multiple scales. However, what they actually offered were syn-
ecdochical accounts of remarkable wild areas or lists of the most abundant
endemic species, to represent the natural history of the whole. The botanist

Manasseh Cutler, for example, had planned to produce a comprehensive inventory of plants in Massachusetts. In 1780–81, he solicited the support of the Harvard Corporation for such a study, proposing "to investigate the botanical character of such Trees and Plants as may fall under my observation, which are indigenous to this part of America, and have not been described by Botanists; also to make out a Catalogue of those which are found growing here, but have been found in other parts of the World, and therefore need no botanical description; and of such as have been propagated here, but are not the spontaneous production of the Country."[29] But rather than a comprehensive inventory, in 1785 Cutler published *An Account of Some of the Vegetable Productions, Naturally Growing in this Part of America, Botanically Arranged.* This work described "some" plants confined to Cutler's neighborhood on the north shore of Boston, listing those species he had managed to find "growing within the compass of a few miles; except a small number that happened to be noticed at a greater distance." At least Cutler acknowledged the limited scope and implications of his research, whereas Jacob Bigelow, one of his successors, argued that the vegetation of the city of Boston should "serve as a tolerable specimen of the botany of the whole New England states."[30]

Conversely, natural histories or surveys provided overviews of the most widespread species. For example, Thomas Caulfield, a governor of Nova Scotia in the early eighteenth century, wrote a detailed report of the natural resources in various sections of the province. His description of Cape Breton Island's vegetation was particularly succinct: according to him it was "intirely a Rock covered over with moss."[31] In a similar rhetorical move, Belknap's botanical survey of New Hampshire offered only "briefly to take notice of such as we are endowed with the most."[32] (And recall Arthur Young's phrase "north of tobacco," which made the presence of tobacco a metonym for the South and its absence a metonym for the North.) While Cutler and Bigelow's representative samples were concentrated in small areas, Caulfield, Belknap, and Young restricted their attention to commonly dispersed species. All four presumed the interchangeability of the selective and the general, the part and the whole. As D. Graham Burnett has written of imperial maps, natural histories of early northeastern America "conjured omniscience" but were more often "merely a handful of views from here to there"—a collection of microcosms or exemplars from which the environmental conditions of broader areas could be extrapolated.[33]

Such extrapolation was necessary partly because of the real physical challenges surveyors faced in New England's most rugged and thinly inhabited tracts, hills, and mountains, such as the White Mountains range in New Hampshire. Although New Hampshire had been colonized in the early seventeenth century, there was little agricultural settlement before the 1760s, and the mountains limited farming and commercial activity.[34] To encourage more settlement, "a company of gentlemen," including Belknap and Cutler, hiked the mountains in the summer of 1784. Most of the men were interested in the area's real estate potential; for Cutler, the tour was a scientific expedition during which he would survey and collect birds and botanical specimens and measure the height of Mount Washington to provide Belknap with materials for revising the natural history section of his book on New Hampshire. A botanical study of the highest elevations in the region could corroborate or dispute leading theories about the climatic similitude of all alpine regions or the essentially unified biogeography of all northern countries across the globe. Barometric measurements could have contributed to an understanding of local weather. The purpose of such fieldtrips was thus threefold: the gentlemen toured marginal areas to contribute empirical data to the encyclopedic project of Enlightenment natural history, to perform their expertise by detecting known or new species, and—most important of all—to assess regional commercial possibilities.[35]

As it happened, the 1784 survey in the White Mountains contributed to these goals only "in part," Belknap told members of the American Philosophical Society, because the formidable "weather while we were in that region hindered us from making some observations which we intended." According to Cutler, "it happened, unfortunately, that thick clouds covered the mountains almost the whole time." The dense cloud cover "rendered useless" the sextant, telescope, barometer, thermometer, and other instruments he had packed. In addition, the cold air at higher altitudes "nearly deprived him of the use of his fingers." By the time the party reached the summit, all the gentlemen were numb fingered: one of them tried to "engrav[e] the letters NH but was so chilled with the cold, that he gave the instruments to Col. Whipple, who finished the letters." Even on hot days much of mountainous New Hampshire was disappointing to them. Although they were momentarily excited by "the appearance of a close-fed pasture" above tree line, on closer inspection the green turned out "to be a mere mass of rocks, covered with a mat of long moss." The "extreme" temperatures and "barren plains" were bad for botanizing. "This is a most wretched place indeed," Cutler huffed in his diary, "miserable huts, on very poor, rocky, rough land, constantly uphill and down."[36]

Together, the hikers were able to identify several arctic-alpine species, as well as "various kinds of vegetables, most of them such as we had never seen before."[37] Nevertheless, Cutler decided that their findings were insufficient. His measurements were dubious and limited by the fact that "all the instruments were unhappily broken in the course of the journey through the rugged roads and thick woods." Moreover, the extent and character of most of the mountains in the range were still unexplored, "the country round them being a thick wilderness." He wrote in a personal letter to Belknap that it was "no small mortification to me not to be able to give a more accurate account of the height of this mountain, after taking so much pains to ascertain it."[38] "Yet till a better account" was possible, Belknap was sure that information from this expedition would at least "prove more satisfactory than any which has yet been published or reported."[39] Cutler reassured Belknap that future hikes would allow for better measurements: "Another tour, I hope, will remove many of our present doubts and uncertainties."[40]

In the meantime, Belknap supplemented his natural history of New Hampshire with predictions of its future state. In early America (and likely in other colonial contexts), regional descriptions emphasized improvement and the shape of things to come rather than the history or current state of nature. This invention of a paradoxical forward-looking tradition was part of the western discovery narrative that long dominated ideas of the New World, from the Renaissance through the twentieth century. It was implicit in eighteenth-century North American improvement literature, including travel narratives, gazetteers, and natural histories, which hardly mentioned an indigenous presence or ignored precontact history altogether (as Morse wrote in his entry on New Hampshire: "There are no Indians in the state. Their former remains of scattered tribes retired to Canada many years since").[41] Elites in settler colonies revised this narrative by emphasizing their expertise and the role of science in uncovering the resource potential of insufficiently exploited lands. In the territorial disputes in Acadia–Nova Scotia, for example, the twinned goals of colonization and improvement encouraged conceptions of economic potential as if the British would prevail in the peninsula. In 1717 a group of merchants expressed their interest in helping the Board of Trade to establish "a Colony of your Majesty's Subjects" in the place that "the french call'd Accadie." According to the merchants, this was self-evidently "a large Country upon the Continent of North America, adjoining Westward to New England, having on the North the River and Gulph of Canada, and is surrounded on the South & East by the Ocean," a large country that was "capable of great Improvements."[42]

Writing in the late eighteenth century, Timothy Dwight urged travelers to adopt a similar attitude in assessing the qualities of a place—to "overlook" anything "rude, broken, or unsightly" in New England's landscape as only a "temporary defect." Dwight had set off on his travels keen to observe the beneficial environmental effects of two centuries of European settlement. He expected to encounter a thoroughly cleared, settled, cultivated, and flourishing landscape. The wilds, wastelands, and neglected or incompletely settled areas that were evident even within long established communities troubled him. The unevenness of agricultural improvement across the region was an uncomfortable empirical fact that he only reluctantly acknowledged. But he believed that these environmental conditions were ephemeral and anticipated further improvements. Accordingly, his descriptions focused on bright spots, opportune places for the formation of new settlements, and the transience of the current scene. Knowing "from experience of the rapid progress of cultivation," he allowed himself, in viewing underdeveloped lands, to be "transported in imagination to that period, in which, at a little distance, the hills, and plains, and vallies will be stripped of the forests . . . and be measured out into farms."[43] Geographical gazetteers or chorographies that placed so much emphasis on improvement were descriptions of prescriptive or ideal future land uses, rather than reflected images of landscapes in their current state.

Colonial or early national histories could seem to be about the status quo or an ever-unchanging natural world, as when Samuel Williams stated that "in most productions of nature, the subject is fixed, and may always be found and viewed in the same situation." But Williams showed that this stasis was deceptive. Through "a steady course of observation," scientists were able to "discover and ascertain the laws by which [nature is] governed," laws that would ultimately help guide land use and predict "the situation [nature] will assume in other periods of time." Claiming that this process was evident in the dramatic shifts of Vermont's climate, he wrote that "instead of remaining fixed and settled, the climate is perpetually changing and altering, in all its circumstances and affections; And this change instead of being so slow and gradual as to be a matter of doubt is so rapid and constant, that it is the subject of common observation and experience. It has been observed in every part of the United States; but is most of all sensible and apparent in a new country, which is suddenly changing from a state of vast uncultivated wilderness, to that of numerous settlements, and extensive improvements."[44] Williams observed the future in the present just as Dwight inferred the face of the country "as it truly appeared, or would have appeared eighty or one hundred years before." Likewise, Volney's description of a fully continental United States—a nation conterminous with North America—embraced this rhetorical strategy of mixing past, present, and prospect.

Historians of the Enlightenment have shown how profoundly stadial theories of progress from savagery to civilization shaped ideas about the history of non-European and colonial environments. These ideas, as Joyce Chaplin writes, "prescribed a static image" of people and landscapes. However, Dwight knew that in order to draw such a static image of the landscape, all its "imperfections must be left out of the account." Depicting the countryside only "as it is in a state of nature, or as it will soon be in a state of complete cultivation"—that is, representing only the beginning and end points of improvement—was a (sometimes necessary) act of gross omission. On the other hand, when early American naturalists depicted the transitional phases between discrete stages of development, they did so in order to emphasize the dynamism of environmental change.[45]

## WORKS IN PROGRESS

The combination of spatial indeterminacy and chronological blurring in eighteenth-century natural histories made them composites, inherently ambiguous documents of landscape conditions. Early American naturalists sometimes endorsed the veracity of topographic surveys or natural history accounts through a rhetoric of accuracy and precision, but they were also aware that these were relative values because geographical knowledge was understood to be indeterminate and subject to revision.[46] If naturalists detected signs of landscape or climatic improvement, they knew that their surveys would be rendered outdated by successive reports documenting progressive changes. This emphasis on improvement encouraged naturalists to imaginatively transcend the constraints of political boundaries and posit geographical concepts in more expansive and adaptable terms, terms more fitting to the labile nature of early America. Naturalists writing about early northeastern America were not naively contradictory or careless in their empirical and descriptive practices. Often they were deliberately, even methodically ambiguous for pragmatic, intellectual, and ideological reasons. By emphasizing the temporariness and ambiguity of their natural histories, early American naturalists demonstrated a fidelity to empiricism and at the same time guarded their work against the vagaries of environmental change. As further developments in measuring instruments narrowed the definitions of precision and accuracy, the range of practical and rhetorical meanings of imprecision, ambiguity, and indeterminacy was inevitably broadened. Perhaps the pressure to treat nature in the terms dictated by these heightened standards of exactitude had raised expectations of what should and could be represented with accuracy. Eighteenth-century naturalists in extra-European or provincial contexts who worked under material or physical constraints were particularly sensitive to these new pressures (and their corollaries) and employed a language of imprecision to evade charges of incompetence, incompleteness, or the imminent obsolescence of their work.

Most important, naturalists in Northeastern America developed a language of empirical imprecision as a means of reckoning with and encouraging ideas about the dynamism of nature. Empirical imprecision may seem to be a paradoxical scientific method, but it reflected naturalists' understandings of historical instabilities, as well as their expectations for imminent change. These writers were particularly focused on how new European settlement through the eighteenth century was transforming the geography, landscapes, and climate of North America. But their insistence that imperfect characterizations were the most faithful representations of unstable environmental phenomena could prompt us to reconsider the relationship between the environmental sciences and environmental change over the longer term. The pragmatism of imprecise description in eighteenth-century America points to a broader cultural history of scientific attempts to represent the radical temporal contingency of environmental change.

# CHAPTER 3

# FARMING AND NOT KNOWING

## AGNOTOLOGY MEETS ENVIRONMENTAL HISTORY

### FRANK UEKOTTER

IN THE MODERN world, ignorance is first and foremost a bad thing. With the omnipresent buzz about "knowledge societies," there is a broad consensus that information is a key resource of modern societies, giving an inherently negative ring to the absence of knowledge. Historians of science have been particularly hesitant to challenge this rationale. Ever since Francis Bacon, the notion that "knowledge is power" runs through the Western philosophy of science, and that put a heavy burden on those who wanted to study ignorance dispassionately. Even more, Baconian science offered a simple remedy for the problem of ignorance: it implied a promise that over time the realm of ignorance would shrink with the advancement of science. Furthermore, with mandatory publication of all results of research—a key requirement of Bacon's philosophy—access would not be an obstacle either. So why ponder questions about ignorance if the generation and dissemination of new knowledge is so much more exciting?

From an environmental history perspective, the rationale for the study of ignorance looks equally dubious. After all, environmental historians have stressed time and again that action, rather than knowledge, was the crucial challenge. The causes of photochemical smog have been clear since the

1950s, and yet the problem continues to torment city dwellers all over the world.[1] The importance of Colorado River water has been apparent since John Wesley Powell's legendary journey, but managing it in a sustainable fashion became subject to enduring conflicts.[2] In the ongoing debate over global warming, the issue of knowing and denying anthropogenic climate change has evolved into an epic global struggle.[3] In all these cases, the problem of ignorance has an obvious response, namely impartial research and free access to the results, and environmental historians were disinclined to move beyond this notion.

In recent years, a number of researchers have challenged these readings and raised the question whether there might be more to the problem of ignorance. Under the banner of "agnotology," they have set out to develop new perspectives on the phenomenon, stressing that ignorance has shown a remarkable resilience even in an age where the legitimacy of scientific research appears by and large uncontested. The first part of this article outlines their approach and describes where the debate within STS currently stands. The second part provides a case study that pushes the debate further: it makes a case for the liberating power of ignorance. More specifically, it describes a situation where ignorance was the result of an interplay among different parties: researchers, advisors, farmers. While academic knowledge production was proceeding apace, ignorance did not shrink, as scientists got caught in an exchange with other groups. In the end, scientists became accomplices to a body of knowledge that in some respects ran counter to the best available expertise. The conclusion will argue that understanding these trends and their underlying dynamics may emerge as an important field at the intersection of environmental history and science and technology studies.

## THE MAKING OF AGNOTOLOGY

The project of agnotology stands in skeptical distance, if not open opposition, to the promise of Baconian science. Thus, it is noteworthy that the field did not emerge out of a theoretical treatise from the philosophy or sociology of science. The most important advocate of the study of ignorance, Robert Proctor, is a professor of the history of science at Stanford University, who came to the topic through his work on the history of cancer research and cancer knowledge. The concept of agnotology thus remains severely undertheorized to this day, and Proctor makes no bones about this fact, depicting his project first and foremost as a wakeup call for scholars to look at the issue more closely. In the introduction to a recent essay collection, Proctor wrote, "The point is to question the *naturalness* of ignorance, its causes and its distribution."[4] After all, it is "remarkable," in Proctor's view, "how little we know about ignorance."[5]

Proctor casts a wide net in his programmatic essay, stressing the diversity of types of ignorance. His remarks include discussions of manufactured ignorance, sanctioned ignorance to protect the right to privacy, foolish ignorance (also known as stupidity), and even virtuous ignorance. However, Proctor is by no means the first to ponder the multitude of social meanings of ignorance. More than half a century ago, Wilbert Moore and Melvin Tumin published an essay on "some social functions of ignorance" in the *American Sociological Review*. They approached the issue even more broadly, noting functions as diverse as the defense of privileged positions, the avoidance of jealousy over unequal rewards, and the preservation of fair competition.[6] However, Moore and Tumin's timing looks exceedingly poor in hindsight: in the context of post–World War 1 America, Vannevar Bush's emphatic declaration that science is "the endless frontier" was clearly more attractive as a research agenda than a critique of the "rationalistic bias" that stood in the way of a sociology of deficient knowledge.[7] It is only in recent years that ignorance has received more than scant attention among scholars.[8]

In order to understand the rise of agnotology, it is helpful to sketch the path that led Proctor toward the proclamation of a new scholarly field. In his work on the history of cancer, he came to devote great attention to the risks associated with tobacco smoking, which led to the publication of two widely cited volumes on cancer research and the anti-cancer and anti-smoking efforts of the Nazi regime in Germany.[9] His insights on the role of tobacco companies in debates over lung cancer eventually led him beyond academia and into the courtroom, where he testified as an expert witness against the tobacco lobby. This activity clearly left a mark on his scholarly work, all the more since he represents a minority within his own scholarly community: Proctor reports that as of 2005, he was one of only three scholars who testified against the tobacco lobby, whereas the industry had hired no fewer than thirty-six academic historians to speak under oath on its behalf.[10] His aversion to the tobacco lobby is plain, to put it mildly. In a recent newspaper article, he speaks of "the greatest health disaster of all times," specifically targeting the "killer machines" that produce cigarettes. Moreover, he stresses that "the best—and deadliest—machines came from Germany," a more-than-tacit reference to the Holocaust.[11] It is clearly a specific type of ignorance that is closest to his heart, namely deliberate manipulation by vested interests. With obvious pleasure, he quotes a statement from officials at the cigarette company Brown & Williamson who noted internally that "doubt is our product." Proctor presents this as a smoking gun.[12]

It goes without saying that the intentional production of cognitive uncertainty holds merit in the field of environmental history as well. Many debates over environmental risks take place under the impression of significant disagreements over the actual extent of the hazard, and industry has an obvious interest in using this situation to its own advantage. The tobacco lobby may be an extreme case in this regard, but it is by no means singular. The climate debate provides a case in point, and it is by all means fortunate that Proctor's volume, edited together with Londa Schiebinger, includes an article on climate science.[13] Every newspaper reader knows that the challenge from so-called climate skeptics is an ongoing one, as a recent outcry from the climate community serves to attest.[14] And yet this type of deliberate ignorance is only one aspect of the overall challenge of agnotology, and the focus on manipulation by vested interests leaves important dimensions in the study of ignorance unexplored.

The following section thus seeks to open a new path toward the study of ignorance. In the case study presented herein, deficient knowledge results from an interplay between different academic and nonacademic groups, where scientific concerns become subject to long and tedious negotiations. With that, this paper underscores an important contention of science and technology studies: knowledge production is a social process. Scientific principles do play a role, but they do not define or determine the long and complicated process of knowledge production. At times, scientific rules even yield to other imperatives—to the need to reduce complexity and to reach decisions within reasonable spans of time, for instance. The following case study describes such a situation and goes on to offer some suggestions in methodological respect. Environmental history is not only an important field for scholars of ignorance, but it may also provide some hints as to an urgently needed theorization of agnotology.

## IGNORANCE AND THE PERILS OF MONOCULTURE: PLANTING CORN IN GERMANY

Agriculture is certainly not the most popular topic within the history of science. In fact, there is good reason to lament its marginal status within the discipline, as much of what Deborah Fitzgerald bemoaned two decades ago still holds true: "For historians of science, agricultural science represents a distinctly 'blue collar' phenomenon, which, like engineering, suffers from neglect partly because of its practical aspects. If the physical sciences are highest in intellectual status, the agricultural sciences are near the bottom."[15] To some extent, environmental historians have helped to correct this imbalance of scholarly interest in recent years, yet it may help in raising the profile of the agricultural sciences to stress that it is particularly rewarding for the study of ignorance. In fact, agnotology may serve as a welcome reminder that knowledge is more than an issue for academia. Scholars have tended to look at the production of knowledge in laboratories and test plots, while giving scant attention to how knowledge travels out to the field, and have rarely pondered questions as to whether something is coming back. Ignorance as it emerges in the following is an issue emerging from the interaction between scientists and other groups.

In making sense of science-based agriculture, it is crucial to acknowl-edge the enormous dynamism of agricultural knowledge. Farming methods have changed dramatically in the modern age, with yields-per-acre and other indicators of productivity rising dramatically. It is equally plain that scien-tific results have played a crucial role in the process, a fact that agricultural science textbooks routinely tout. But agricultural production is, at its core, a closely interconnected system with a huge variety of factors, making for a wide range of unexpected and poorly understood interactions. Knowledge production and ignorance thus went hand in hand, and the latter was a chal-lenge for researchers, advisors, and farmers: it allowed, and even encour-aged, certain types of behavior. One of the most dramatic and consequential innovations in twentieth-century German agriculture, the introduction of corn, provides an instructive case in point.

In the nineteenth century, corn was a marginal plant in Germany. Ger-man farmers planted corn on only forty-five hundred hectares in 1893, mostly in the south.[16] In contrast, corn has emerged as a crucial part of fac-tory farming in the twenty-first century: wherever animal production ex-ists on a grand scale, corn is an indispensable part of the diet for the poor creatures that are supposed to grow as fast as they can on their way to the slaughterhouse. In regions with heavy factory farming, corn is cultivated annually on at least a third of the arable land; at times, almost half of the land is covered by the plant.[17] In fact, the stage may be set for yet another boom, as advocates of bioenergy are placing high hopes on corn, drawing on its enormous capacity to generate organic matter for fermenting. All the while, plant scientists proclaim the promise of even higher yields: with more research on heterosis, productivity might jump to thirty tons per hectare. At the moment, the genetic potential allows yields in the range of fifteen to eighteen tons per hectare.[18]

As one of the most important crops on the globe, corn was anything but an unknown plant in the late nineteenth century. Originally a native of Mexico, corn (biological name: *Zea mays*) came to Europe as part of the Columbian exchange, finding a new home in the agricultural economies of southern Europe. Due to its tropic origin, the plant requires a relatively warm climate, and it took decades of breeding efforts to make corn accept the harsher climate of Central Europe. However, corn had a crucial biological ad-vantage: it is a C-4 plant, meaning that it can assimilate light to four carbon atoms. Other Central European crops conduct photosynthesis differently, as-similating light to only three carbon atoms (C-3 plants). In other words, corn can use sunlight with superior efficiency, and that gives the plant an inher-ent natural advantage. And yet it took time and specific conditions for this advantage to come into play. While corn became a crucial part of the African diet, Germans, like most Central Europeans, refused to eat corn in great quantities.[19] German corn production thus focused on its merits as animal feed, an important market given the fact that animal production emerged as the key source of revenue for German farmers in the late nineteenth century. However, most farmers preferred a different solution for the lack of animal feed within the confines of Germany: rather than experimenting with a new crop, they bought feed from abroad. In the years before 1914, Imperial Germany emerged as the greatest net importer of animal feed in the world.[20]

However, the context was different after World War I. Due to the collapse of world markets and the burdens imposed on Germany after the war, it was no longer possible to solve the need for feed through globalization. All of a sudden, increasing feed production on German soil became an urgent necessity, and autarky became a key theme of agricultural research. For one thing, interest in grasslands increased enormously, with the first association for the promotion and intensification of grassland research founded in Straubing, Bavaria, in 1919.[21] But soon attention went beyond the familiar types of grass toward the strange neophyte that promised to produce enormous quantities of plant matter. As corn became a plant of great interest in the 1920s, the call was for experts inside and outside academia to provide information on cultivation, harvesting, and many other factors that together constitute proper plant use. Knowledge production found an institutional home in 1925 when the German Agricultural Association (Deutsche Landwirtschafts-Gesellschaft) founded a special committee on corn.[22] Committee work had been a mainstay of scholarly work within the German Agricultural Association since its foundation in the 1880s, yet something was peculiar about the Committee on Corn (Sonderausschuss Mais): it was the first committee that focused on a specific plant. Traditionally, the work of the German Agricultural Association had centered on specific parts of agricultural work: fertilizer, pest control, agricultural implements, and so forth. However, the unusual characteristics of corn called for a synthetic approach that comprised all phases of cultivation from seed production to feeding, and the new character of the Committee on Corn mirrored that fact.

It takes a broad view of the trajectory of agricultural knowledge in the modern era to understand the challenge that researchers, advisors, and farmers faced in the Committee on Corn. As in most academic fields, specialization had been the dominant trend in the agricultural sciences since the late nineteenth century. Whereas universities commonly had only one chair for agriculture since the creation of the first institute for agriculture at the university of Halle in 1863, the trend toward specialized professorships for agrochemistry, plant breeding, and veterinary medicine had become irreversible by the turn of the century. The Committee on Corn thus bucked one of the most powerful trends in modern science in that it had to draw together widely divergent fields of expertise: its members had to find ways to produce the new seed, to study technologies for harvesting and processing, to adjust crop rotations, to find ways to conserve and store corn for the farm year, and so on. No system of corn production could work without attention to these widely disparate challenges, and this meant that the usual process of knowledge production, bound by disciplinary binders, would not be sufficient. Characteristically, the Committee on Corn decided to go on a field trip to Romania and Hungary as one of its first items of business. By studying an actual system of corn production, the researchers were hoping for clues on how it all might fit together.[23]

When it came to biological potential, the report was unambiguous. "If we cultivate and conserve corn properly, it can produce more feed on a given plot of land than any other plant," the excursionists noted, stressing the importance "for our German fatherland" when it came to "autonomous food production with no input from other countries."[24] However, the German agricultural science network was slow to react, clearly a reflection of the cognitive challenges that the neophyte raised. Four years after its foundation, the Committee on Corn had only one trip report, two public meetings, and three essays to show.[25] It took the peculiar conditions of the Nazi era to instill a notable upswing in corn production: acres planted rose from 3,697 hectares in 1933 to 59,394 hectares in 1938, clearly a reflection of the unprecedented extent to which Nazi policies could dictate food and plant production in Germany. After all, autarky was at its core a policy to prepare Germany for the next war, and the Nazis were willing to take decisive measures toward that goal, leading some to boast about a "movement" that was "only beginning."[26] However, that statement had clearly more to do with the Nazi penchant for "movement" than with the reality in German fields, as the forced nature of that boom makes perfectly plain. Hans Buß, chairman of the Committee on Corn, noted in a 1936 essay that research had achieved "clarity in all important technical questions," yet that statement merely mirrored the peculiar challenge of corn production in Germany, as knowledge production was not simply a matter of adding up pieces of information.[27] It was not enough to resolve isolated issues if they could not be integrated into a unified system, and that synthesis was clearly lacking in the 1930s. When the Nazi autarky regime collapsed, farmers were swift to abandon corn production.

Corn galvanized interest among farmers again around 1960, and this time, it was farm practitioners who were pushing the issue. With the rise of industrial-style animal production, farmers were searching for a simple way to produce great quantities of feed, and corn was an obvious choice. However, researchers were no better prepared in 1960 than they had been in the 1920s and 1930s. In retrospect, the innocence of the academic community is almost touching in some cases: a researcher from the agricultural university of Stuttgart-Hohenheim published a synthesis on corn cultivation in 1960 that made no mention at all of the crop's potential for the Federal Republic of Germany, treating corn as an important crop for a wide array of countries from the United States to South Africa, but not for Central Europe.[28] Prospects for guidance from the research community were dim, as a satisfactory synthesis for the different branches of knowledge was nowhere in sight. In fact, the challenge of corn production had actually grown with the advent of highly productive hybrid varieties. Hybrid seeds had shown a clear edge in terms of productivity, but they were sterile, meaning that the farmer could not reproduce the seed by himself; rather, he (and it was almost always a "he") had to buy new seed from a company each year. Hybrid seeds had conquered the American corn market in the 1930s, with enormous repercussions for the relationship among farmers, extension agents, and corporate America.[29] With that, the threshold for farmers was even higher: not only did they have to master the peculiarities of an unusual plant, but they also had to accept close dependence on seed companies.

The growing interest of farm practitioners inspired a notable upswing of knowledge production. Researchers clearly recognized the need to investigate the neophyte more closely and supply the farmers with instructions and advice. Yet the boom was notably selective: instead of embracing a broad synthetic approach, research remained confined to established fields of academic interest. Three disciplines stood out in particular. One of them was plant breeding, a discipline that was among the first to develop a distinct specialized profile within the agricultural sciences in the late nineteenth century.[30] Seed companies as well as state-run research institutes underscored the trend toward a separate field of research, and it is by no means surprising that in the postwar years, they were focusing narrowly on seeds: their key goal was to adapt the plant to the Central European climate in general and to regional specificities in particular. Another booming discipline was economics, a field that generally received more attention in the postwar years. The third booming field was agricultural technology, as engineers worked on a fully mechanized production chain that, once in place, allowed the farmers to produce corn without touching the plant one single time.

It requires no lengthy explanations of the complexity of farming to note the exceeding selectiveness of this research boom. Quite a number of important themes received scant attention at best: crop rotation, pest and weed control, soil erosion, the health of the soil, and the balance of organic matter. And yet this precarious knowledge base was good enough to sustain an enormous expansion of corn cultivation: the plant became the industrial farmer's favorite plant, and cultivated acres skyrocketed. In fact, corn eventually pushed all competing plants to the margins, essentially monopolizing the market for animal feed production. When agronomists compiled statistics for farming in the northwest German region of Westphalia and Lippe in 2003, the hegemony of corn was evident. While corn claimed a total acreage of 172,499 hectares, grass came to only 13,419 hectares. A number of other plants that went into animal feed accounted for a paltry 2,281 hectares.[31]

It is not difficult to identify mixed feelings about this boom. In 1966, an article in a widely read farming periodical noted "that corn, unlike other plants in cultivation, is quite new for us, and we should not ignore that we have to collect experiences, both individually and for the entire region."[32] Four years later, the same author cautioned against "cultivation at any price," stressing the need to check "whether conditions are suitable at all" for corn.[33] In an overview of "up-to-date cultivation of cereals," Gustav Aufhammer noted that corn should not be planted more than two times in a row, a recommendation that stood in stark contrast to the realities of German fields, where corn soon emerged as the emblematic plant of monoculture.[34] In fact, it is by no means surprising to find mixed feelings among experts, as the ecological risks of corn monoculture were plain.

However, this skepticism soon became entangled in the power relations of the farm sector. While autarky was an accepted paradigm for all parties in the interwar years, the postwar boom went back to the wishes of farmers, putting experts in the position of having to follow suit or else lose their audience. No group felt the ensuing tensions more strongly than farm advisors. Scientists could retreat into the ivory tower if they wished, but advisors risked losing their audience if they remained silent. On the one hand, farm advisors were in an ideal position to witness the numerous problems of corn monoculture, as they occupied the space where specialized research met with the inherent complexity of farming practice. On the other hand, they were forced to make decisions and recommendations even when scientific knowledge was deficient: no farmer paid much attention to people who were trying to duck difficult questions with calls for more research. Farm advisors were also aware of the fact that they would be held accountable for errors, and given the shaky knowledge base that the boom of corn was resting upon, skepticism was probably the most natural reaction.

One might summarize the cognitive situation in the 1960s as follows: there was little hope that knowledge production along existing lines would provide a solution for the gaps of knowledge. Given the specialized and fragmented nature of research, researchers could provide individual bits of information—say, on the best seeds or the most efficient machinery—but they were in no position to supply a holistic vision that took the many aspects of corn cultivation into account. As a result, the knowledge system that underscored the expansion of corn cultivation grew "from the ground up": it developed out on the farms and in the fields, in an unplanned, improvised way, with bits and pieces coming together in an ad hoc fashion. Given these origins, it is by no means surprising that this knowledge system was far from holistic.

The fragmentation of knowledge was closely related to the fragmentation of groups of actors. A key innovation was the hiring of external service providers (so-called *Lohnunternehmer*, an oxymoron as it depicts service providers both as entrepreneurs and as recipients of wages) who took care of plowing, harvesting, and other important tasks. Of course, farmers lost a good deal of control and knowledge about their fields in the process, but for the farmers of the postwar era, who were busy mastering the complicated business of industrial-style farming, delegating time-consuming activities was what counted more. Characteristically, a handbook on corn cultivation mentioned the help of these service providers as an advantage, rather than a liability.[35] The rationale for the division of labor clearly took precedence over the rationale of knowledge, as service providers were primarily responsible for doing their job, and not for monitoring ecological problems on the fields of their customers.

Ignorance was thus built into the knowledge system of corn from the very beginning. While seeds, machinery, and economics received huge attention, issues like erosion or crop rotations were marginalized—not because people felt that they did not matter, or for lack of information about their importance (after all, crop rotation was the issue that gave birth to the agricultural sciences!), but because the hegemonic knowledge system had other priorities. Corn monoculture offered great advantages in terms of farm economics and machinery, and these economic and technological rationales ultimately overwhelmed competing ecological and agricultural rationales. After all, there was a way to make up for deficient knowledge: excessive resource use. The extreme wastefulness of corn monoculture in terms of agrochemicals and energy was not a mere "coincidence" or "externality"—it was built into the knowledge base from the very beginning, as the resource input would have been far smaller with a greater attention to the cycles and interconnections on the farm. In the absence of systemic thinking, farmers stuck to "rules of thumb," simple remedies that often presupposed a significant deal of brutality. When it came to pest and weed control, a narrow reliance on the chemical approach soon rendered all deeper thoughts obsolete. Atrazine in particular became the quick fix that farmers sprayed indiscriminately, and textbooks offered lavish thanks to the chemical industry, noting that the herbicide "has contributed greatly to the expansion of corn cultivation in recent years."[36] After three decades of excessive use, atrazine showed up in the groundwater, eventually leading to its ban within the Federal Republic of Germany in 1991.[37]

Plant nutrition provides an even more drastic case in point. Corn de-vours nutrients, but there was never a systematic attempt to adjust fertilizer doses exactly to its needs. Instead, farmers drew on the fact that corn was not prone to overfertilization, quite unlike traditional grains in Central Europe. In the awkward logic of industrial agriculture, that was even an advantage, since factory farming produced great quantities of manure: one could simply dump excrement on cornfields, and corn boosters were jubilant about the "ideal crop for liquid manure [*Güllefrucht*]," noting that farmers could dump double the usual amount of excrement on corn fields "without damages."[38] Revealingly, textbooks were expecting a completely different fertilizing re-gime, even noting that manure would become "dispensable" with sufficient doses of mineral fertilizer.[39] However, the sheer need to dispose of the grow-ing tide of animal waste soon swept away the ideas of the experts.

To be sure, this practice was open to criticism from more broadminded scholars: at a meeting of the German Agricultural Association in 1988, a speaker sardonically remarked that farmers in some regions saw fertilizing with liquid manure "primarily as a form of 'waste removal [*Entsorgung*],' and not so much as a way to supply nutrients."[40] But excessive doses of nutrients at least ensured that plants did not suffer from fertilizer deficit, and that argument ultimately carried the day. The brutal approach took precedence over the sophisticated one. Wasteful use of resources allowed a degree of ignorance that, while staggering from a purely scientific viewpoint, was pre-cisely what farmers needed, as it was hugely profitable from a short-term per-spective. Thanks to ignorance, farmers were free to act in a way that would have been irresponsible, and arguably impossible, with greater attention to the complexity of agriculture. In short, wasteful use of resources was the functional equivalent of deficient knowledge.

At this point, the merits of a dispassionate view on ignorance become apparent. While conventional readings would stress the gaping holes in the knowledge base, suspecting that knowledge production must have gone ter-ribly wrong at some point, agnotology notes that ignorance fulfilled a social function. To be sure, this merely supplements the view of the costs; agnotol-ogy is about explaining things, not excusing them. After all, it is clear that the long-term costs of this approach were staggering, especially when one considers external effects like groundwater contamination. In fact, even the handbook on corn cultivation (*Handbuch Mais*) made no bones about the hazards of corn monoculture in its 1990 edition, at least for certain regions with precarious conditions: "The logic of 'more corn—more animals—more liquid manure—more income,' while self-evident in economic terms on first sight, clearly has led us into a dead-end street."[41]

## IGNORANCE WAS STRENGTH

One could summarize the trajectory of corn in Germany with three words: ignorance was strength. It was only through ignoring certain issues that corn monoculture became an enticing prospect for German farmers. Focusing narrowly on the short-term economic advantages—the huge amounts of animal feed, the chance to dispose of manure, and the chance to optimize the load for costly farm machines—farmers glossed over troubling issues that cautioned against an excessive reliance on corn. However, "ignorance is strength" is, if anything, a loaded phrase: together with "war is peace" and "freedom is slavery," it is a key slogan of the totalitarian state of Oceania in George Orwell's *Nineteen Eighty-Four*.[42] And yet it would not be completely off the mark to argue that the knowledge system that sustained the boom of corn after 1960 was totalitarian in its own way. Of course, no farmer was legally compelled to plant corn, but from a short-term economic perspective, it was the cheapest way to produce animal feed and dispose of manure. Any farmer who opted to plant corn cautiously risked being outcompeted in the overflowing market for farm products. In other words, it was a good idea not to ponder deeper thoughts about erosion, soil health, or pests, as doing so would run the risk of weakening the farmer's competitive edge. For farmers who wanted to survive the cost-price squeeze of the postwar years, ignorance on certain issues was indeed a matter of strength.

In Orwell's novel, "ignorance is strength" epitomizes the method of "doublethink" that the totalitarian regime embraces as a way of suppressing dissent. In a nutshell, "doublethink" seeks to reconcile ideas that collide as a matter of common sense, presenting them as unified and thus "harmless" thoughts. Awareness of "freedom" and "peace" no longer leads to inconvenient conclusions that could undermine the dominant regime. The concept is quite appropriate here, as farmers, advisors, and even researchers displayed a mind-set reminiscent of "doublethink": on one level, they knew about the risks of corn monoculture; but on another, they were forced to think in a manner that blocked out the tricky issues. However, STS scholars do not need to embrace Orwell here, as there is a wealth of approaches that discuss and explain the discursive formation of scientific knowledge. Since Ludwik Fleck, the idea that scientific progress is essentially a matter of communication has gained currency and has almost become conventional wisdom in the twenty-first century (at least within the STS community). In this regard, agnotology may offer an important addendum: the making of scientific knowledge is a social process, and so is the development of ignorance.

Environmental historians can learn a lot from agnotology. However, it is rewarding to turn the question around by way of conclusion and reflect on whether environmental history has something to offer to scholars of ignorance. After all, the production of ignorance surrounding corn was notably different from the making of ignorance about tobacco smoking as outlined by Proctor. There was no powerful lobby that pushed corn monoculture out of a vested interest. If anything, there was a powerful lobby behind the *established* grains. Seed manufacturers as well as the producers of farm machinery were forced to hastily switch to corn when farmers began planting it in ever greater quantities. The example points to the need to stress social interactions in the production of ignorance, as farmers, advisors, and scientists came together in an improvised coalition that, while delicate from a cognitive viewpoint, allowed corn to occupy ever greater space on German fields.

However, interaction does not presume an equality of these groups, and the story at hand is revealing as to their relative weight. In the interwar years, when scientists were pushing corn, the project was an utter failure once the exceptional context of Nazi autarky policy was gone. In contrast, corn became part of the German landscape once farmers emerged as the prime movers in the postwar years. Agricultural scientists could do little more than voice their objections in journals, and farm advisors were in an even more precarious situation as they sought to find their path between expert warnings and farmers' demands. In fact, advisors deserve attention not only as they seek to mediate between different fields of expertise but also as indicators of trouble within knowledge. They will likely emerge as a crucial group for social studies of ignorance, as advisors are often forced to make decisions in spite of severe constraints.

These constraints include lack not only of scientific evidence but also time. One of the underlying aspects of the example at hand is that decisions on crops had to be made under a rigorous temporal reign. The farm year implied firm deadlines for decisions on plants, seeds, pest control, and so forth, as a delay of decisions would mean a subprime harvest at best or a lost year at worst. In other words, the choice of corn was frequently a "forced decision," where farmers simply went down a certain path because deep inquiries into academic matters required investments of time that they simply did not have: ignorance was a way to live with an immutable business cycle. Time also emerges as a crucial parameter in a second respect, as the example points to the need to adopt a long-term perspective on the development of ignorance. It took more than one generation to move from the tepid experiments of the 1920s to the feverish boom of the 1960s, as farmers and advisors adjusted to the gaping holes in the working knowledge only over time. Ignorance that results from social interaction calls for a view on the *longue durée*.

Finally, it might seem that environmental history calls for an amendment to the typologies that have become so prominent in the field of agnotology. The example shows that there is an empowering, liberating power of ignorance. After all, this is not the kind of ignorance that the tobacco companies bestowed on the general public: the parties that embraced the narrow knowledge base of corn monoculture were *directly implicated* in the production of ignorance. Knowledge production was a multi-actor enterprise in the case at hand, and scientists were far from defining the key parameters. Without the farmers asking for an easy, hassle-free way to produce feed for animals, there simply would not have been any production of ignorance. It showed that it was easier to do certain things if one did *not* inquire about certain issues—at least for a certain time and in a specific context, the brutal "race for survival" that ruled among farmers after 1945. As Moore and Tumin noted so correctly more than sixty years ago, "ignorance must be viewed not simply as a passive or dysfunctional condition, but as an active and often positive element in operating structures and relations."[43] There is a type of ignorance that makes the ignorant strong, in some respects, for some time, and environmental historians are well advised to inquire about similar situations more frequently—not least of all because this ignorance rarely seems to play out to the advantage of the environment.

# CHAPTER 4

# ENVIRONMENTALISTS ON BOTH SIDES

## ENACTMENTS IN THE CALIFORNIA RIGS-TO-REEFS DEBATE

### DOLLY JØRGENSEN

The choice that someone would make about whether to go with complete removal or partial removal, it's just not clear. It's obviously nice if you come up with something that's obviously the optimum decision in all circumstances, but in this case, the choice that someone would make or the preference that they would have depends on how they value the various kinds of impacts.

—Brock Bernstein, lead author of *Evaluating Alternatives for Decommissioning California's Offshore Oil and Gas Platforms*

IN HIS ORAL presentation in June 2010 of a report commissioned by the California Ocean Science Trust, Brock Bernstein highlighted a conundrum facing Californians interested in establishing a policy for the decommissioning of obsolete offshore oil and gas production platforms: not only did the individual characteristics of each platform affect which disposal option was most favorable, but value judgments played a vital role in the decision-making process. His presentation slide titled "Desired Option" said it all: "Choice of option depends heavily on preferences." If a person valued ecosystem integrity, strict legal compliance, clear ocean access, and limiting potential state liability, they would favor complete removal of the multistory steel structures currently standing off the coast of California. On the other hand, a person who valued reductions in air emissions, retaining existing biological communities, limiting costs, providing recreational fishing, and limiting impacts on water quality would favor converting the structures into fish habitat as artificial reefs.[1] Both sides could claim (and rightfully so) that they were doing what was best for the environment; yet deciding what was best was a value judgment.

Bernstein's astute observation about the role of values was revolutionary in the debate about what to do with California's offshore oil structures at the end of their lives. Before Bernstein's report, the controversy over rigs-to-reefs—the conversion of decommissioned offshore structures into artificial reefs to serve either as nature reserves to protect fish or as recreational fishing reefs—was framed as a question of acquiring better science. In October 2001, when Governor Gray Davis returned Senate Bill 1, which would have permitted rigs-to-reefs in the federal waters off the coast of California, to the California State Senate without his signature, his veto statement placed his decision squarely within the context of a scientific controversy: "There is no conclusive evidence that converted platforms enhance marine species or produce net benefits to the environment. . . . It is premature to establish a program until the environmental benefits of such conversions are widely accepted by the scientific and environmental communities."[2] It thus might seem natural to examine how science functioned in this environmental controversy, similar to the approach of Michael Egan in this volume. The two pieces previously written about the California rigs-to-reefs debate separate science from political/social factors, implying that one side in the issue was more scientifically grounded than the other.[3]

On the contrary, this chapter proposes that while both the proponents and the opponents of the plan mobilized scientific knowledge as a component of their arguments, it was not the root cause of the friction, nor could it provide a resolution. As a whole, the debate centered on different visions of nature, rather than on scientific knowledge. Scientific knowledge in this case reinforced competing, incompatible enactments of nature. Rather than argue about the validity of the science used by each side or the external motivations various actors had for getting involved, this chapter attempts to expose the two basic enactments of nature present in the debate. These enactments were based on a different set of values, all of which can be considered "environmentally friendly," yet those values contradicted each other.

## MULTIPLICITIES OF NATURE

Classic environmental histories of the environmentalist movement tell stories of pro-environmentalists fighting against anti-environmentalist interests that typically opt for economic gain over environmental preservation. But every history is not this simple. In the debate over whether or not former offshore oil installations should be converted into artificial reefs in California, both sides claimed to be doing what was best for nature. How can environmental history explain a case like this where both sides claim to be pro-environmental? Is it simply that one side is operating from "bad" science or has external motivations?

By applying the science and technology studies (STS) concept of enactment, this paper explores how both sides can be genuinely pro-environmental while advocating exactly the opposite approach. Enactment of nature, per the work of STS scholars John Law and Annemarie Mol, involves multiplicity; that is, there are different practices that make manifest different versions of nature and propose courses of action for dealing with it.[4] Rather than thinking about multiple viewpoints of one object, Law and Mol propose to see the knowledge of objects in practice as producing multiple versions of the object. This is not to say that nature does not exist, but only that we know it through the versions of nature we produce.

Mol's groundbreaking study of atheroscleroses of the legs revealed the various ways in which the disease was constructed by practices—ranging from technical medical practices to daily routines of patients. In her telling, the identity of atheroscleroses is "fragile and may differ between sites"; the disease has a complex present, not just a past construction.[5] In the conclusion of her study, Mol advocates further exploration "into the diverging and coexisting enactments of the good. Which goods are sought after, which bads fought?"[6]

This is the call that resonates most with this particular study. How did actors enact what would be beneficial for California's coastal nature or oppose that which they believed would be harmful? Like Mol's analysis of how a disease is created by practice, this case tries to uncover how California's offshore nature was enacted through practice. Through the way they understand the oil structures, sea life, and humans, the actors on both sides created seascapes.

In this case, two versions of Californian waters were enacted by the actors: one that incorporated manmade structures to increase fish stocks (proponents of rigs-to-reefs) and the other that returned the ocean floor to a pre-oil-development condition (opponents of rigs-to-reefs). The proponents wanted to keep the obsolete offshore oil structures as reefs, whereas the opponents insisted on removal. Both sides based their position on creating a better environment. Because the enactments of California's offshore nature called for opposite actions, a full-blown environmental controversy emerged.

Much work in STS has centered on how controversies are brought to closure. In the subfield of technology studies, Wiebe Bijker's work on the development of the bicycle exemplifies this approach. Bijker traced the ways in which various actor groups interpreted the safety bicycle and how design changes, as well as redefinition of the problem, eventually led to closure in the technological controversy.[7] The closure of scientific controversies has received much attention, particularly the role of politics and ethics in the making of science.[8]

Research on the closure of environmental controversies has followed this pattern. For example, in Leland Glenna's recent study of a controversy over supplying New York City with water from a far-reaching watershed, Glenna argues that differences in the actors' positions were due to varying ideas of ethics and justice; when New York City addressed the underlying competing theories of justice of the watershed residents, the conflict was resolved.[9] Although this may be a valuable approach, Glenna makes two assumptions that this chapter avoids: (1) that closure of environmental controversies is likely with the right negotiation and (2) that different viewpoints are based on differing ethical positions.

Against the first, this chapter follows the lead of Mol, who believes closure is in fact a rare event: "Differences aren't necessarily bridged; they may be kept open—with suitable hard work. They need not be overcome, be it by agreement or force, they may just keep going."[10] Even contradictory enactments may be stable—that is, they may not be easily changed because they are built on strong practices and beliefs. This can result in the continued existence of incompatible enactments, thus no closure.

As for the second, the rigs-to-reefs controversy will show that rather than a question of ethics—that there are philosophically more desirable positions—there are multiple enactments in play that are based on different value structures, each with environmental merit and yet incompatible with each other. As we will see in the course of this chapter, the positions of both the proponents and the opponents have a history, and both are based on scientific knowledge as well as personal values. The enactments—the way offshore rigs are understood as an object—depend on the value preferences. An investigation of these enactments is not simply a postmodern philosophical exercise; differing enactments can thwart closure of environmental controversies, an issue vital to environmental policy making and politics.

## THE NATURE OF RIGS-TO-REEFS

Offshore oil and gas structures have a life span based on the petroleum reserves under them. When the petroleum either runs out or becomes noneconomical to extract, the structures become obsolete and, under existing US federal and international rules, must be removed. But every structure need not be taken to shore. An alternative reuse of some structures as artificial reefs, called rigs-to-reefs, has been available under US law since 1985. The conversion process can involve toppling in place, partial removal of the upper structure while leaving the lower portions in place, or relocating a structure to a designated reef planning area. Most often the reef is made

Figure 4.1. Aerial view of the platforms Ellen and Elly, eight miles offshore of Long Beach, California. Photo by Bob Wohlers © 2012.

from the huge steel legs (called jackets) that support the working platform rather than any part of the platform or equipment that contacted petroleum directly. The idea is that the colonizing organisms like mussels, barnacles, and corals will attach themselves to the steel, creating an artificial reef. Small fish soon discover the reef, which offers a source of both food and protection from predators. Larger fish likewise are attracted to their food sources that inhabit the reef.

Obsolete offshore installations were one of the materials of opportunity listed in the National Artificial Reef Plan of 1985, opening the way for the development of state-level rigs-to-reefs programs.[11] The states of Louisiana and Texas developed artificial reef plans in compliance with the National Artificial Reef Plan in order to create state-level rigs-to-reefs programs.[12] These two states had the vast majority of the US offshore petroleum facilities in 1983: 4,056 out of the total of 4,094.[13] From the beginning of the programs to 2006, approximately 10 percent (238 total) of all obsolete structures in Texas and Louisiana were converted into reefs instead of being removed.[14] Under these two state programs, half of the money the company saved by choosing the artificial reef option is donated to the state along with the structure; the state then assumes all liability for the reef structure. In both cases, the rigs-to-reefs programs have received no public funding—the monetary donation pays completely for the program costs.

In the Gulf Coast, the concept was almost always touted as a "win-win" solution for industry, the environment, and fishing interests; even environmentalists supported the plans. Recreational fishermen and divers were particularly strong and vocal advocates of rigs-to-reefs. The discourses surrounding the Gulf Coast rigs-to-reefs programs stressed the beneficial habitat for large game fish such as red snapper, the betterment of a Gulf largely devoid of life naturally, and the potential coral growth on the structures.[15]

## WILL CALIFORNIA HAVE RIGS-TO-REEFS?

Twenty-eight oil and gas production platforms also stood off the coast of California when the National Artificial Reef Plan of 1985 allowed states to create rigs-to-reefs programs. But advocates of the program did not appear right away in the state. When California had its first regular offshore wells decommissioned in 1988, no mention was made of a rigs-to-reefs conversion option. Rigs-to-reefs finally entered the discussion in the early 1990s, when Chevron was planning the decommissioning of four platforms (Hazel, Heidi, Hilda, and Hope—often referred to as the 4 H platforms).

Chevron submitted a decommissioning proposal for the 4 H platforms in 1993. Chevron was one of the leading companies in Louisiana's artificial reef program, having converted seventeen platforms by 1994, so it was quite familiar with the rigs-to-reefs concept. According to Chevron representatives, when the idea of a rigs-to-reefs option came up in discussions in February 1993 with state regulatory agencies, it got "a chilly reception," and the agencies specifically requested that Chevron *not* include a rigs-to-reefs option in the decommissioning analysis.[16]

That might have been the end of the rigs-to-reefs discussion, except that the sportsfishing organization United Anglers of Southern California (UASC) became interested in keeping the structures as reefs to enhance recreational fishing. In early 1994, the group approached Chevron, suggesting that it cut the jackets thirty to fifty feet below the water line in order to preserve the existing marine habitat and relocate the removed sections as part of an artificial reef complex in Ventura County. UASC then approached the California Department of Fish and Game (DFG) with the idea, suggesting that DFG work with Chevron to convert the structures into reefs.

Because the decommissioning process was well under way, Lee Bafalon, a senior land representative for the Chevron USA production office in Ventura, was not pleased with what he considered "an 11th hour" proposal, believing that all the right permits could not be acquired within the planned schedule.[17] Although the DFG agreed to review UASC's plan, Chevron was proved right. The schedule was too tight, and the final permits were approved in December 1994. In late 1995, the platforms were removed to shore for scrap recycling as originally planned.

Enthusiasm for rigs-to-reefs was slow in developing. The lead agencies for offshore oil production in California, the US Mineral Management Service (MMS) and the State Land Commission (SLC), cosponsored a workshop in September 1997 on future offshore decommissioning. The workshop included industry representatives, scientists studying platforms as fish habitat, environmentalist groups, and regulatory agencies. The immediate result of the workshop was the establishment of an Interagency Decommissioning Working Group (IDWG), but this group ranked rigs-to-reefs as a low priority with no action required from agencies; "industry" was given the responsibility for the item.[18]

If rigs-to-reefs was to become a reality in California, industry and other interested parties like UASC would have to act. Industry took the initiative with California Senate Bill 2173 (1997–98), introduced on February 20, 1998. This bill, introduced by Senator Bruce McPherson with the backing of the oil industry lobby, would have established a rigs-to-reefs program, including a Marine Protection Endowment Fund to preserve and enhance marine fisheries and habitat with funds from reef donors.[19] When the bill came before the Senate, it failed to pass.[20]

A sitting member of the Natural Resources and Wildlife (NR&W) committee, Senator Dede Alpert (Thirty-ninth District, San Diego, a Democrat with a pro-environmental background), latched onto the Marine Protection Endowment Fund aspect of the rigs-to-reefs bill as a way to fund marine environmental programs and decided to introduce a bill in the next session with similar goals, SB 241 (1999–2000), "California Endowment for Marine Preservation."[21] The bill required owners and operators of decommissioned facilities that were permitted as artificial reefs to deposit money into the endowment, which would dole out the money to marine preservation projects.[22] The Santa Barbara–based Environmental Defense Center (EDC) and legal counsel Linda Krop immediately voiced their displeasure with the bill, and the senator agreed not to issue any public statements about the bill until it could be discussed by the legislature.[23]

Alpert commissioned a Select Scientific Advisory Committee on Decommissioning to review the current status of scientific knowledge about the marine ecosystem effects of various decommissioning options, including rigs-to-reefs. The committee issued its final report in November 2000. The report ended up being used primarily by the opposition as evidence of the lack of scientific certainty about the benefit of rigs-to-reefs, particularly the main conclusion that while "local, short-term effects" of the options could be estimated, the "longer-term regional effects" could not be predicted based on current scientific knowledge.[24]

SB 241 went through several contentious committee hearings and was passed by the Senate. However, the bill ended up being placed in the inactive file in August 2000 instead of going to the Assembly for a floor vote because of last-minute changes requested by Governor Gray Davis's office that might have resulted in a veto.[25]

Alpert took a second shot at the legislation, introducing SB 1 (2000–2001). This bill listed decommissioned platforms as potential artificial reef material but required an evaluation of any such material by the relevant agencies and had a much more elaborate endowment structure.[26] The text's length had drastically increased, from thirty-seven lines in McPherson's 1998 bill, to about three hundred in Alpert's 1999 version, to nearly nine hundred lines in SB 1!

The debate on SB 1 was hot and heavy in both the legislative chambers and among the public, with several hearings and significant media coverage. In spite of fierce opposition from environmental groups, trawler associations, agencies, and legislators from the Santa Barbara area, where most of the offshore facilities are located, the bill eventually passed both the Senate and the House. But when the bill was presented to the governor, he refused to sign it, and there were not enough votes to override the veto.

## ENACTING NATURES

In the hearings and public comments on rigs-to-reefs, the proponents and opponents were unwavering. Both sides claimed that they were proposing the environmentally friendly course of action. The same people appeared multiple times, and the arguments remained consistent over the course of the debate—from the beginning of the proposal to convert the four Chevron facilities into reefs through the 2001 hearings. The most vocal supporters were sportsfishing organizations, particularly UASC, backed by the oil industry and the fisheries scientist Milton Love. The opponents rallied around Linda Krop of EDC and trawling groups (who met regularly with EDC and other environmental groups in the Santa Barbara area as part of an anti-oil coalition).

While the scientific controversy cited by the governor in his 2001 veto statement plays into the various points of view, looking at the ways nature was envisioned through its component objects provides a much more fruitful way to make sense of the conflict. Differing values led to competing understandings of the components of the offshore ecosystem. Turning to the arguments used by both sides from 1994 to 2001, we can reconstruct the enactments of nature at work.

The following sections attempt to disentangle three elements of the enacted nature: technology (offshore oil platforms), environment (fish), and culture (humans). It is often hard to separate the three, but such an investigative strategy allows us to examine nature as a composite construction. The way that the groups understand all three elements and the roles they play in current and future configurations of nature contributes to the proponents' and opponents' enactments of the California seascape.

## REEF OR RUBBISH?

The technological artifact—the offshore oil structure itself—plays a central role in the construction of the two enactments. The structure could be viewed as either a nature-enhancing device or a nature-destroying one, depending on what the actors valued. How did the actors incorporate this artifact into their enactments?

The proponents argued that existing offshore oil structures were integrated, productive components of nature. When divers and sportsfishermen discussed the structures, they invariably pointed to the existence of marine ecosystems on the standing production structures. For example, the UASC position was that "in the decades since offshore oil production began in California, these oil rigs have developed a rich ecosystem around them. They are, in fact, living reefs of extraordinary biodiversity, harboring many species of marine life, some of which are threatened or protected species."[27] Oil industry representatives echoed the sentiment. In one interview, George Steinbach, decommissioning project manager for Chevron, said, "These platforms are already artificial reefs. There is significant marine life on and around these platforms."[28]

The supporters relied on experiences from the Gulf of Mexico programs, which had been active since 1986. In the media coverage, USAC and oil industry representatives often invoked the Gulf of Mexico's programs as a model for California.[29] Proponents picked up on discourse on Gulf of Mexico

Figure 4.2. A school of opaleye (*Girella nigricans*) swim around the bivalve-encrusted legs of a standing offshore platform in California waters. Photo by Bob Wohlers © 2012.

rigs-to-reefs that called the rigs oases of life: "Just as a spring in the middle of a desert provides the infrastructure for a biological community, a drilling platform performs the same function in the ocean."[30]

Images of underwater life around the platforms supported this view. In a television news report, CNN featured footage by undersea filmmaker David Brown showing colorful anemones and starfish on the platform legs with fish darting around. Brown commented during the broadcast, "I am always in awe of nature's ability to take our stuff, our mess, and create something beautiful from it. All those hard edges get softened and all that drabness turned into color and life."[31] Legislative committee members were treated to videos showing the abundance of creatures living on or around the standing platforms.[32] In fact, one trade journal article credited Alpert's introduction of SB 241 to her seeing underwater marine life from a Coalition for Enhanced Marine Resources (CEMR) submarine.[33] One of the key supporters of rigs-to-reefs, Daniel Frumkes of UASC, even contrasted the visual aesthetics of the structure above and below the water: "You look out there now and you see it above the water and it's an eyesore. But you look at it below the water and it's gorgeous."[34]

From this position, proponents argued that removing the rigs amounted to killing sea life: "Completely removing these reefs will necessarily kill millions of pounds of sea life."[35] According to one article, Love believed "it would be immoral to pull out the vibrant pink pillars and hang the millions of marine organisms attached to them out to dry."[36] Rather than destroy these existing ecosystems, advocates wanted to promote "preserving the ecosystems associated with the platforms."[37]

On the other side, for much of the local population of Santa Barbara, where almost all of California's offshore oil production takes place, oil platforms are understood as a source of pollution. The legacy of the Santa Barbara oil spill of 1969 looms large in the public perception of the structures. The 1969 spill is considered one of the catalysts of the US modern environmental movement and has remained a poignant event for southern Californians. As a *New York Times* article from 1985 began: "To many residents along the scenic coast here, the 20 or so oil rigs that rise out of the sea a few miles off shore are ugly reminders of a disaster—an oil spill that in 1969 blackened miles of ocean and beaches."[38]

Media coverage of the rigs-to-reefs controversy often invoked the spill, such as the voice-over for a video broadcast on the cable news channel CNN: "In 1969, for example, Platform A spilled nearly 4 million gallons of oil. At the time, it was the biggest U.S. oil spill in history and helped launch the modern environmental movement. Susan Rose, Santa Barbara County Supervisory Board: We don't look kindly to oil off our coast, and it certainly has done a lot of damage to us over the years. We'd like those rigs removed."[39] In the Senate floor debate on Alpert's first bill, the 1969 spill was invoked in clear terms by Senator Tom Hayden, the chair of the NR&W committee, who opposed the bill: "This is a bill that tests our institutional memory. I don't know how many were impacted by the oil spill in 1969 in Santa Barbara, the worst until the *Exxon Valdez* collision a few years later. But the, the impact of that was to put oil drilling off the coast in the national consciousness and to establish protections for the people of California in coastal communities."⁴⁰ In the various comments that invoke the 1969 spill, we see an association of oil structures with oil spills. The structures are viewed as undesirable because of these oil spills.

Because the structures were tainted by oil, the opposition considered the steel structures themselves current and future sources of pollution. As the Pacific Coast Federation of Fishermen's Associations put it: "The offshore oil platforms and their operations . . . have altered the natural environment and destroyed fishing grounds."⁴¹ The EDC invoked this position in a letter sent in 1999 to the NR&W committee, making statements such as "certain components of the platforms themselves contain toxic materials and substances that may corrode or leach into the environment."⁴² The idea was reiterated in a 2001 letter saying that rigs-to-reefs "will pollute the ocean due to corrosion of the platform structures as well as the leaching of toxic drilling muds and cuttings."⁴³ The platform materials that would be converted into reefs were called "debris" and "industrial junk," and actions to convert the materials into reefs were termed "ocean dumping."⁴⁴ According to Krop, "The materials are not compatible with marine life."⁴⁵ The Southern California Trawlers' Association stated in no uncertain terms: "It's our belief that offshore rigs are essentially junk with a purpose. Without the purpose, they're just junk."⁴⁶

The structures were also seen as an aesthetic blight on an otherwise beautiful coastal view. Linda Krop put it this way in one interview: "We always think of Santa Barbara as paradise. Yet a lot of visitors come here and one of the first things they say is, 'These things are ugly. How did you guys ever let them do that?'"⁴⁷ Several letters sent to the governor in reference in SB 1 cited aesthetic concerns with the final legislation, including Santa Barbara House representative Hannah-Beth Jackson, who called the rigs "unsightly."⁴⁸ The Sierra Club California representative complained that "most of the citizens who live onshore of offshore platforms definitely do not want to see lighted oil platforms along the California coast long after they have stopped producing crude oil. This is certainly not in line with long held expectations that when oil was no longer produced from a platform, that platform would no longer be seen from shore."⁴⁹

In the opposition enactment, the platforms were foreign substances that contaminated the ocean, as well as the pristine shoreline view. These technological artifacts did not belong in nature.

At a basic level, the role of the technological artifact—the oil platform— was polarized in this debate. One side understood the structures as foreign intruders, the other side as an assimilated part of the ecosystem. Both of these positions can be considered pro-environmental because the structures could legitimately be viewed in either way. If one valued nature as it was before the structures were added, then the opposition enactment made sense; if one preferred to focus on the biological habitats created by the structures, then the proponent enactment was favored. This first component of the enacted natures reveals how fundamentally different the proponent and opponent enactments were.

## FISH OR JUST FISHY?

The second component of the two enactments of nature is sea life. The actors all invoked fish, both around the platforms and in the coastal area in general. In line with their thinking about the rigs as part of the ecosystem, proponents interpreted the fish life swimming around each standing rig as valuable individuals, whereas the opponents turned to the issue of regional fish productivity.

For proponents, research on fish ecology around the platforms by Milton Love, a researcher at the University of California at Santa Barbara, added scientific weight to the anecdotal evidence about rigs serving as fish habitat. Love was regularly interviewed for newspaper articles and testified at the Senate committee meetings. His studies focused on rockfish populations and showed higher concentrations of rockfish, particularly the bocaccio species, around the platforms than on the natural rocky reefs in the Santa Barbara Channel. He argued that the platforms served both as a protective nursery for juveniles and as shelter for adults since they would be safe from trawlers' nets around the platforms.[50]

Scientific evidence of fish using the platforms was highlighted in correspondence with the Senate committee. For example, Dick Long, the president of the San Diego Oceans Foundation, wrote in support of SB 1, "The eggs, spores and larvae of the many plants and animals associated with an oil platform habitat move throughout the Southern California Bight and 'seed' areas far away. . . . The drilling platforms are home to several threatened species."[51] Alpert referred to these same arguments regularly in her support of the bill. For example, in a personal letter to Governor Davis asking him to sign the bill, she wrote, "Fishes currently using the platforms as habitat are also preserved, which in the case of declining species, like *bocaccio*, is significant."[52]

The prospect of removing the existing rigs meant removing a vibrant ecosystem and killing the local fish. Love was quoted as saying, "If someone said, 'Well, there's this rocky reef and we're just going to dynamite it, haul it away, and cut it up for scrap,' people would go nuts. And yet that's what's being done."[53]

The primary supporters of the rigs-to-reefs idea, however, showed that they were not necessarily concerned with increasing the numbers of just

Figure 4.3. A copper rockfish (*Sebastes caurinus*) in the reeflike area at the bottom of a standing California platform. Photo by Bob Wohlers © 2012.

any fish. Comments such as "Our idea is controversial, but we think it will benefit the fish. In the long run, it would provide more fish for the trawlers and the sportsman," by UASC president Tom Raftican, reveal the extent to which their vision of the sea was tied to species that are used by humans. [54] Jim Paul, the president of UASC, made this clear during the discussions with Chevron about the 4 H platforms: "They're such great fishing spots, we don't want to see them disappear."[55] In the Gulf of Mexico discussion, a similar emphasis on sport-catch species was also present.[56] Since the idea behind the sportsfishing groups' interest came from the Gulf experience, this should come as no surprise.

The proponents saw the local fish population around each rig as valuable, perhaps even more so than natural reefs. Because offshore installations boasted populations of fish and sea life desirable to recreational fishing and diving communities, they were seen as worthwhile habitats. The proponents therefore measured the positive effects of a potential rigs-to-reefs program by its preservation of these local pockets of fish life.

The opposition, on the other hand, did not recognize the individual fish living on and around the manmade reefs as valuable. Instead, they honed in on the idea of regional fishing stocks as the measure of what is good for nature.

The opposition regularly cited the conclusion of the Select Scientific Advisory Committee on Decommissioning report of 2000 that there was no conclusive evidence of the benefit of rigs-to-reefs to regional fish stocks in letters to the Senate committee and the governor.[57] The report also discussed the potential environmental benefit of local pockets of fish, but in the opposition enactment of nature, the local fish population was unimportant—regional biomass was the only thing that mattered.

Although supporters often pointed to the great number of fish swimming around standing rigs, the opposition questioned the ability of the fish to recognize good habitat. During a TV interview, Krop summarized this view: "There's no evidence that they [platforms] provide habitat. You may see fish out there, but you see sea gulls at a landfill. You know, you see birds on telephone poles."[58] This position was often bolstered by reference to the "production versus attraction" issue, a general scientific debate about the ability of some artificial reef habitats to actually produce more fish rather than simply attract fish away from alternative habitats. The Select Scientific Advisory Committee on Decommissioning report pointed out that scientists had reached no consensus about whether rigs-to-reefs attracted fish so they appeared more plentiful or actually increased their total numbers.[59] The opposition hooked on to this scientific difference of opinion and presented artificial reefs made from obsolete platforms as attractors, rather than producers, of fish. Zeke Grade of the Pacific Coast Federation of Fishermen's Associations argued, "People seem to think that with rigs you get instant fish, but we don't know if they're simply attracting fish from other places."[60]

In this view, artificial reefs had the potential to attract fish away from productive natural reefs, and if the artificial reefs were open to fishing, a net decrease in the regional fish population might result. According to the president of the Southern California Trawlers Association, "This creates the 'shooting gallery' effect—allowing sport anglers easy access to fish they would otherwise have to search for on hard-bottom, reefs or other high-relief habitats, thereby unduly stressing valued fish populations."[61] The opposition thus argued that if reefs were made, they should not be for fishing: "There should be no connection between fishing and artificial reefs. In fact, if artificial reefs are created, they should be delineated as preservation zones not subject to fishing pressures."[62] This ran directly counter to the main proponents, sportsfishermen who wanted to make accessible fishing reefs out of the rigs.

Instead of being concerned with individual fish residing at particular habitats, the opposition enactment of the California seascape focused on regional fish populations. Because scientific evidence could not show a regional ecological value of a rigs-to-reefs program, it should not be part of nature. Moreover, rigs-to-reefs was presented as a detriment to the regional ecology because of the ability of artificial reefs to attract fish away from other habitats.

The fish then play a different role in the two enactments. For the proponents, local fish populations evidenced both by science and by sight were worth preserving; for the opposition, only the total regional fish stock mattered. The enactment was different depending on what was valued—local versus regional ecosystems.

## Promises Kept or Promises Broken?

The enactment of nature in this case involved more than just the technological artifact (the platform) and the environment (the fish)—it also involved people. What should be the role of people in the marine environment? And more generally, what was environmentally friendly behavior?

The background of the actors tells us much about how they envisioned humans within nature. The driving proponent groups were sportsfishermen, with the backing of recreational diving groups. These same groups had led the charge for rigs-to-reefs in the Gulf of Mexico.[63] For these groups, humans participated actively in nature, coming face-to-face with sea life through fishing and diving.

The supporters brought up the ability of humans to modify nature, both for bad and for good, and stressed that humans had an obligation to right past wrongs. According to Frumkes, "We have overexploited our resources. If man has done the negative and we have a way to kind of compensate, I don't think that's inappropriate. We can do that by increasing the amount of good places for marine life to live."[64] A diving group that wrote to support the bill argued that the diver's view of the underwater world allows them to "see the damage that may result from carelessness and neglect," and artificial reefs helped mitigate those negative effects.[65] In these statements, humans are made responsible for counteracting previous environmental damage, and the rigs-to-reefs program provides an avenue for doing that.

The leading protestors were environmentalists from the local Santa Barbara area. For them, human activity should be separated from nature— nature needs protection. They say that politics makes strange bedfellows, and such was the case here, as environmentalists joined forces with trawlers. The trawlers did not take the same "hands-off" approach as the environmentalists did—they in fact advocated that they should have full access to the sea floor—but for the trawling interest groups, nature had no room for the oil industry, and this made them convenient partners for the environmentalists.

Throughout the debate, the opposition cited the current requirement that the oil companies remove all objects from the sea floor during the decommissioning process. Both environmentalists and trawlers believed that any change in the legislation would amount to "corporate giants wriggling off this hook."[66] When talking about the existing requirements, the trawlers' association often spoke in terms of "promises" such as "We were promised that we'd gain it back when production from offshore platforms was exhausted. We expect the federal government and the oil companies to keep that promise";[67] and "When they put these platforms in, they promised to remove them completely. Now, they are hoping to sell everybody on a bunch of junk science for not taking them out."[68] In trawlers' opinion, the oil companies thus had an obligation to take the structures out, regardless of the potential benefit of leaving them. The oil industry was considered an interloper in California's nature.

For the proponents, people were responsible for the vibrant local habitat created by the offshore oil platform—they had created it, and they had an obligation to retain it. For the opponents, the oil industry had made a promise to return the ocean floor to its preindustrial state, and they should stick by that promise because the rigs are sources of pollution and do not have regional effects on fish populations. Both of the positions are justifiable, but they are based on different environmental values. The problem is that they are mutually exclusive positions—you cannot both take out the structures and leave them.

## THE DEBATE GOES ON

The development of mutually exclusive enactments of nature in this case led to an unresolved controversy. Even after the defeat of SB 1 in 2001, several proponent organizations, including UASC and California Artificial Reef Enhancement (CARE), a nonprofit organization supported by Chevron, continued to have media events, including an underwater platform tour in 2003 and a rigs-to-reefs conference in March 2007. Newspaper articles regularly appeared on the topic, as special interest groups continued to press for legislation. The 2001 legislative defeat did not close the controversy.

In February 2010, Assembly Bill 2503 (2009–10), "Ocean Resources: Artificial Reefs," was introduced in the California Assembly. This act, like its predecessor SB 1, allowed the conversion of rigs into reefs under appropriate conditions (i.e., permitting and environmental benefit analyses) and established a fund to handle the money contributed by the oil companies. The main difference was that a portion of the money is due up front when the application is made (rather than when decommissioning is complete), which may have provided an incentive to legislators in these troubled financial times. The bill was passed unanimously by the Assembly on June 1, passed the Senate, and was signed into law by Governor Arnold Schwarzenegger.

Between 2001 and 2010, a plethora of scientific articles were published on California's platform ecologies to address the lacunas in knowledge that were identified in the 2000 Select Committee report.[69] Yet in spite of ten years of work and the publication of "better" science, the support and opposition actors used exactly the same arguments for and against the 2010 bill as they had a decade before. In fact, the wording is almost identical in many cases.[70] This reveals the conviction of the actors in their enactments, which correspond to their environmental values. Science only bolstered arguments on both sides; it did not make them.

When Brock Bernstein presented his committee's report about offshore installation decommissioning in summer 2010, he admitted that the actor groups held different environmental values and that these led to different opinions on rigs-to-reefs policy. In contrast to Governor Davis's 2001 veto statement, which called for no action until consensus within the environmental and scientific communities was reached, Bernstein's research recognized that such a consensus was not possible. As we have seen, the sides in the debate constructed enactments of the California seascape through understandings of technology, nature, and humans that called for opposing actions. And nothing in the nine years since SB 1's defeat had changed those enactments.

Even though rigs-to-reefs is now allowable under the law in California, the opposition may not consider the issue closed because it runs counter to their enactment of nature.[71] Since each application for a rig conversion requires a public permitting process, the opposition may still be successful at promoting their enactment and requiring removal of the structures.

When looking at environmental histories, particularly histories of policy development, attention must be given to the ways in which the actors understand nature, especially how they enact it through their practices, based on different value sets. In this case, the concept of enactment allows us to see why California's rigs-to-reefs debate developed the way it did and why a middle ground was not found. Such an enactment-based analysis might also be extended to examine why other controversies have been successfully closed when the enactments were either compatible or values shifted.

# CHAPTER 5

# THE BACKBONE OF EVERYDAY ENVIRONMENTALISM

## CULTURAL SCRIPTING AND TECHNOLOGICAL SYSTEMS

### FINN ARNE JØRGENSEN

THE ICONIC FACE of Norwegian recycling is a small round hole in the wall. More than a billion beverage containers pass through this hole every year, bringing vast amounts of glass, aluminum, and plastic back into the recycling loop. Bottles and cans are made from raw material and are then sold—or leased—to bottlers, who fill them with beverages and send them through distributors to grocery stores. From the stores they end up in shopping bags that go to private homes where the beverages are consumed. This is where the recycling loop comes to a critical point; the empty bottle must return to the appropriate facility in order to get recycled or reused.

The Norwegian beverage market represents one of the highest-performing recycling systems in the world. An overwhelming majority of all beverage containers in Norway are reused or recycled, remaining in the loop. Instead of throwing the bottle in the trash, most Norwegians bring the empty bottles back to a grocery store, where they feed the bottles through the hole in the wall. From there, the beverage containers are sorted and transported either to a bottler, where the bottle gets washed and reused, or to a recycling facility, where aluminum is melted down to aluminum ingots, glass is crushed, and plastic is shredded for use in other products. Large amounts of energy and raw materials are conserved through the reuse and recycling of beverage containers.

In the case of container recycling in Norway, the entire product loop hinges on the hole in the wall, or more precisely the Reverse Vending Machine (RVM). The RVM is a high-tech machine that receives, identifies, and processes empty beverage containers and is most often placed in grocery stores where consumers can return their bottles and cans when shopping for groceries. The users of the RVM have delegated many types of work to this machine. Most obvious are the physical tasks that the machine performs in handling and counting bottles. More subtle are the interactions that take place when the RVM meets the consumer-user returning empty bottles.

How can we understand the success of this environmental system? Scholars have long asked *why* we (or why we don't) recycle.[1] Beverage container recycling is one of the most widespread forms of consumer recycling in the Western world, and Norway has a particularly high recycling rate of more than 90 percent. Recent scholarship on waste and consumption has demonstrated that general recycling motivations have changed significantly over time.[2] Originally mostly a question of resource reclamation, consumer recycling has today become an activity with a markedly green tint, a way of protecting the environment and conserving resources.[3] While decidedly a powerful narrative, "greenness" is a relatively new way of interpreting recycling. Focusing solely on the symbolical aspects of recycling as a modern environmentalist activity is also somewhat limiting, since the material movements and transformations that take place are equally critical.

What I want to do in this chapter is instead look at *how* Norwegians recycle and how they know what is the appropriate action in a recycling situation. If we are to understand recycling as an environmentalist activity, we need to examine the innermost workings of the technical systems that have been set up to support and require recycling and how these systems have come to be in a historical context. Some of these systems technically have nothing to do with environmentalism, yet they are critical to the efficacy of environmentalist actions. In other words, environmentalist actions and technological systems are intimately connected, but the connection between these cannot be taken for granted.

This chapter explores the shifting relationships among consumers, containers, and environmentalism at two critical points in the history of Norwegian beverage container recycling. At the first point, the industry-run beverage container return system came under pressure from disposable containers in the 1960s. Strong resistance against littering led to a new, state-mandated deposit system based on refillable glass bottles. The second point came in the 1990s, when this deeply culturally engrained system was rearranged to accommodate a parallel recycling system for disposable aluminum and plastic containers. The events and controversies at both of these points contributed to creating the highly successful Norwegian infrastructure for everyday environmentalism.

This chapter considers the ways in which consumers are enrolled into the system as recyclers and how technologies give consumers conceptual tools to interpret their own actions as environmentally friendly. How does the RVM as a technological artifact transport and transform meaning between its different users? In this perspective, the RVM not only performs physical tasks for different users; it also becomes a mediator that connects concrete practical activities with values and ideologies, local actions with global resource chains, and different actor groups that may or may not have common interests.

## EVERYDAY ENVIRONMENTALISM: CONVENIENT AND UNCONTROVERSIAL

At certain times and places, some environmentalist actions can seem so culturally engrained as to have become part of the social fabric. In Norway, beverage container recycling is one such action.[4] In 2009, consumers returned 305 million cans and 91 million disposable PET bottles, providing 5,500 tons of aluminum for new cans and 3,700 tons of plastic.[5] In total, 99.2 percent of the reusable glass and plastic bottles, 92 percent of aluminum cans, and 90 percent of disposable PET plastic bottles were returned.[6] These are extremely high numbers—in comparison, Americans recycled 27.8 percent of glass bottles, 45 percent of aluminum cans, and 23.5 percent of disposable PET bottles in 2006.[7]

A large part of the existing environmental history literature deals with the making of modern environmentalism.[8] By looking at the rise of the environmental movement, activist organizations, and their influence on policy and society, environmental historians have attempted to grasp the emergence of a new social contract between humans and nature. Many classic works have focused on environmentalism as a social movement, as a philosophy, or as a value statement, but there have been all too few studies of everyday, mundane environmentalist actions. A certain tension seems to appear when scholars and others look at the concrete manifestations of consumer environmentalism.

However, the most widespread environmentalist action today is consumer recycling of packaging, including paper, cardboard, plastics, bottles, and cans. The packaging leftovers of the billions of consumer products we purchase every day would end up in landfills or as litter if consumers did not return them to recycling points. I call these actions *everyday environmentalism,* a form of environmentalism that runs shallow and wide. The actions of everyday environmentalists are small and in themselves insignificant; source separation of household waste, occasionally biking to work instead of driving, using low-energy light bulbs, purchasing carbon credits when flying somewhere, and so on. Everyday environmentalism does not require drastic sacrifices and lifestyle changes. Rather, it is a form of environmentalism that is compatible with consumer society: convenient, comfortable, and uncontroversial.

Yet I would argue that we should not underestimate the significance of consumer recycling. Recycling paper and returning empty beverage containers are two of the most pervasive "green" consumer activities in the Western world, but in the face of global climate change and a rampant consumer society, many have argued that these are insignificant and next to useless activities compared to bilateral agreements, global emission reductions, and so on.[9] It is true that these isolated actions are insignificant. At the same time, they can take place in large technological systems that harness and coordinate the aggregated effects of consumer environmentalism. The little things add up. However, environmental history as a discipline is poorly equipped to properly evaluate such forms of faceless and anonymous environmentalism, without great heroes and institutional archives.[10] In cases such as these, approaches from science and technology studies (STS) examinations of knowledge, values, and action can be a valuable contribution to an environmental history of recycling.

## SYSTEMS, SCRIPTS, AND ENROLLMENT

Scholars in the diverse and expansive STS field have long been concerned with how knowledge, values, interests, and power are embedded into machines and technology and how these machines interact with society.[11] STS provides us with a set of tools that can help in studying the flow of functions and meanings in this recycling process. In particular, the concepts of *large technological systems, scripts,* and *enrollment* can help us explore the largely anonymous environmentalism of the masses. *Large technological systems* direct our focus toward the technical infrastructure that makes certain actions possible and effective, while *scripts* help tie together the systems and consumer actions and values through *enrolling* users in the technological system.

It is precisely the question of what happens when delegating work and responsibilities to machines that spurred Bruno Latour's classic and highly influential article "Where Are the Missing Masses? The Sociology of a Few Mundane Artifacts," where he posits that mundane, anonymous technologies are the missing pieces that keep society together.[12] These technologies are not simply passive bystanders while social interactions take place between a strictly human set of actors; they are the very things that make the social interactions possible, as well as actors in themselves. Aligning myself with Latour's central point, I will argue that the extremely high beverage container recycling rates in Norway cannot be explained solely from moral and ethical values—in other words, environmentalism as a philosophical outlook—among Norwegian consumers. Rather, this everyday environmentalism exists within the configurations of technologies, organizational forms, materials, and users that channel and enable it.

The concept of scripts gives us an analytical entry point to understanding the interactions that take place around beverage container recycling, while constantly reminding us that we can't meaningfully separate the symbolic and the utilitarian content of any given technology. In a widely cited article, sociologist Madeleine Akrich defines scripts as "devices installed by designers to control the moral behavior of their users."[13] In other words, scripts become a kind of user instructions embedded in the technology itself or in the social packaging of the product.[14]

In order to understand evolving historical configurations, we need to consider both the design of the RVM and the relationship between its many designers and users. Scripts are a tool for identifying and analyzing how designers "define actors with specific tastes, competencies, motives, aspirations, political prejudices and the rest," as well as designers' assumptions "that morality, technology, science and economy will evolve in particular ways."[15] Design historian Kjetil Fallan argues that "the inscription of meaning in an artifact is by no means limited to its 'technical content'" and that script analysis thus can be a "highly valuable tool in the quest for a better understanding of how a product's utilitarian functions, aesthetic expressions, social meanings, and cultural identities are constructed."[16] The instructions given by scripts perform a variety of tasks and may "measure behavior, place it in a hierarchy, control it, express the face of submission, and distribute causal stories and sanctions."[17]

The effectiveness of scripts cannot of course be immediately decoded by looking at a particular artifact, particularly because users are not passive recipients of the intentions and imperatives that designers aim to embed in products. The function and meaning of any given product are not complete when it leaves the hands of the designer. In the words of Latour, "the fate of facts and machines is in later users' hands."[18] Scripts can be misunderstood, ignored, adapted, or subverted through practical use, over extended periods of time.[19] Historical studies are thus well suited to analyzing the creation of scripts and the investment of meaning and functionality into technological products. Environmental historians, in particular those concerned with consumer culture, are well advised to carefully consider the complex and changing relationships among designers, consumers, technologies, and commodified products on one side, and environments, natures, and our ideas and values about nature on the other.

## "WHAT ARE WE SUPPOSED TO DO WITH IT?" BOTTLE-CENTERED SCRIPTS IN THE 1960S AND 1970S

The history of Norwegian beverage container recycling provides us with considerable insight into the complex and shifting relationships among various actors, technological systems, and the ideas and practices of environmentalism. STS concepts like scripts, systems, and enrollment can help us tease out the critical components of these configurations. Brewers originally created the Norwegian bottle return system as an infrastructural and economically motivated tool.[20] As early as 1902, consumers had to pay a small deposit on each bottle, which was refunded when they returned the bottle to the point of purchase.[21] Bottles were expensive to purchase, so it made economic sense for the brewers to rent out the bottles to consumers rather than sell them along with the beverage. Such setups were common all over the Western world. In some cases, such as the pre-1930s soda bottles embossed with "Property of Coca-Cola Bottling Co.," this relationship among bottlers, bottles, and consumers was physically and symbolically inscribed on the bottle itself. Many bottlers also embossed the name of the place of bottling on the glass, further inscribing the bottle into a particular local context.

During the first half of the twentieth century, bottle return habits became second nature to Norwegians. Almost all glass bottles came back to the bottlers and could be washed and reused twenty to thirty times or more before they were retired. The success of this arrangement and the affordances of the reusable glass bottle—a thick-walled and relatively heavy 35cl or 70cl container, short and brown for beer and long and transparent for soda (which also came in small 20cl bottles)—shaped the entire Norwegian beverage industry in ways that still influence the market. The sizes had been regulated by law since 1912 and allowed different bottlers to reuse each other's bottles.[22] The high price of new bottles made it rational for them to have a cooperative relationship.

Grocery stores served as both the point of purchase and the return point. As such, they became an important interface between bottlers and consumers. Bottle returns became intimately connected with grocery purchases, practically and symbolically tying the actions of purchasing and returning together. The deposit was a simple mechanism that encouraged consumers to return the bottle. We can classify it as an *economic script,* using financial incentives to motivate consumers to return their bottles. A consumer could not choose whether or not to accept this script—in order to buy a bottled beverage, the deposit had to be paid. Of course, the consumer could choose not to return the bottle after use but would then face the economic loss of the deposit value.

While the deposit was a powerful script, it operated within a larger system. Entire theses have been written about economic incentives and deposit systems, but they tend to disregard the system context necessary to make bottle returns work in practice.[23] The bottle return system as it existed in the 1960s depended on manual labor and control of bottles upon return. While labor intensive, the system functioned fairly well, with high return rates and a general acceptance of this way to handle bottles. Consumers knew what to do with empty bottles, and they knew what happened to the bottles afterward. The flow between consumer and bottler was relatively transparent; bottles even carried physical scuff marks—the ring of white scratches around the thickest part of the bottle, where the bottles touch each other as they go down the conveyor belts in the bottling facility—that grew more pronounced the more times they had been washed and reused. The entire reuse system had been designed around the standard bottle types and deposits as an economic script that motivated consumers to return bottles.

Two new types of beverage containers put the Norwegian beverage container system under considerable pressure in the 1960s, threatening to unravel the configuration that made bottle-centered scripts so efficient. First, the introduction of the throwaway bottle on the Norwegian market in the early 1960s threatened to destabilize the entire system. Such "no deposit, no return" bottles had become very common in the United States in the 1950s but had not yet found a place in Norway. For instance, a 1956 book described disposable containers as bottles "that are sold together with the beer," a novel idea in the Norwegian mind-set. The book conceded that canned beer had potential but required such large sales that it would probably never be profitable in Norway.[24] Motivated by potential economic gains, a few of the larger breweries wanted to begin using disposable bottles. Since these bottles did not have to be returned to the bottling facility for cleaning and refilling, they allowed the bottlers to set their sights on a national, rather than a regional market.

Some bottlers began using the 35cl "stubby" bottle, a short and thin-walled glass beer bottle designed to be thrown away rather than refilled, in 1960. Since the stubby did not carry a deposit, consumers had no economic incentive to return it to the grocers. Nor did the grocers accept such bottles that the customers tried to return, since they received no compensation from the bottlers for handling them. After use, the bottle became waste, something that had to be disposed of, completely disconnected from the established web of relations of bottlers, bottles, and consumers. In a domestic setting, consumers could throw it out with the household waste, though city sanitation departments generally did not appreciate fragile glass containers mixed in with regular trash. But since bottled beverages increasingly became tied to a mobile leisure culture, much consumption took place on the road, in public spaces, and in nature, where trash containers were not always at hand. As a result, bottles—whole or broken—began to litter cities, beaches, roadsides, and forests.[25]

While many Norwegians obviously embraced the new disposable containers, even more pushed back at the changing bottle scripts. The stubby bottle attracted public ridicule in newspapers and in popular culture. For example, comedian Carsten Byhring made fun of the fact that Norwegians simply did not know what to do with nonreturnable bottles in a popular show in the fall of 1961. Posing as a hobo, he lamented the new disconnect between himself and the bottle flow; no longer part of the interdependence among consumers, stores, and breweries, he was now all alone in the world, without any responsibility for even the smallest bottle.[26] Angry consumers soon demanded a stop to the disposable bottle, mostly due to the littering problem. The Norwegian government could not flat out ban these containers under the current laws but managed to negotiate a voluntary agreement among the bottlers not to use disposable containers in 1965.[27] The stubby bottle soon disappeared from the Norwegian market, but it had become clear to all the involved actors that the existing beverage container system and the internalized container return habits it depended upon could not be taken for granted. The bottle-centered return script lost its efficacy with the devaluation of the physical container.

The grocers had somewhat different interests than bottlers and consumers did. They had not minded the disposable bottle since it meant less work for them. After the agreement to keep disposable containers out of the market, though, they realized that they were stuck with receiving empty bottles in their stores. The second new bottle that appeared in the 1960s—the one-liter glass soda bottle—only exacerbated their bottle problem. Thanks to this innovation, made possible by the introduction of the screw cap, consumers could now buy large bottles instead of small single-serving bottles, since soda would no longer go flat between servings. Bottlers and consumers liked these bottles, which carried a deposit and could be returned for refilling like any other bottle. For grocers, though, the thick-glassed bottles were heavy and unwieldy.[28] The one-liter bottle fit in the regulatory system, but not in the physical system that handled the material bottle flow.

The grocers' approach to their particular bottle problem became key to the technological rearrangement of the bottle return system that took place in the 1970s. While a practical arrangement for consumers, grocers found the practicalities of dealing with empty bottles challenging and had begun seeking technologies that could help them in as early as the 1950s. This started a systematic, technologically mediated reconfiguration of the Norwegian beverage container recycling system. Several mechanical RVMs existed at this time, both from the Swedish company Wicanders and from the Norwegian company Arthur Tveitan AS. These machines could receive bottles, count them through simple mechanical sensors, and give consumers a token for each bottle, but since these machines were custom built around the physical size of the bottles, they could not handle the new one-liter bottle. They could also easily be tricked by nondeposit containers of the same height, such as ketchup bottles.

A new and improved RVM launched in 1972 aimed to solve these specific problems. In 1971, a group of grocers commissioned the new RVM from two brothers: Petter Planke, an experienced salesman with many contacts in the grocery industry, and Tore Planke, an engineer with a background in cybernetics and mechanical engineering. Together they developed the Tomra I, a high-tech reprogrammable RVM that relied on optical bottle recognition with photocells. I will not go into a detailed origin story of the RVM, its inventors, and its relation to other beverage container return technologies.[29] For the purpose of this chapter, it is sufficient that we have a clear idea of what the machine does and how it does it.[30]

When consumers met the prototype RVM, they saw a metal plate on the wall, with a bottle-sized hole in it, large enough to handle one-liter bottles. They could place the bottle in a small box in front of the hole and then press a button to start a conveyor belt that transported the bottle into the machine.[31] The final version of the machine used an automatic sensor to start the conveyor belt when a bottle was inserted—the button confused many users. The RVM would register the deposit value and print the total on a small paper slip. The bottle ended up on a storage table, where the grocer could move the bottles into crates at a convenient time. This setup was much simpler than the older machines, where each bottle type had a separate hole and the consumer received a token for each bottle. Still, a one-page user instruction had to be posted on the wall to help consumers operate the machine, as we can see in figure 5.1. One of the things that the directions instructed consumers to observe was that the bottles had to be clean and empty, though in fact the first generation of the machine could not tell the difference.

The first version of the Tomra RVM accomplished two main tasks. First, it simplified grocers' work significantly and provided them with better control of the bottle flow through their stores. Second, by making the actual bottle return more convenient, it removed a major barrier to technology-centered consumer returns. The old machines created much frustration when they couldn't handle the new bottles; the new ones could easily handle all the bottle types on the Norwegian market.

Tomra's new RVM soon won a central place in people's everyday routines. One no longer simply returned the empty bottles; one returned them through the RVM. A simple *practical script* of feeding the bottle through the hole in the wall was thus established. In this process, the personal exchange relationship between the grocer and the consumer changed, becoming one

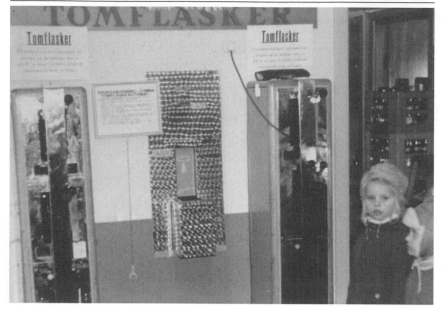

Figure 5.1. The prototype Tomra I in the middle and the old Tveitan bottle machine on both sides. The openings in the Tveitan machine have been blocked, as the machine is no longer in use. Image courtesy of Tore Planke.

between the machine and the consumer instead. This transition was almost frictionless, since the RVM did not add anything fundamentally new to the relationship. Instead, it was a clear case of delegation, where the machine simply took over some of the grocer's physical labor.

Part of the success of the machine comes from the inventor's expressed intention of making recycling so convenient "that there was no resistance."[32] By having only one opening for all bottles and using one receipt for all returned bottles, Tore Planke inscribed into the machine a vision of recycling as an act that required convenience rather than ideological fervor as the primary motivator. The success also tied into the establishment of a state-mandated bottle bill in 1974, requiring bottles to carry either a deposit or a large tax.[33] After the failure of the bottle-centered deposit script, the Norwegian government saw this as the best way to maintain high return rates. The previously mentioned littering discussion played a central role in the making of the bottle bill, but protectionist concerns in the beverage industry were equally important. By requiring all bottles to be part of a deposit-refund system, Norway in effect closed its borders to mass imports of foreign beverages. No matter the motivation, the bottle bill firmly secured the deposit as an economic script.

State-mandated bottle deposits and high-tech RVMs thus became a new configuration to regulate and manage the flow of bottles in Norway. Bottle handling became automatic and standardized, rather than manual and dependent on increasingly fragile cooperation from the bottlers. Manual bottle handling could only survive for so long in the face of rising beverage consumption, higher demands for efficiency, and international competition. Through economic and practical scripts, the reusable bottle entrenched its position as the morally and materially "correct" bottle for the Norwegian market, barring disposable and nondeposit containers from the market.

## "RETURN EVERYTHING—ALWAYS!" SYSTEM SCRIPTS AND UBIQUITOUS BEVERAGE CONTAINER RETURNS IN THE 1990S

The 1990s saw a considerable reconfiguration of the Norwegian beverage container system, primarily to accommodate disposable containers. As a result, the bottles themselves became less important as carriers than as scripts. Instead, an RVM-centered return system directed attention to consumer actions and to the larger social and environmental context of recycling. In the twenty years after Tomra's RVM first gained a foothold in the Norwegian market, RVM-centered bottle returns became ubiquitous in Norway. Internationally, Tomra had some ups and downs, but the company's position was virtually unchallenged at home. Almost every Norwegian grocery store had a Tomra RVM by the early 1990s.[34] The RVM had an equally strong position in the everyday routines of consumers.

A new generation of environmentally conscious consumers embraced bottle recycling as a way of protecting the environment in the late 1980s. Framing bottle returns as recycling for environmental reasons, not just for economic reasons, became a new script added to the existing configuration. Bottle recycling became a way of "thinking globally and acting locally," the credo of the Norwegian environmental movement in the 1980s. The success of the refillable bottle, however, created new problems for the bottlers when disposable containers again came under consideration in the early 1990s.

The beverage industry had a significant interest in using lightweight disposable PET plastic bottles and aluminum cans, and they had strong support from the grocers when pushing for disposables in the late 1980s.[35] However, the success of the existing reusable glass bottle system meant that such disposable containers had to change their image as being environmentally harmful. Norwegians generally considered aluminum cans to be somewhat exotic containers (often purchased on vacation trips to Sweden, which implemented a deposit system for aluminum cans in 1984) that had no proper place in the Norwegian bottle flow. The recent upsurge in environmental concerns meant that disposable containers faced even tougher resistance. Many feared that an aluminum recycling system—especially a successful one—would threaten the existing bottle reuse system.

A consortium among aluminum producers, grocers, bottlers, and Tomra designed a new recycling system for disposable containers at the beginning of the 1990s.[36] Tore Planke—the inventor of Tomra's RVM—was the key architect behind the new system, dubbed Resirk, which plays on the Norwegian word for recycle: *resirkulering*. He designed it around a new generation of RVMs with online capabilities to provide more control and prevent fraud. The Norwegian parliament finally allowed the Resirk system to start up in 1999 after a drawn-out legal controversy. In this system, all disposable containers carry a differential tax to be adjusted annually based on recycling rates. In other words, the higher the recycling rate, the more profitable disposable containers could be. Foreign beverage containers that were not part of the Resirk system would have to pay the full tax. Resirk then had a huge incentive to make the recycling system as effective as possible to benefit its members. Resirk was set up as a nonprofit organization to handle the administration of the new deposit system. The deposit would follow the beverage container through the entire life cycle of the container, from producer to bottler to grocer to consumer and back again. In this way, the producer would lose money unless the container was returned and recycled. Resirk paid out a small handling fee to grocers, as well as to the recycling depot as compensation for handling the empty containers.

The consortium set up the new system as complementary to, not competing with, the old glass bottle return system.[37] Resirk aimed to minimize the public distinction between different container types. Tomra specifically designed a machine that could handle all containers, whether plastic, aluminum, or glass. The Tomra 600 launched for Tomra's official twenty-fifth anniversary in 1997—before Resirk got final approval, but it was clearly designed to be part of the Resirk plan. By using the same hole in the wall for all containers, Resirk emphasized the similarity of the containers, as well as making recycling more convenient for consumers. Resirk also subsidized grocery stores that wanted to upgrade their RVMs to ensure that as many stores as possible adopted the new technology. By choosing to use one machine to handle all beverage container types, even ones Resirk was not responsible for, the consortium blurred the distinction between old and new containers. The established glass bottle return script was strong, so Resirk wanted to tap into existing recycling habits.

Resirk put considerable effort into encouraging beverage container returns through the RVM. They turned to advertising as a new form of beverage container script. Through a series of high-profile magazine and television ads, Resirk developed an effective rhetorical strategy connecting the local, everyday action of beverage container returns to the global environment. I have chosen to call these *epistemological scripts,* as they deal with the transfer of knowledge and meaning through the RVM and the system, conveying a specific message about the connections among the beverage container, the consumer's actions, and the global environment.

Their ads consistently feature a hole in the wall—the RVM—through which the consumer returns a container. Through this visual strategy, the machine became interwoven with the recycling activity. Resirk's ads thus reiterated the central position of the RVM in the system. Most of the illustrations in the ads featured a picture of a person about to put a beverage container into the RVM—seen from inside the machine itself.

The pinnacle of this advertising was perhaps a 2006 ad/music video featuring the rapper Ravi and Norway's grand old man Odd Børretzen. Through a catchy song, the ad gives listeners a step-by-step guide to how to recycle:

> One—you get out the bottles and the cans.
> Two—not just some, but they all go in the bag.
> Three—you find the deposit automat.
> Four—you take a little breather before the finale.
> Five—then you take the bottle or the can, you'll put it into the wall and
> then it will return.

As a sparkling new can magically appears in the blond cashier's hand, Børretzen asks, "How does it feel then?" and Ravi replies, "It feels so good!" The song thus provides not only clear instructions for how to recycle but also for how to feel about it; recycling is a simple, convenient, and feel-good activity. Returned cans get a new life instead of being wasted.

A new campaign appeared in print in magazines and newspapers all across the country in 2007 and stated that "the globe has a fever!" Released during the Easter holidays—when all good Norwegians are supposed to go skiing—the ad features images of climate change, with ski slopes turned into barren deserts. The ad intones that "we all need to contribute to reduce energy use and emissions. Recycling beverage containers is a small, but important effort for a big cause. Recycling cans and bottles is very environmentally friendly compared to making new aluminum or plastic. Do something small for something big: return everything during Easter!"[38] Having firmly established the *how* of recycling by tapping into the preexisting glass bottle return scripts and through several high-profile advertising campaigns, Resirk now also aimed to say *why* consumers should recycle. With climate change on top of the global agenda, it made sense for Resirk to use it as the basis for a campaign. The RVM could connect individual recycling activities—every single bottle—to global environmental change. The critical point here is that the *why* and the *how* were intimately connected: without the technical infrastructure of RVMs and the willingness of consumers to return their bottles through it—or, as Carsten Byhring's abandoned 1960s hobo might have put it, to *entrust* their empty bottles to a technological system—bottle returns would have been an atomized and individual activity where all the little bits did not add up to an effective environmental system.

## SCRIPTING ENVIRONMENTALISM: AWARENESS VERSUS ACTIONS

What do advertisement campaigns like this say about the responsibility of consumers and producers? The many-layered scripts that appear in the Resirk ads can best be compared to the even more iconic "Crying Indian" public service advertisement, launched by Keep America Beautiful (KAB) on Earth Day 1971.[39] In the famous one-minute video, the actor Iron Eyes Cody paddles down a river in full Native American garb, past an increasing flow of litter. Like Resirk, the Crying Indian attacks littering, but KAB placed the blame solely on consumers: "People start pollution. People can stop it." While the authenticity of this environmental message has been heavily debated, one cannot doubt the powerful role the ad played in raising environmental awareness among the general American public. A close eye on scripting of environmental messages teaches us that there is a significant difference between environmental awareness and environmental action. The Crying Indian presents the viewer with an imperative "don't do it" but does not say what consumers *are* supposed to do with the litter. Instead, KAB left it up to the consumer to figure this out, as an individual and atomized activity.[40] Resirk, on the other hand, consistently tells the consumer exactly how to properly dispose of your containers and, just as important, assures the individual everyday environmentalist that even these little actions matter.

There are generally two takes on the fundamental mission of KAB, which like Resirk was founded and funded by packaging companies and organizations.[41] The first and most positive story states that KAB was established to educate consumers in how to properly dispose of used packaging products. The second and less rose-colored story argues that KAB aimed to direct the responsibility for littering away from the packaging companies and over to the consumer, particularly to avoid environmental taxes and fees such as deposits. While both stories can be said to be valid, it is a fact that KAB actively lobbied against bottle bills.[42] KAB was founded in 1953, at a time when American beer breweries in particular were moving away from refillable bottles and over to disposable containers.[43] Between 1965 and 1975, soft drinks followed suit, increasing from 10 to 70 percent disposable containers. In the new disposable beverage age, KAB and its corporate backers rejected the idea of deposits, since these implied responsibility and involved packaging producers in the practical return system. KAB approached littering as a social problem. As a result, KAB ended up promoting awareness-raising campaigns. In other words, KAB rejected both economic and practical scripts and instead adopted only the epistemological script. For Resirk, however, littering was more of a technical problem, and it focused instead on the concrete actions that consumers had to do in order to return their bottles. The key insight provided by comparing these two organizations is that neither technological infrastructures nor voluntary recycling initiatives turned out to be sufficient. Consumers must be recruited into the recycling system, and their actions must be convenient and infused with meaning.

Different types of scripts enable the successful functioning of such sociotechnical infrastructures. *Economic scripts* make bottle recycling worth it, whether they are the differential taxes that motivate businesses or the deposits that motivate consumers to recycle. The wallet has repeatedly proven to be a surprisingly strong motivator and also helps recruit unlikely cleanup personnel, such as kids and hobos collecting empty bottles to provide some extra cash. *Practical scripts* address the specific activities consumers need to perform in order to return their bottles. These need to be convenient and nonintrusive, in order to remove possible sources of resistance. Finally, *epistemological scripts* help consumers connect their own actions to global environmental issues. They facilitate the transfer of meaning and knowledge through recycling activities. In other words, they make recycling "feel so good."

My analysis in this chapter suggests that effective environmental action needs to be scripted. By that I mean that one needs to have a clear focus on *how* consumers can do the correct thing. One might argue that the environmental issues the world faces today can't be solved at such a microlevel and that nations and international organizations need to take responsibility rather than individual consumers. I wonder, though, if it is only by breaking down environmental problems to such individual, yet systemized actions that we can get to the larger issues. International environmental treaties and agreements are certainly important, but they have to be tied to a close focus on mutual responsibility and accountability among consumers, businesses, and governments. The Resirk story shows that a business consortium built a remarkably efficient recycling infrastructure, but they only did so because the Norwegian government held them accountable for the efficiency of the system and because consumers expected a certain way of returning beverage containers that Resirk could tap into. Resirk thus co-opted the strong economic and practical scripts that made the glass bottle return system so successful rather than competing with them and added an epistemological script on top. The economics of the situation were such that the potential littering problem caused by disposable containers could not simply be legislated away by the government. Nor could littering be solved without active participation from all the involved actors—bottlers, grocers, distributors, and consumers. The RVM and the larger technological recycling system that it enabled became the thing that tied together all these actors and interests.

I have argued that the work that the RVM has performed as a functioning technology and as a mediating agent is the glue that kept the Norwegian beverage container system together as a technological infrastructure, an everyday consumer routine, and a green knowledge system through scripts. It is not my intention to conjecture regarding counterfactual history, but it is very likely that without the RVM, this recycling system would have collapsed in the 1970s under the pressure of new container types, new materials, and new consumer preferences.

Theoretical and methodological insights from STS can help us think not only about the design and function of technological artifacts such as RVMs and their interactions with users but also about the connections between consumer recycling and environmentalism, how environmental choices and goals can be embedded in effective sociotechnical systems. Without the RVM and the technological system, recycling would have simply been a symbolic activity signaling concern for and awareness about environmental issues. Using the STS method of identifying and analyzing scripts can help environmental historians understand the investment of meaning into technologies, and the everyday actions of consumers can add up to become effective environmental infrastructures on a large scale.

PART II

CONSTRUCTIONS OF
ENVIRONMENTAL EXPERTISE

# CHAPTER 6

# THE SOIL DOCTOR

## HUGH HAMMOND BENNETT, SOIL CONSERVATION, AND THE SEARCH FOR A DEMOCRATIC SCIENCE

### KEVIN C. ARMITAGE

The history of every Nation is eventually written in the way in which it cares for its soil.

—Franklin Delano Roosevelt, "A Presidential Statement on Signing the Soil Conservation and Domestic Allotment Act," March 1, 1936

ADDRESSING THE 1941 meeting of the American Society of Agronomy, Hugh Hammond Bennett, chief of the Soil Conservation Service (SCS), pleaded for scientific solutions to the nation's agriculture problems:

Shall science allow silt, the product of erosion, to continue to shoal our streams and turn them into swamps and marshes where malarial mosquitoes may breed and multiply? Shall science allow erosion to continue to strip off layer after layer of productive soil, leaving behind nothing but raw unproductive subsoil, so that man cannot produce enough on it to make a decent living? . . . Shall science allow erosion to continue to cut gullies that disgorge soil and subsoil material, and sand, rock and gravel to cover fertile lowlands, and at the same time pour out water with maximum speed into flood-swollen streams? . . . Shall science allow erosion to continue to blow up giant dust storms and dissipate the wealth of thousands of farms and homes?

As daunting as such problems remained, "they are only some of the challenges of erosion to science."[1]

Bennett clearly held that science must be socially responsible, a servant to the public interest broadly construed. Yet even if "science" could produce solutions to seemingly intractable problems such as erosion—and Bennett was sure it could, claiming that "the soil erosion problem can be conquered. Science has proven that it can be done"—left unanswered was the problem of bringing science to farmers.[2] Bennett's supreme faith in science was tempered by his modest hopes for how the public would respond to its findings. Nor was this an inconsiderable obstacle. Bennett noted that the entire history of American development was based upon forgoing "restraint in the use of our natural resources." Indeed, the erosion problem was simply the logical extension of a history of exploitation: "Having cut down our trees, killed off the buffalo, and plowed up our grasslands of the plains, we began to exploit the soil." Progress itself was defined as "not how long you can keep a thing, but how quickly you can economically scrap it."[3]

Challenging belief in an economic cornucopia was not the only issue; rural people distrusted science and conservation. Farmers, often uneducated, suspicious of outsiders, and working under an economic system that demanded maximum production, at times resisted implementing the new methods of conservation agriculture. Bennett allowed that "some farmers were somewhat dubious or distrustful at first. This was, as they saw it, a new kind of agricultural endeavor involving something of the experimental."[4] Education was key; some farmers were, frankly, "untutored and . . . inept."[5] Bennett saw the fundamental problem the Soil Conservation Service faced as needing to "penetrate the wall of a traditional, exploitative, erosion-inducing farming system and introduce an ever-widening wedge of scientifically sound, modern, conservation agriculture."[6] Economics made the wall that much more difficult to breach: "In some cases, this new arrangement may not entirely fit the farmer's pocketbook."[7] Bennett and his SCS needed to solve agriculture problems and to find ways to implement its findings among wary agriculturalists.

Bennett's challenge exemplifies a fundamental tension that has informed the entire history of environmental reform: how to translate expert knowledge into common practice. How can experts inform the broad public? Can a genuinely democratic social movement derive from the elite knowledge of experts? The quandary was how to reconcile democratic and expert knowledge.

Science could help solve environmental dilemmas. It also made many of those problems apparent in the first place. For example, Bennett's understanding of the erosion crisis and its solutions derived from his scientific —and more specifically ecologic—understanding of the world. Though such problems as gullied hillsides might be readily apparent, seeing and understanding the consequences of sheet erosion or the leaching away of soil nutrients depended upon a scientific understanding of agriculture. Only scientists had the expertise to confidently diagnose these complex problems. "We sometimes hear farmers . . . say 'Erosion is not a problem in our locality.' In some places it is not a very serious problem . . . but in many other localities where it is believed not to be operative at all, much damage is actually being done."[8]

Bennett's undertaking, then, demanded the deployment of technical capability and the translation of expertise into the democratic sphere. To examine this complex process we need to turn to a set of theoretical tools that help explain how Bennett attempted to create what I call a science public —citizens who unite to overcome, through the use of science, a problem common to them. A science public is created through framing a complex issue and then establishing a public that uses science to bring about ameliorative action. In the example of Hugh Hammond Bennett and the Soil Conservation Service, we can examine how the environmental movement has framed technical expertise, attempting to make science a public and democratic practice. We will also find that in the constant reconfiguration of frames, a science public can easily return to larger cultural narratives that threaten its considerable achievements.

## THE MAKING OF A SOIL DOCTOR

Hugh Hammond Bennett came upon his life's calling naturally. Born near Wadesboro, in the agriculturally rich Piedmont region of North Carolina, he grew up on a 1,200-acre cotton plantation that his father operated. As a child he helped his father plot level terrace lines. His job was to dig small holes that would mark successive points, which in turn would denote the plow line for turning up soil for terraces. A homemade "horse" or wooden bipod substituted for a transit when plotting the lines. It was difficult and cumbersome labor. When he asked his father why they went to such trouble, his father supposedly snorted, "To keep the land from washing away!"[9] Bennett later recalled that this lesson impacted him more greatly than any he learned in school.

Upon graduating from the University of North Carolina, Bennett accepted a job with the Bureau of Soils, then a division of the Department of Agriculture. The Bureau of Soils embarked on county-based soil surveys in 1899. A turning point that Bennett referred to repeatedly throughout his professional life occurred in 1905 when he and his colleague W. E. McClendon investigated declining crop yields in Louisa County, Virginia. There they inspected two adjacent pieces of land, one with forests that grew from mellow, loamy, friable soil that, even in the Virginia heat, was moist and could be dug into with bare hands. The adjacent cultivated hillside consisted of rock-hard clay. Bennett surmised that both plots were originally in similar condition but that erosion had washed away the fertile, nutrient-rich soil of the cultivated hillside.

This insight was crucial for two reasons. The denuded clay hillside was not gullied with erosive channels. Yet Bennett was sure erosion was the cause of the missing soil and theorized that rather than eroding in a gully or rill, overland water had removed the soil in a sheet of more or less uniform runoff. After a rain this soil would show up in local creeks as muddy water, but the lack of gullies hid the source of the erosion. Bennett coined the term *sheet erosion*—now an accepted geologic concept—to describe this phenomenon. Second, it struck Bennett that the erosion problem affected the entire rural society, not just the individual farmer. The problem facing the Bureau of Soils, then, was not simply to dispense technical advise to agriculturalists, but to consider the social and economic dimensions of the erosion problem. Bennett made little headway on the erosion issue until he seized opportunity when it came via the conservationist presidency of Franklin Delano Roosevelt.

FDR's profound attention to agriculture and conservation provided Bennett with two openings. Most generally, Roosevelt championed scientific research as a way to inform government policy. FDR increased funding for both basic research into fundamental scientific problems and applied research that sought to advance policy goals such as conservation and public health. In 1933 he established the Science Advisory Board. Though that board quickly became moribund—its duties replaced by the National Academy of Sciences—it positioned science as central to the administration's goals. Moreover, if the structure of the advisory board did not work, science itself was not jettisoned but rather remained indispensible to a government dedicated, as FDR put it in his 1932 address at Oglethorpe University, to "bold, persistent experimentation." One bold experiment was to create the Soil Erosion Service, dedicated solely to overcoming the erosion problem, directed by Hugh Hammond Bennett.

Despite this promotion, Bennett had some grievances with the new agency, beginning with its name. He disliked the name Soil Erosion Service because it emphasized the problem of erosion rather than solutions. Bennett favored a plain designation—the Soil Conservation Service—a name that sounded like a typical New Deal agency but was comparably optimistic, positive, and indicative of action. The name also more accurately connoted the many interrelated methods of erosion control that Bennett wished to institute. Two years later, the agency did change its name to Bennett's preferred designation. This change was not merely bureaucratic or cosmetic. Bennett had a keen sense of needing to sell how his agency would use science to serve the public interest. In doing so he was engaging in what many theorists of social action term *framing*.

## MAKING FRAMES

Beginning with Erving Goffman, social scientists have employed the idea of analytical frames or framing processes to analyze social action. Goffman defined frames as "schemata of interpretation." They are the cognitive models people use to interpret events. Frames enable people to "locate, perceive, identify and label" events within their lives and in the world at large, thus rendering meaning, organizing experience, and guiding action.[10] Frames, in other words, condition how people respond to social life. The social construction of reality—never a static phenomenon, but a historically grounded process in constant, dynamic reconfiguration—helps explain how individuals respond to social phenomena. Along with resource mobilization and political opportunity, framing occupies a central position in the analysis of social movements. As Robert D. Benford and David A. Snow, leading theorists of frame processes, note, "collective action frames are action-oriented sets of beliefs and meanings that inspire and legitimate the activities and campaigns of a social movement organization."[11]

Advocates of social change have three fundamental framing tasks and one overarching constraint. First, they must diagnose a malady. Quite simply, they need to convince citizens of a problem. The framer's second task is to develop a prognosis, which poses the central question raised by any predicament: What is to be done? What action might alleviate the malady? Motivation comprises the final framing task. This is the call to arms to engage in ameliorative action. The frame must contain compelling accounts for the need to act. Framing is thus a process of diagnosing problems, proposing solutions, and encouraging people to act. Crucially, frames are not mere marketing but must comport with obdurate reality. This is the frame's essential constraint. People will only be motivated to act if the frame rings true. Scholars term this resonance—how well the frame comports with lived experience—its potency.

Bennett's framing of the soil erosion problem was complicated by his faith in scientific remedy, for framing scientific solutions to problems entails particular difficulties. The troubles begin with diagnosis. Although science has credibility among the public, it often points to uncomfortable truths, hidden causations, and unsettling conclusions. Like a great poem, science can force people to see the world anew. While science is a stunningly powerful diagnostic tool, many of its fundamental insights—that space and time are related or that humans share common ancestors with all other forms of life—are deeply upsetting of commonly held norms. For the scientifically untrained, science can seem an alien and antagonistic force instead of a tool for problem solving.

Superficially, diagnostics, given the dramatic background of the Dust Bowl, would appear to have been Bennett's easiest task. Yet this is not the case. Bennett needed to convince farmers that erosion and other forms of land degradation were not natural but resulted from their own actions. If natural processes such as drought caused erosion, then not only is the prognosis beyond human control, but the motivation for ameliorative action is greatly lessened. One day the rains will return. Furthermore, soil erosion did not derive from social injustice, at least as justice was traditionally and widely understood. Thus the most common and potent form of diagnostic framing—the fight for social justice—could not be the central focus of his diagnostic frame.

Prognosis develops from diagnosis. The identification of a cause tends to shape the types of responses considered reasonable. In the case of using science to help frame social understanding, prognosis often flows from the structure of the diagnosis itself. Indeed, the logic of the scientific frame often points to a scientific answer—even when science helps detect a social malady. Like a physician prescribing a medicine, the soil doctor diagnosed the problem and prescribed the solution—a technical rather than a social remedy. Prognostic framing, then, often reveals broader ideologies. Those who propose systematic, structural change might well be at odds with reformers, even as both factions recognize the indispensability of each other. Bennett's chief prognostic task was to overcome fear that his solutions might not work. New farming methods by definition depart from the tried and true. Farmers had everything to lose and were rightly wary of being guinea pigs in an experiment conducted upon their lives and livelihoods.

The resonance of a frame depends upon two further factors: credibility and relative salience. The credibility of those who articulate a frame is an important factor in determining how publics receive a frame. Credibility does not imply simple credentialing: one need only to think of Aristophanes's cliché, the out-of-touch intellectual divorced from mundane reality, to understand that credentials do not guarantee credibility. Scientists need to demonstrate their credibility as much as any other social actor. Instead of credentials, credibility refers to the consistency and empirical reliability of the frame. It must comport with reality. Salience refers to how well a frame fits the targets of mobilization. Are the frames congruent with the lived experiences of the people they intend to motivate? Or are the frames too abstract and distant from the lives of people to motivate action? This last pitfall is particularly important to science, which features abstract, often mathematical thinking, specialized vocabulary, and ostensibly esoteric concerns.

Successful science frames must use the idioms of science in a manner that resonates with prosaic experience.

For the most part, frames are not purposefully manufactured but unconsciously adopted as part of everyday or specialized communication. Frames such as economic individualism caused farmers to be, in Bennett's words, "more or less unconscious of the (erosion) process." Bennett thus targeted his diagnostic framing to farmers, hoping that they would accept his understanding of erosion—and act upon his suggestions. The task of the frame, then, is to enable different kinds of communication. When the targets of framing processes adopt the frame, they are enrolled in a new cognitive understanding of their situation. Enrollment—a key concept in framing studies—defines the ongoing process of persuasion by which actors adopt a frame. Once people adopt a frame, they rethink what kinds of action matter to them and accept tacit theories about how the world works and why.

The process of enrollment always occurs in a context. When analyzing science, that context is often what scholars have termed a technological frame. Developed by Wiebe Bijker, the technological frame is "the concepts and techniques employed by a community in its problem solving."[12] Bijker emphasized that the technological frame is meant to include the interaction of a variety of actors, as well as material conditions. Both social and material, the technological frame accounts for both how social environments condition the development of technology and how technology conditions social interaction. Technological frames are never totalizing, even when stable for a long time; like other frames they are in constant reconfiguration because people participate in them to varying degrees. The key insight of the technological frame, one that helps explain Bennett's actions, is that it includes material reality. Frames adjust according to facts on the ground as well as to changes in worldview.

## NOT ARMCHAIR SCIENTISTS: FRAMING SCIENCE

Bennett attempted to penetrate the very powerful frame of liberal economic individualism and replace it with a frame that recognized nature's limits. Bennett's first target for his framing efforts was not the broad public but fellow scientists. Bennett assumed that his reports on the "erosion phase" of Orangeburg sandy loam or a report on the soils of Fairfield County, South Carolina, that identified a new soil type, "rough gullied land," would spur federal action against erosion. Instead indifference met these reports, due to more than bureaucratic inertia. The dominant idea among government scientists working in conservation was that soil erosion was simply not a problem. Professor Milton Whitney, an admired soil scientist, argued in *Soils of the United States* that "the soil is the one indestructible, immutable asset that the Nation possesses. The soil . . . is the one resource that cannot be exhausted; that cannot be used up."[13] Bennett later recalled that upon reading this report he "didn't know that so much costly misinformation could be put into a single brief sentence."[14]

Bennett sought to counteract misinformation by making science useful. In 1921 he published *The Soils and Agriculture of the Southern States,* in which he framed wise soil use in the language of efficiency and as a promising opportunity for using science in the service of enhanced agriculture. Farming methods "are constantly being improved," wrote Bennett. "From time to time old methods must be abandoned and new or better ones adopted in their place." Such efficiencies derived from recognizing ecological realities; both "scientific agriculturalists" and "practical farmers" could benefit from using soils "more in accordance with their adaptations and requirements." Bennett thus neatly sewed the recognition of ecological realities to improved efficiency. Such a prognosis fit a prominent rhetorical strain within conservation that emphasized efficient use of resources and also brought science into the realm of service for the farmer.

By the mid-1920s Bennett expanded his outreach beyond the scientific community to include the general public. He published in popular and farm magazines such as *American Game* and *Country Gentleman.* From late 1925 through 1926 he contributed a column to *Farm Journal.* A breakthrough came in the late 1920s. The great Mississippi River flood of 1927—the costliest in lives and property the nation had ever seen, with 246 lives lost and over $400 million in damages—helped focus the nation's attention on erosion.

Despite the widespread attention given to the flood, securing the first federal appropriations for erosion research required some canny political framing. A. B. Connor, director of the Texas Agricultural Experiment Station and a colleague of Bennett's, approached his congressman, James P. Buchanan, regarding funding for research on erosion. As expected, Buchanan demurred, explaining that federal funds should only be used for items such as defense. Connor, who described himself to Buchanan as "both scientist and farmer," replied that protection of the soil defends the nation as much as military expenditures. How could the congressman vote $35 million dollars for a battleship but completely ignore erosion control? His interest piqued, the congressman consented to have Bennett testify before his committee. The result was a $160,000 appropriation "to make investigation . . . of the causes of erosion . . . and to devise means to be employed in the preservation of the soil, the prevention or control of destructive erosion, and the conservation of rainfall."[15]

Bennett seized the appropriation and the lessons in political framing. Connor and Bennett were able to convince Congressman Buchanan to appropriate funds by connecting scientific research to national defense. The diagnosis held that soils were a security issue. Not only did this task place erosion in an arresting new light, but it also connected scientific labor with patriotism, national purpose, and the general welfare. Defense is almost by definition motivational, and by employing this frame Bennett placed science in the center of diagnosing and prognosticating solutions to issues of broad public concern.

Not surprisingly, Bennett employed this frame for the rest of his public career. The title of his 1942 soil conservation jeremiad, *This Land We Defend,* explicitly embraced the security frame—one that was surely especially potent given the backdrop of World War II. "Land is [the] most important of all man's material possessions," argued Bennett and his coauthor, William Clayton Pryor. After reiterating Roosevelt's "Four Freedoms," the authors added to them something they believed was even more basic: security in healthy landscapes. "We dare not forget," counseled Bennett, that "the source of peace, security, and freedom" is found in "the preservation of the land."[16]

If soil represented security as much as battleships, it followed that those who worked against erosion were soldiers for the common good. Indeed, erosion might be thought of as an invading enemy: "If a foreign nation should invade this country and destroy ninety thousand acres of land, who doubts the nation would hasten to spend twenty billion dollars to redress the wrong?"[17] Bennett employed this frame widely. During one radio address he lauded the "farmers in all parts of the country" who were "taking an active interest in conservation work" as nothing less than soldiers working for "soil defense."[18] Technicians working in agricultural research stations were also bathed in military metaphors: "in the early days of battle"—before farmers were educated to the erosion menace—"field technicians of the [Soil Erosion] Service were the shock troops."[19] The frame was particularly strong as a motivator, because security is integral to national well-being. This frame associated conservation with widely shared values, thereby enhancing the meanings of both conservation and the security frame.

Though Bennett relied upon the security frame, he deployed it in conjunction with other frames that also connoted national pride and patriotic duty. "I have too much confidence in the patriotism, the inventive genius, and the good common sense of Americans," declared Bennett, "to believe that we, as a nation, are going to stand by and let our soil—the most basic of all our natural resources—slip away into the rivers and the oceans."[20] Allowing the destruction of the soil was a deeply shameful practice that entailed bad economics, bad science, and bad citizenship. Conservation, by contrast, embodied core American ideals: "We have chosen to maintain across the United States a domain of productive land, breathing life and producing vitality to sustain independence and the love of liberty and freedom in the people of the United States. . . . We have chosen to keep the source of health and vigor intact, a bulwark against whatever storms may blow."[21] Prudent and patriotic, conservation, as framed by Bennett, tapped into both the pragmatic and idealistic side of America's sense of shared values.

The security frame not only inculcated conservation with patriotic values but also cast it within the realm of necessity—and pointed toward radical rethinking of the human relationship with nature. Just as security motivated the soldiers' sacrifice, conservation required understanding nature's limits and adapting to them. "Man may for a time attempt to impose artificial and economic consideration on the use of land," wrote Bennett, "but ultimately the inexorable influence of physical laws will prevail and man will come face to face with the utter necessity of cooperating—not competing—with nature in the use of land."[22] Recognizing the limits of nature was a consistent theme of Bennett's speeches. Addressing the Engineering and Human Affairs Conference at Princeton University in 1946, he argued that "we cannot dig deeper into the Earth and find new productive soil. We cannot pump it from wells, plant it with seeds, or dig it from mines. We must keep what we have or do without, for when soil has been washed or blown into the oceans it is not recoverable."[23] Irresponsible use of land by farmers was thus "deserving the wicked names 'soil mining' and 'soil robbery.'"[24] The idea of soil mining is an important diagnosis because mines are exhaustible and are frequently exhausted—values opposite those that the conservation security frame promoted. Conservation was security, a fact that necessitated the clear-eyed recognition of ecological limits.

## AN ACTION AGENCY: THE SOIL CONSERVATION SERVICE

Bennett was keenly aware that framing alone, no matter how effectively conservation was tied to national purpose, could not succeed without resonance—the comportment of the frame with the lived reality of the farmer. Confident that farmers would respond if approached properly, Bennett understood that the very structure of scientific outreach must not impinge upon the autonomy of the farmer but become a useful tool that the farmer chose to adopt. If conservation science was imposed from above, the patriotic frame would be lost, and the necessity frame would shift to a hard lesson in state power rather than nature's limits.

In this sense, Bennett was trying to create what the philosopher John Dewey, writing at the same time as Bennett, termed a *public*. A public is a group of citizens who, affected by the actions of others, band together to organize their collective response. In organizing themselves the public forms a "state" that will protect its interests and establishes officials who comprise the government. The efficacy of the state is judged by how well it performs its function of attending to the public interest. Publics, then, precede the state but may often be organized in response to state action. For Dewey, the creation of publics was a central feature of a functioning participatory democracy.

As always with Dewey, this abstract theorization was meant to help clarify real-world conditions. And in this light Dewey found the public in poor shape, for reasons that bear directly upon Bennett's project. Most generally, Dewey found the public struggling to define itself against modern institutions. Mass society had helped produce a public that was too large and characterized by being simultaneously homogenized and diffuse. Such a public could not generate sound collective judgment. Citizens lived in a world of proximity but not togetherness. This problem was particularly acute in relation to technology. Citizens faced a mind-bogglingly complex world defined by technical expertise and science—so complex that even experts "find it difficult to trace the chain of 'cause and effect.'" Issues before the public were "so wide and intricate, the technical matters involved are so specialized, the details are so many and so shifting, that the public cannot for any length of time identify and hold itself."[25] Bennett echoed these concerns. Stopping erosion would be "very difficult" because it required the "widespread cooperation of soil scientists, engineers, practical agriculturalists, the extension agencies of the nation, bankers, merchants, and railroads."[26] No wonder the public could not identify and hold itself.

Dewey's solution was communication; the proximate but diffuse country needed to recognize common concerns and the associations articulated through collective knowledge. Communication enabled the articulation of mutual interests and the consequent ability to engage problems shared by the populace. But communication must be artful. It must pierce the shell of conventional, mass thought. It must break traditional frames of meaning (such as the "natural" economic order) to enable the intelligent convergence of genuinely shared interests. In the language of frame theory, the communication must diagnose ills, propose solutions, and resonate so that a public may be formed. Bennett's problem was to communicate in a manner that would help create a public receptive to scientific solutions to erosion.

Due to widely held competing frames such as the technological quick fix, such communication was fraught. A science public might degenerate into the promotion of technology as a universal remedy. Bennett worried that the public might trust that "an engineering method of attack—one employing a single implement of combat—is the complete and final answer to the erosion problem."[27] The issue came to the fore while planning federal action for soil conservation. An allocation of $5 million from the National Industrial Recovery Act to establish terraces for erosion control was to be allotted to states according to their cultivated acres. The Special Board for Public

Works announced that the program "provides for the practice of terracing, which agricultural engineers have found to be the most effective means of controlling erosion."[28]

Upon hearing this news, an agitated Bennett approached Assistant Secretary of Agriculture Rexford Tugwell. Though Bennett supported terracing, he noted that "in the long run" terraces had caused "more harm than good."[29] The problem was the uniform, top-down approach in the proposed program. Bennett knew that terraces could be effective only in conjunction with mutually supporting practices such as strip cropping, contour plowing, grassed waterways and crop rotation. Furthermore, terraces themselves needed to be carefully situated and maintained—how would this program inform farmers on proper use of terraces? Terracing was not a universal remedy to the erosion problem and if sold as such would harm enrollment and burden the creation of a science public.

Bennett succeeded in significantly altering the proposal. The revised plan limited work to ten-acre plots where "terracing, strip cropping and seeding to permanent pastures are to be the principle control measures employed on the crop land, with possibly some tree planting on the steepest and most severely washed slopes."[30] As head of the Soil Erosion Service, Bennett wielded considerable sway over the implementation of the program. He directed a staff of twelve and, more importantly, had an abundant source of labor: Civilian Conservation Corps (CCC) enrollees. Initially, only a few CCC camps engaged in soil erosion work. That number expanded greatly as fighting the Dust Bowl became a national priority. As recorded by CCC historian Neil Maher, from March 1935 "until Congress terminated the CCC in 1942, the Soil Erosion Service supervised approximately 30 percent of all corps camps nationwide."[31]

The Dust Bowl crisis opened other opportunities for Bennett to frame his work. In 1935, Congress was debating a bill to make the Soil Conservation Service (Bennett's preferred name) a permanent agency in the US Department of Agriculture (USDA). Called to testify for the bill on April 2, 1935, Bennett knew beforehand that a major dust storm was blowing toward Washington, DC. During testimony Bennett detailed the intricate facts of soil erosion. He belabored points, adding nuanced and specific answers to questions meant to elicit a concise response. The delaying tactic served two goals. First, Bennett needed to emphasize the facts of conservation and the causes of the Dust Bowl. In 1934 the *New York Times* could still editorialize that the "explanation of the storms is quite simple . . . the soil from the West is drier than usual."[32] The way farmers used the soil was of secondary concern. Second, Bennett knew the storm, and the dramatic impact it could have on the Senate, was on its way. Eventually the sky darkened; one senator wondered if a rainstorm had descended upon the capital. Many senators walked over to the window, where they witnessed not the sodden thunderclouds of a rainstorm, but a thick blanket of dust settling over the capital city. For once nature cooperated with Bennett. The bill passed congress unanimously and was signed into law by FDR on April 27, 1935.

Having abundant federal resources only began Bennett's work. He still needed to tie the science of soil conservation to frames that would encourage the rural populace to become a public that acted upon his recommendations. Bennett insisted that his frame include the health of rural society, as well as applied science. Conservation meant the stabilization of rural America, as well as the preservation of the soil. Bennett explained the idea in *This Land We Defend:* "Soil Conservation means more than stopping erosion on hillsides. It means conservation of people as well as conservation of resources of the land." And it worked: conservation "has stabilized the people *on* the land as well as the land itself."[33] Conservation meant not just efficient use of resources, but secure rural families contributing to a productive farm economy. This framing made conservation personally relevant and emotionally potent. It resonated.

Bennett's powerful diagnostic frame could not by itself enroll the rural community. He gained credibility by illustrating his ideas through demonstration projects. Demonstration enrolled farmers in Bennett's frame because it allowed farmers to see for themselves: "The results of an applied program of erosion control, taken from a large number of demonstration projects in various distinctive problem areas throughout the country, will illustrate the relation of this kind of work to the alleviation of rural difficulties caused by excessive soil erosion."[34] After witnessing such demonstrations, "farmers themselves decide what they want to do to improve their land and water resources, and how they go about doing it."[35] Demonstration gave pragmatic credibility to science and thus allowed the frame of democratic science to stick: "Soil Conservation Districts are, of course, essentially democratic mechanisms," argued Bennett. "They are the farmers' own governmental units—of, for and by the farmers."[36]

Bennett understood that farmers had to uphold their participation for his program to work. The SCS thus required farmers to sign a "cooperative agreement" that, as Neil Maher explains, "required them to contribute personally to the work being performed by [CCC] enrollees and perhaps more important, to maintain the improvements resulting from such work for at least five years."[37] Farm work itself thus became part of the conservation frame. Bennett explained that "when the farmers know what to do they will act. They must have assistance and demonstration of methods" but will then proceed on their own.[38] According to one investigative body, the program was very popular among farmers; it enjoyed "near universal cooperation. Farmers say that it is practical and they take to it immediately."[39]

Land-use capability surveys were another tool put into the hands of farmers. The survey was a map, usually color-coded, that described "no more than about five land classes. These classes range from land of such favorable quality as to require no special treatment for proper cultivation and adequate protection, through land that requires a variety of treatments . . . to land that should never be cultivated under any circumstances."[40] Armed with information, farmers could become a science public and implement their own "design for farming."[41] Such practices worked well and had an important political effect: they helped rally rural support to the New Deal and its conservation programs.[42]

The creation of a science public through the structure of the SCS and Bennett's rhetoric helps explain its success. Farmers enrolled in a plan of scientific expertise, but they implemented the reforms. Encompassed within a technological frame, expertise was not encased in an ivory tower, but outside, active, and created in conjunction with constituents. Bennett explained that SCS employees who made land capability surveys "are not armchair scientists. They go out on the land to study it. They study the kinds of soil, the drainage, the degree and direction of slope; the condition of the land; and its use, as for pasture, woodland, or cultivation." The frame worked because it used science to solve practical difficulties. Bennett explained that SCS recommendations "have not been worked out on paper by office experts, but on the earth itself."[43]

The practical uses of science pointed to a final frame that Bennett effectively employed. If science could show how to wisely use nature's bounty, then perhaps the object of scientific inquiry—nature—should be quantified in terms of its economic benefits: "Virtually all life rises from the soil; man depends on it. As the soil produces man prospers; as the soil fails, man fails. These are truths—truths so basic, apparently, that they are sometimes overlooked." The frame was intended to evoke practical concern for the nonhuman world. The key was linking the services of nature to human prosperity.

Bennett calculated the damage done by soil erosion in monetary terms: "In dollars alone, the cost of soil exploitation is terrific . . . the estimated annual bill which America pays in one way or another for erosion is $3,844,000,000." The "itemized statement" for this staggering invoice included $3 billion for erosion that took "away three billion tons of soil material." The remaining $844 million derived from "the direct cost to farmers of the country, through reduced farm income and forced abandonment of land left in ruin by erosion." Like so many costs in a capitalist system, the public paid for these "externalities," the many related expenses capitalists impose upon society. For example, Bennett cited the silting damage to reservoirs that cost the public "$63,000,000 annually." Flood damage to roadways and navigable waterways accounted for another annual public expense of $309 million. Moreover, these expenses contribute to rural poverty: "poor land means poor crops. . . . Low yields mean low income and low incomes mean a low living standard, unpaid taxes, reduced trade in the community, impaired school facilities, often malnutrition, hopelessness, and migration."[44]

Bennett was stressing a frame that many contemporary scientists, environmental economists, and environmentalists are now making: nature (the "ecosystem" in contemporary language) provides humanity with fantastically great but ultimately calculable services. The frame means to bring those natural services into the calculations of economically rational actors. If under capitalism people must ultimately privilege economic rationality, then nature's services must be calculated if they are to be valued. If soil is understood to have great monetary worth, citizens will treat it as the valuable commodity it is. Bennett thus framed the history of the exploitation of the natural world in economic terms: "we didn't pay back to the land the riches we took from it . . . we robbed Peter to pay Paul."[45] Conservation, then, was the beginning of not just repaying a debt but of assuring that wealth would be available in the near and long-term future. Economic individualism—the most powerful frame in America—demanded conservation. The same motivations that push people to get rich should also make them into conservationists.

The success of the frame depended upon expanding economic thinking—and economic self-interest—in ways that consider the natural world. Given the abandoned, dried-up farms of the Dust Bowl, this frame surely had a great deal of resonance. Yet that resonance was only as deep as a sheet of water that carries away rich soil as it washes over a hillside. Once technology or better weather overcame, at least temporarily, an immediate crisis, the older frame of economic rationality that disregards nature's value returned to the fore, begetting the next crisis. Does the economic frame reinforce those habits of exploitation that brought about the soil crisis?

Questions regarding the usefulness of the ecosystem services frame—a topic of intense debate in contemporary circles—might be illuminated by examining the unanticipated ends of Bennett's frames. Bennett's diagnostic frame focused on specific harms at the expense of the social causes that induced them. Surely part of the reason farmers exploited the soil was due to frames that ignored nature's limits. Too, in many cases the implementation of new, smarter farming techniques—contour plowing, crop rotation, gully control, and the like—achieved their goals: farms returned to productivity. Once productivity returned, farmers returned to an economic frame that demanded maximum production. As Donald Worster, in his classic study *Dust Bowl*, makes clear, "a capitalist-based society has a greater resource hunger than others, greater eagerness to take risks, and less capacity for restraint." By ignoring the underlying political economy of resource exploitation—

by treating the symptom, not the cause—Bennett and others reinforced the quick technological fix, the hubristic frame that supposes humans can transgress nature's limits. As Worster explains, "We are still naively sure that science and technique will heal the wounds and sores we leave on the earth, when in fact those wounds are more numerous and more malignant than ever."[46]

This unwitting accommodation to existing mythologies did not happen easily. Bennett continually stressed the limits of the natural world. "Today's necessity for public action," thundered Bennett, "is the outgrowth of yesterday's failure to look more carefully to our land." Furthermore, "Unless the United States goes ahead vigorously, persistently, and speedily to defend and conserve the soil and to make *far-reaching adjustments in our complex land economy*, national decadence lies ahead."[47] Despite this attempt to question fundamental assumptions, most farmers heard precisely what they wanted to hear: that new techniques enabled business as usual. Any message that questions basic assumptions is susceptible to this kind of interpretation. The immediate problem, erosion, can be redressed though technological solutions. Thus those technological frames will have greater resonance than more abstract language about underlying factors. Indeed, the prognosis— new farming techniques—is likely to displace attention to the underlying economic factors.

The question of motivation haunts diagnostic frames. A diagnostic frame that indicts an entire economic system is likely to fail. Like "science" it can be too abstract to motivate the formation of a public. Prognosis becomes greatly problematic. This makes prognostic frames that provide immediate, tangible solutions much more appealing—more people will understand and embrace them. Even against the background of the Great Depression, questioning capitalism without an immediate prognostic frame (such as joining a labor union) was not likely to engender much support. Bennett's occasional insistence upon recognizing and working with nature's limits had little chance against the lure of better technique producing greater profits. He was greatly successful in bringing conservation science to people who desperately needed it; he was not successful in fundamentally altering the complex land economy. To achieve that end, a science public would have to persist. Bennett's career demonstrates that instead of a reckoning with nature's limits, conservation can be made over into land-exploitation technology and the agricultural sciences into the servant of capitalism.

# CHAPTER 7

# COMMUNICATING KNOWLEDGE

## THE SWEDISH MERCURY GROUP AND
## VERNACULAR SCIENCE, 1965–1972

## MICHAEL EGAN

Canaries are sometimes used in mines to investigate the presence of methane gas. Let us imagine that a mining accident does happen. The intention is to send rescue staff down into the mine, but there is suspicion that the staff would run a serious risk of gas poisoning or suffocation. A great number of birds are sent down into the mine, and people wait in vain for them to return. In that situation, would we expect the start of an endless academic discussion on whether the birds had been killed by the gas or by some other cause? And would people actually be sent down into the mine, the justification being that even though the birds might have been killed by the gas, they are just birds and humans might fare better?
—Carl-Gustaf Rosén, *Svenska Dagbladet*, September 16, 1965

ACCORDING TO ITS members, Carl-Gustaf Rosén's editorial in the Swedish daily *Svenska Dagbladet* officially marks the formal creation of a rather informal and unofficial group of scientists who immersed themselves in the early science and politics of mercury pollution in Sweden and challenged the traditional patterns of social and environmental inquiry.[1] Frustrated with the inertia surrounding both the process and the conclusions of addressing mercury along more formal channels, Rosén and a group of younger scientists—from a variety of disciplines, who had left their independent research and turned their attention to mercury problems—forged a new kind of hybrid scientific praxis, engaging in interdisciplinary research while simultaneously entering the mainstream debate and arguing vociferously for more radical responses to mercury pollution. Their place in Swedish science was a liminal one as they formed something of a specialized fourth estate, studying and publicly challenging the findings and reports from their more established colleagues and from various scientific bodies within the Swedish government. Coming from disparate backgrounds, these young activist-scientists came to refer to themselves (not at all self-consciously) as the "Mercury Group," as they developed working relationships and pushed their findings into the mainstream media.[2]

This chapter means to investigate the social dynamic that fostered this particular response to the Swedish mercury problem while paying special attention to the manner in which technical information was communicated to the public and policy makers. It seeks to marry—or at least reconcile—two intellectual avenues of inquiry. On the one hand, this chapter provides a historical analysis of the contentious science and politics of knowledge creation and authority, raising themes that are germane to science study literature. On the other, it adopts as a fundamental anchor the longtime staple of environmental history: the notion that nature (or the physical environment) is more than just a passive spectator in the history of human activities. Mercury's evolution to unwelcome ecological hazard offers an intriguing blend of human and natural partnerships of the sort that make environmental history an important avenue for historical and environmental inquiry. Indeed, if a coherent and consistent argument holds the following narrative together, it is that the mercury cases transformed the role of science and scientists in Swedish environmental regulation, marking their integration into public debate and policy making in unrivaled form, stressing the intricate nature of the coproduction of environmental knowledge and regulation. On the one hand, it ensured swift action on subsequent environmental action, but the closed and centralized nature of Swedish politics also created some space for scientific dissidents like the Mercury Group. The Mercury Group's particular brand of vernacular science creation, therefore, provides intriguing points of communication for telling especially novel stories about nature, to use William Cronon's wonderfully apt phrase.[3]

The larger global history of knowing and regulating mercury pollution can be reduced to a complex struggle for epistemic clarity in and between science, policy, and the public. In Sweden, it took on the very specific challenge of creating a vernacular science designed to inform and galvanize a nonscientific audience into action. This process involved an interesting and increasingly public debate between government and industry scientists who downplayed the gravity of the mercury problem, on the one hand, and, on the other, the young, critical members of the Mercury Group who challenged government reports and pushed for more clarity, public participation, and action. In addition to communicating through government documents and scientific journals, the Mercury Group engaged mainstream media outlets. Though hardly unique to Sweden and mercury, efforts to establish a network of translating technical science into a vernacular language that could be interpreted and understood by citizen groups concerned about the environment constitutes one of the most significant and underestimated developments in twentieth-century science. The Swedish case study also provides an intriguing example of the urgency for accessible information as it pertains to environmental problems. Over time, epistemic clarity reveals itself; in the case of pernicious environmental pollutants, time is frequently a luxury that cannot be enjoyed. Findings, policies, and public education all need to have been done yesterday, and experts are pressed to reach conclusions very quickly, often with incomplete data and knowledge.

The scientific, regulatory, and public responses to mercury came in two overlapping stages. In the first, the initial discovery of mercury in the Swedish landscape was made by conservationists who observed reductions in bird populations and a remarkable number of bird carcasses around the country. Agricultural fungicides' subsequent identification as the source of the problem and their fairly rapid control eliminated mercury's threat in this capacity. The second stage involved the discovery of mercury in Sweden's water systems as a result of industrial practices that involved mercury use in factories and a concomitant concern about the apparent bioaccumulation of mercury as it moved up the food chain. With regard to aquatic mercury, which followed the agricultural case, scientists were the first to identify mercury, because of their experiences with mercurial fungicides, and it was their studies that demonstrated how mercury concentrated in human bodies through the consumption of mercury-tainted foods. Controlling the mercury pollution was similarly fairly straightforward. The real source of consternation, however, stemmed from efforts to warn the public about the potential hazards associated with fish consumption from contaminated freshwater sources. Whereas farmers were quickly presented with alternatives to the hazardous fungicides, the fishermen's plight was less easy to resolve, because mercury persisted in the water. Regulating the aquatic mercury case was made even more complicated by the fact that the fishermen were not responsible for the initial source of the pollution. Further, building on the understanding of mercury's threat to human health as it developed during the agricultural case, scientists were much more worried and vociferous in their assertions during the second debate, with the Mercury Group front and center.

## SETTING THE STAGE: MERCURY IN SWEDEN

The Mercury Group's involvement in the Swedish mercury problem developed after the discovery of mercury in fish in 1964, but by that point mercury had already produced a significant scare throughout the country and was already a mainstream issue. Mercurial compounds were effective fungicides because of mercury's significant antiseptic qualities, which the ancient Greeks had even recognized. In modern agricultural circles, their application warded off wheat bunt, barley smut, snow mould, net blotch, rice blast, and various forms of leaf stripe, which rendered seed unusable. Mercury-based fungicides were developed in Germany in 1914 to coat seeds during storage; in 1915, Upsulun, a Bayer product, was the first mercurial seed disinfectant put on the market. By 1920, Saccharin Fabrik had introduced Germesan, I. G. Farbenindustrie had brought out Ceresan, and the British company Imperial Chemical Industries was selling Agrosan. Their adoption in European and American agriculture was rapid and widespread. These early fungicides consisted of an inorganic form of mercury—phenylmercuric acetate—but during World War II, a new generation of organic mercury dressing agents—whose active ingredient was methylmercury dicyandiamide, which was especially toxic to vertebrates—was introduced in liquid form.[4] In spite of the hazards, their ease of application and universal effectiveness made the new, liquid alkylmercury fungicides nearly universal treatments all over the world. The Stockholm-based Casco Company developed Panogen, the first of these commercial formulations, which quickly dominated the market and was readily adopted by Swedish farmers. By 1950, Swedish farmers were using considerable quantities of Panogen to treat fungus-infected seed and to prevent fungal growth on healthy seeds. "Using Panogen," one observer commented, "became as routine in farming as plowing."[5]

In the early 1950s, ornithologists and members of the Swedish Conservation Association (Svenska naturskyddsförening) remarked on a noticeable reduction in the populations of seed-eating birds from traditional nesting and breeding habitats, while also encountering more bird carcasses around the countryside; by 1960, predatory birds that lived on mice and other small rodents that ate seedgrain were also found in reduced numbers. Farmers and conservationists also observed increasing numbers of wild birds flopping helplessly on the ground. Through a quirk of history, the State Institute for Veterinary Medicine was the first scientific body to systematically analyze the problem. Game birds and animals in the countryside were traditionally the property of the crown, which ensured reserves of game for royal hunting. After World War II, the crown's authority had morphed into a sound justification for ensuring an effective means of conservation, but in keeping with the old laws, game carcasses and the carcasses of predatory birds had to be sent to the state-run Veterinary Institute for examination.

As early as 1950, the State Institute for Veterinary Medicine's Karl Borg had examined a rook sent to him from Skåne; after considerable analysis and deliberation in his Stockholm lab, Borg concluded that it had died of mercury poisoning. Later, in 1956, he received a number of wood pigeons, and in 1957 a merlin, all of which had died from high levels of mercury. Noting the beginnings of a disturbing trend, the Veterinary Institute initiated more routine mercury analyses on dead birds.[6] Having access to all the samples allowed Borg and his colleagues to make effective comparisons and compile valuable data. The Veterinary Institute's research returned findings indicating elevated residues of mercury in the birds' systems, particularly in the liver and kidneys.[7] As a result, mercury's discovery and its link back to methylmercury as a seed-dressing agent rapidly became a persuasive working hypothesis for the dead birds.[8] Subsequent studies showed a drastic increase in mercury levels in birds after 1940 and the introduction of mercury into the agricultural environment.[9] Planting methods invariably left a proportion of the treated seed uncovered before germination, while seed tended to spill during transport to farms, surplus seed was frequently dumped, and rodents attacked sacks of seed.

The Swedish chemical industry and its agricultural experts dismissed allegations that its fungicides were responsible for the spread of mercury poisoning throughout the country's avian populations, but all indications suggested that the high mercury content was linked to agricultural fungicides and the treatment of seedgrain.[10] Lesions were found during preliminary examinations, and because the birds' illnesses paralleled the planting season for grain crops, seasonal parallels also suggested that fungicides were the primary culprit; whereas the first hatch of baby birds in the spring— coinciding with the planting of treated seeds—often died, survival rates increased among second and third hatchings.[11] Indeed, according to later estimates, "seed-eating birds and rodents managed to consume at least 1 of the 80 metric tons of alkylmercurial fungicides sown on Swedish crops between 1940 and 1966."[12] In 1964, concerned that Panogen might cause further damage to bird populations, the Swedish Plant Protection Institute recommended cutting in half the amount of mercury in seed dressing and stressed that fungicides should only be used to treat seed that was already infected. To reduce the impact on bird populations, the Plant Protection Institute also suggested that mercury only be used during the fall planting to avoid breeding season.

During the investigations into the source of mercury in birds, scientists began considering the repercussions if mercury used in agriculture should find its way into freshwater systems. According to one account of the Swedish response to mercury pollution, "not much imagination was needed to realize the potential hazard to human health of the mercury in fish."[13] In 1964, Alf Johnels, a zoologist at the Swedish Museum of Natural History, and Torbjörn Westermark, a nuclear chemist at the Royal Institute of Technology, began taking samples of fish from several bodies of freshwater in Sweden. In short order, they made a disturbing discovery. Speaking to *Dagens Nyheter* in February 1965, Johnels claimed, "We have not found as high a percentage of mercury [in fish] as has been found in birds. But the variations we have established within the different areas give us reason to suspect that added mercury is caused not only by [agricultural mercury use] but also by industrial waste."[14] This constituted a whole new problem and implied that mercury was becoming a universal environmental hazard throughout Sweden.

In response, the Ministry of Agriculture's Royal Commission on Natural Resources convened a scientific conference on mercury on September 8, 1965, in the Parliament Building. Two hundred fifty scientists, politicians, and industrialists whose stakes intersected with the mercury problem attended; the event consisted of eight lectures, followed by extensive discussion and debate.[15] The conference was headed by the minister of agriculture, Eric Holmqvist, and Sven Aspling, the minister of health and social affairs. Their interest was to gather information on mercury in order to ascertain how government should act, bringing together a wide variety of government agencies and university scientists from numerous disciplines. The scientific consensus indicated that the agricultural use of methylmercury seed dressings was responsible for the decrease in the bird populations, but the science was not perfect, and numerous questions remained. Hans-Jörgen Hansen, of the Veterinary Institute, was one of many who counseled against any knee-jerk reactions. "Our studies of mercury-impaired game are like studying the part of an iceberg that is above the water," he warned. "About the part below the water, we only know that it is incomparably larger than that part that is above the water."[16] More science was needed before definitive answers could be given and actions taken. While the Plant Protection Institute made its 1964 recommendations compulsory, its reluctance to implement an outright ban stemmed from the absence of an effective substitute for farmers; Panogen and other alkylmercury fungicides, such as Betoxin, had demonstrated their value by increasing crop yields by as much as 10 percent.[17] Similarly, the National Poisons and Pesticides Board refused to stop licensing mercurial compounds the day before the conference, citing the need to strike a balance between risk and need. The board's chairman, pharmacist Rune Lönngren, asserted that they had reached the conclusion "that the time had not come for a total ban." This committee of experts' reluctance to stop licensing compounds made control and enforcement of a ban particularly difficult, practically and politically. In a statement to the media, Holmqvist declared that he saw no reason to oppose the position of experts within the Poisons Board.[18]

## SCIENTIST ACTIVISM: THE MERCURY GROUP

The relative inaction during and after the conference riled the younger, more concerned members of the Mercury Group. Summing up the conference in an editorial a week later, Carl-Gustaf Rosén noted, "The mercury problem was due to negligence in some quarters, and twisted propaganda from others." Rosén also charged that the Poisons Board had sabotaged "the very purpose of the conference" by deciding against a mercury ban before the conference. "It is a particularly strange way of acting," he mused, "first deciding and then taking advice." Nevertheless, Rosén's commentary was more constructive than derisive: "From what is known and from what has been said, it is evident that a solution has to be found rapidly." He concluded with numerous recommendations, both scientific and policy oriented, as an agenda that imperatively demanded action.[19] "We started a debate," recalled Rosén many years later; this debate was initiated by his editorial in *Svenska Dagbladet*.[20] Shortly thereafter, the Swedish Conservation Association, the Swedish Ornithological Society, and the nature conservancy committee of the Royal Academy of Science began collaborating, agreeing to sign a document to send it to the government to push for resolution to the mercury problem. They recruited Rosén to write the document, which turned into a long paper documenting the different mercury compounds and their respective threats to human and environmental health. "This was information the public and most policymakers did not have," he explained.[21]

Another source of concern for the members of the Mercury Group was the clear indication that mercury accumulation became increasingly concentrated as it moved up the food chain. That birds of prey—which ate rodents that had eaten methylmercury-dressed seeds—had such high levels of mercury was a clear indication of its potential harm to humans. In the aftermath of the first major mercury disaster—in the 1950s at Minamata, Japan, where a fishing village suffered catastrophically from subsisting on fish that had been exposed to high levels of methylmercury from a chemical plant that had dumped its waste into the bay—mercury's magnification served as another source of social concern over its continued use. In addition to the potential hazards of eating pheasant or duck, scientists, government agencies, and the public worried about more ordinary foods like eggs, chicken, and meat. The early results only elicited more concern; in 1964, *Dagens Nyheter* had covered the hazards of mercury in poultry eggs in considerable detail, and Austria and Denmark refused to import Swedish eggs because of their high mercury content.[22] The National Institute for Public Health attempted to downplay the alarm, stressing that there was not nearly enough information about the toxicity of mercury to jump to quick conclusions. This was still the official tactic at the time of the September 1965 Natural Resources Commission conference, where the Public Health Institute's Gunnel Westöö reiterated that Swedish eggs were not a health risk.[23] Then and shortly thereafter, however, three early and prominent members of the Mercury Group, Hans Ackefors, Carl-Gustaf Rosén, and Robert Nilsson, objected vehemently, suggesting that "simple arithmetic" combined with World Health Organization experiments with rats indicated that even two eggs a day could be quite hazardous. In 1965, studies showed that Swedish eggs taken from rural farms contained an average of 0.029 parts per million mercury, roughly four times more than eggs from other European countries but, the National Institute for Public Health was quick to point out, still lower than the World Health Organization's recommendations.[24]

Westöö, an analytical chemist, further obfuscated the debate at the conference by showing methylmercury contents in Swedish agricultural products but then denying that mercury came from seed-dressing agents. She inferred, instead, that the high mercury levels might be attributed to naturally occurring mercury in Swedish rocks and soil, as well as industrial mercury.[25] In contrast, however, Rosén worried that mercury-treated seeds served as the beginning of a chain that linked mercury to human bodies, traveling through seeds into the plant, to hen fodder, to hens' eggs, and finally to peoples' breakfast tables.[26] "We cannot wait and see how [the mercury problem] develops," he claimed. "To sit back, as . . . Gunnel Westöö . . . has done, content with the fact that the mercury level in our most common food stuffs has not yet reached the highest level allowed is to turn a blind eye to the realities of the problem."[27] Further studies proved Rosén right. Whereas hens fed grains grown from untreated seeds showed 0.008 to 0.015 parts per million mercury in their eggs, eggs from hens fed with grains from methylmercury-treated seed contained considerably higher rates of mercury, 0.022 to 0.029 parts per million.[28]

Another feature of the Natural Resources Commissions' conference that irritated the Mercury Group was the manner in which science and academics were poorly situated to engage in public debate. "The scientists who presented their reports on the effects of mercury compounds on Swedish fauna at the conference at the Parliament on Wednesday 8 September," Rosén wrote, "did so in a very restrained tone, which is usual in academic circles. This has obviously misled nonprofessionals to underestimate the evidence of the findings."[29] To counteract this, Rosén and Ackefors took their efforts to the media. In an interview with Barbara Soller, they challenged Public Health assertions that eggs and game birds were safe to eat. Pointing out that even the World Health Organization (WHO) standards left little margin for error, they cited studies with rats that showed heightened levels of mercury poisoning with rather modest doses from eggs. Countering another Public Health claim that Swedes could safely eat a square meal of pheasant once a week without risk, the Mercury Group showed the math. Since half of the pheasants in Swedish studies contained more than 1 mg/kg of mercury, this "square meal" could only include 25 g of pheasant in order to keep within the WHO guidelines.[30] In addition to a front-page story in *Dagens Nyheter*, the two were interviewed on national television. Recalled Ackefors: "TV was rather new in Sweden in 1965, and Carl-Gustaf and myself were interviewed. There was only one TV program at that time. Between two and three million people saw that interview that Friday night." Rosén provided further context: "We realized that to arouse the general public we had to have something dramatic, so that's why we focused in on eggs. And we decided that we should do this 'matter-of-factly,' but we should get people scared. . . . We exaggerated things, but we stuck to the facts. We presented things that would highlight the seriousness." Just as important, the Mercury Group understood the tension among science, statistics, and the public. Göran Löfroth, who joined the Mercury Group soon after its inception, observed: "The public can't realize concentrations. Parts per million don't excite the public." Rosén agreed: "But if you say two eggs a day, you would have exceeded the level recommended by the World Health Organization: they understand that."[31]

Mainstream media certainly developed an interest in the story. The number of articles in the media increased steadily, growing from 73 articles in 1964 to 138 in 1968. But something else happened, too: science became a more prominent feature of the Swedish mercury debate. Lennart J. Lundqvist notes that by 1965 "scientists [were] the actors most often referred to in the press." In 1964, roughly 34 percent of mainstream newspaper articles on mercury made reference to scientific findings or authorities; in 1965, that number had risen to 63 percent and then to 67 percent by 1966; it peaked at 78 percent in 1967, before dropping slightly in 1968 to 77 percent and to 62 percent in 1969 as new environmental legislation came to the fore and policy makers became more prevalent in the media reporting.[32] The Mercury Group's members were among the catalysts for ensuring that scientific knowledge was central to the larger political discussion. "We understood [that we needed] to manipulate the general public," Rosén recalled. "Because unless you have the general public stand up and cry, the politicians won't do anything."[33]

Reflecting forty years later, Löfroth claimed: "The turning point was when Gunnel Westöö showed that the major part of mercury in fish was methylmercury."[34] Westöö's discovery stressed the severity of the hazard—methylmercury was widely accepted as one of most hazardous forms of mercury—and brought about sweeping changes in Swedish policy.[35] This prompted government authorities to change their course within a few short months of the September 1965 conference. In 1966, the National Institute for Public Health recommended that commercial fishing bans on certain species of fish be imposed in lakes, rivers, and coastal waters where the concentration of mercury in the particular species exceeded 1 mg/kg wet weight. The new safety standard was accompanied—eventually—by a media and public awareness campaign to encourage the public not to consume freshwater fish more than once a week.[36] Determining a limit of 1 mg/kg was derived from the recent studies of mercury in Swedish fish and married with older data concerning fish consumed at Minamata. Whereas methylmercury levels in Swedish fish rose as high as 8 mg/kg, the fish eaten by Minamata victims contained between 27 and 102 mg/kg. Assessing an average of 50 mg/kg for the fish that had poisoned the inhabitants of Minamata, the National Institute for Public Health justified its safety limit:

> According to general pharmacological considerations, a reduction of the
> levels of mercury in fish to about 1/10 of those found at Minamata, i.e., to
> 5 mg/kg, should prevent the appearance of manifest cases of poisoning,
> even with daily consumption of fish. At such an intake, however, one must
> assume an increase in the mercury levels in certain parts of the human
> body. It can be reasonably assumed that with a further reduction of the
> mercury level to 0.5–1 mg/kg, no toxic effects will appear. According to this
> evaluation, fish with a higher mercury concentration than 1 mg/kg should
> be considered unfit for daily consumption by human beings.[37]

This rationale, however, was based on a critical misreading of the Minamata reports. In January 1967, Löfroth, who would later head the Swedish Natural Science Research Council's Working Group on Environmental Toxicology, discovered that the Japanese data addressed dry weights for fish while the Swedish standard was a wet weight, with none of its natural water removed.[38] Accounting for the discrepancy between wet and dry weights, the corrected Minamata mercury levels were actually 5–20 mg/kg, which suggested that the new Swedish limit was quite high. A few weeks later, at a Ministry of Agriculture–sponsored conference on mercury in fish, Stig Tejning of the Institute of Occupational Medicine in Lund and other concerned health officials pushed for making 0.5 mg/kg the maximum acceptable limit of mercury in fish consumption, but authorities, intent on keeping the Vänern lake fishery afloat, resisted. Representatives of the fisheries industry were concerned with a limit of 1 mg/kg, arguing that since no cases of mercury poisoning had been reported from eating Swedish fish, even that standard was too stringent and an affront to their livelihoods. Moreover, fishermen typically ate more fish on average than the general public, and no cases of mercury poisoning had been reported among their number.[39]

The National Institute for Public Health acknowledged its error in inter-preting the Minamata data, but rather than reducing the safety limits to 0.2–0.4 mg/kg, as its previous calculations would have required, it altered the method of determining a safe level of mercury, arriving—again—at the standard of 1 mg/kg.[40]

## THE POLITICS OF KNOWLEDGE: SCIENCE, MERCURY, AND ECONOMY

More than science was at work in determining safe levels, of course. A somewhat removed member of the Mercury Group, Nils-Erik Landell, a psychiatry student, was working at the Public Health Institute when he came across a draft of a toxicological evaluation of mercury in fish: "I saw it on [my manager's] table, and he had written [the safe limit of mercury content in fish at] 0.5 milligrams per kilogram of wet weight. The next day, the paper was still there on the table, but now I saw that he had rubbed it out and it was now 1.0 milligrams per kilogram. And I asked him why . . . and he said in Lake Vänern, the biggest lake in Sweden, the fishermen had pointed out that the fish had a concentration of 0.7, so he had to raise it to 1.0. And I understood that the evaluation of toxicology was not so sharp as it should be, but it was illustrative of the pressure from different companies and economic interests on the scientists."[41] No big surprise, likely, but for young scientists driven by an overriding social concern, this warranted more public discussion and debate.

And this is the other side of the coin: the inner conversations indicated that scientific knowledge composed only one piece of a much more complicated regulatory puzzle, one in which "political and administrative considerations were allowed to intrude into the scientific realm."[42] Preventing mercury from entering the water and imposing acceptable limits of mercury for fish consumption involved walking a political and economic tightrope. In the balance lay a significant public health disaster, on the one hand; on the other, the potential destruction of the Swedish freshwater fishing industry loomed equally large. And if scientific agencies like the National Institute for Public Health, with no regulatory power, needed to account for economic issues in their recommendations, the pressure on regulating agencies must have been even more severe. After the National Institute for Public Health delivered its 1 mg/kg recommendation to the National Veterinary Board and the National Board on Social Welfare, which had the capacity to enact legislation on the fisheries, no new regulations were implemented.[43] Because of the speed with which the Institute for Public Health had been required to produce data and its reliance on the Japanese case, the decision-making agencies considered the recommendation insufficiently persuasive and found it difficult to justify action on regulating mercury levels in fish. According to Lundqvist, "the situation was one where the [National Institute for Public

Health] recommendation could be interpreted as an admission that there was indeed a risk, but where the responsible authorities did not find it necessary to take any drastic measures to protect the public from that risk."[44] Rather than adopting the National Institute for Public Health's recommendation, the National Veterinary Board distributed information regarding the risk of mercury and its relation to fish consumption to Public Health Committees throughout the country. The letters cited the National Institute for Public Health's "provisional" safety limits, with further information about methylmercury in fish, risks associated with mercury poisoning, the mass poisoning at Minamata, and the differences in mercury levels in fish from the sea and from freshwater systems. The letter closed with reassurance that scientific research was ongoing and that all measures would be taken to ensure public health.[45] More forceful letters were sent to local communities where especially high levels of mercury in fish had been found, warning against daily consumption of fish.[46]

As more research and findings were conducted and disseminated through 1967, however, the National Veterinary Board and the National Board on Social Welfare eventually came to accept the National Institute for Public Health's original recommendation that fish with mercury content of more than 1 mg/kg were not safe for human consumption. In November 1967, the Veterinary Board and the Social Welfare Board blacklisted all bodies of water with fish that had been found to contain mercury levels greater than 1 mg/kg and published an accessible recommendation for fish consumption, which assured the public that they could eat sea fish without risk, that they should completely avoid fish from blacklisted bodies of water, and that they should eat fish from inland or coastal waters with mercury levels lower than 1 mg/kg not more than once a week.[47]

But even as the regulators came to accept that mercury posed a significant environmental hazard, critics considered their efforts both late and unsatisfactory. The Mercury Group had called for much more stringent controls than those posited by the National Institute for Public Health several months earlier and had been critical of the methodology that the Public Health scientists had adopted. Incensed by what they called "guesses" at thresholds for safety that were based on ill-conceived calculations of misread data that had no scientific support, scientists lashed out.[48] Tejning, who noted that evaluating "average" fish consumption was problematic, because some consumers—notably small children and pregnant women—were at greater risk, accused the National Institute for Public Health of bowing to nonscientific considerations.[49] Which they had: in early 1968, in the aftermath of the mercury debate, a Public Health spokesperson conceded that "even a scientific agency had to look at possible ramifications to the fishermen and the nation's economy."[50] But really, the National Institute for Public Health was stuck. It had clearly realized that mercury posed a significant hazard, "serious enough to offset the usual practice of arriving at safety recommendations," and its findings and recommendations were the sole mechanisms for legislative action.[51] As a result, it had moved quickly, only to face vitriol from fishermen, scientists, and regulators. This marked an important boundary for scientific knowledge and its authority; what emerged was a complex interdependence between science and state, wherein Public Health scientists tended to assume the role of highly skilled experts retained to provide legitimacy to government policies, even if bowing to nonscientific constraints undermined their scientific authority.[52]

## THE COPRODUCTION OF KNOWLEDGE: CRISIS DISCIPLINES AND POSTNORMAL SCIENCE

Their critics were working under similar restrictions: under the same time constraints and having to navigate through and across multiple disciplines and areas of expertise. Indeed, the Mercury Group's efforts were consistent with numerous scientific and environmental efforts in other parts of the world after World War II. According to Julie Thompson Klein, "in the latter half of the twentieth century . . . heterogeneity, hybridity, complexity, and interdisciplinarity [have] become characterizing traits of knowledge."[53] Not only were the scientific actors in the Swedish mercury debates based at different university and government agency homes, but there was also a considerable disparity in their disciplinary training and methodologies, their collaboration and communication concentrated on a shared focal point, mercury. This suggests the relative truism that environmental hazards arise irrespective of the rigid parameters of academic disciplines and that new problem-based fields need to be developed to address them. And time is of the essence. The urgency associated with procuring usable scientific information for environmental problems has radically transformed the interface among science, society, environment, and policy. In a 1985 article on the development of conservation biology, Michael Soulé discussed the precarious nature of what he called "crisis disciplines," where, he claims, "one must act before knowing all the facts." Soulé argues that "crisis disciplines" require more than "just science." In fact, they are "a mixture of science and art, and their pursuit requires intuition as well as information."[54] Putting it another way, Brian Wynne and Sue Mayer assert, "Where the environment is at risk, there is no clear-cut boundary between science and policy."[55] A considerable amount of work conducted by the young members of the Swedish Mercury Group involved precisely this kind of intuitive analysis, while also navigating policy channels in the dissemination of their findings.

In my reading, the production of scientific knowledge necessary to understand the problems surrounding methylmercury in the Swedish environment takes place in the incipient stages of what Jerome Ravetz has called "postnormal" science, where knowledge "is uncertain, values in dispute, stakes high, and decisions urgent."[56] From nuclear fallout to global warming, scientific communities have been pressed into action to weigh in—quickly—on the issues of the day. Let me stress *quickly*. In advocating a "third wave" of science studies, which examines the boundaries between experts and the public, H. M. Collins and Robert Evans observe that "the speed of political decision-making is faster than the speed of scientific consensus formation."[57] Indeed, the National Institute for Public Health was forced to use data from the Minamata disaster, rather than starting its own tests and experiments to exhaustively determine the highest zero-effect dosage of mercury, in order to make speedy recommendations. Time was of the essence. The project of this postnormal science—a derivative of Thomas Kuhn's paradigm-based normal science—is not to collect and present definitive knowledge, but rather to function within a highly complex network of policy-making interests, best described by Bruno Latour's notion of "coproduction," which marries the production of knowledge with the production of social order.[58]

The post–World War II environmental crisis has transformed the practice of science and its place in environmental politics. Consistent among the various issues that we might refer to cumulatively as the modern environmental crisis is that scientists of various specialties have been called in after the ecological problem was already widespread throughout human and physical environments. This reactive science—where uncertainties in the knowledge are significant—is necessarily reduced to part of a larger conversation in which it holds no special authority over other participant groups, in large part because "it is impossible to separate facts from the value commitments, themselves often controversial, that underpinned their production."[59] At the same time, however, a major thrust of the Royal Commission on Natural Resources' 1965 mercury conference was to involve independent researchers who were not directly tied to industry or government. These independent researchers were able to direct the scientific consensus away from agricultural and industrial interests in favor of more general public and environment health. And a balancing act exists in how governing bodies use science. On the one hand, policy makers recognize the need for reliable knowledge to support their administrative and political authority. The Royal Commission on Natural Resources' conference sought to develop knowledge that was sufficiently trustworthy to generate faith in subsequent policy decisions, even if some agencies had already decided what actions they intended to make. On the other hand, social decisions—risk assessment, political and economic interests, local knowledge—remain the key determinants in environmental governance, that gray area between electoral politics and administrative rule making. Branches of scientific knowledge, therefore, join several parties in a fantastically complicated dance in which environmental decision making takes place.[60]

# CHAPTER 8

# SIGNALS IN THE FOREST

## CULTURAL BOUNDARIES OF SCIENCE IN BIAŁOWIEŻA, POLAND

### EUNICE BLAVASCUNAS

THERE ARE MULTIPLE ways to know nature, particularly nature that is forested and "wild." But which experts do people trust when those experts speak about nature, about the ontology of the forest? Which compositions of plants and animals belong there? And why might it matter to scholars of science technology studies and environmental history if those experts produce facts in the new borderlands of the European Union in post-Socialist Poland?

In ecological promotions for tourists and nature lovers, the Białowieża Forest's most frequently associated trait is its primeval character. In such portraits, roots of alders reach their way out of rust-colored bogs. Towering spruce grow to their upper height limit to create perches for eagle owls and three-toed woodpeckers or splay as huge root plates next to majestic oaks. The preeminent symbol of this forest, the European bison, belongs to this portrait of ancient woodland. Its restitution from near extinction after World War I created a reverential status for the relict ungulate and its "original" forest dwelling. This forest image can been seen in the Białowieża National Park, founded in 1921, often described by nature activists as too small (16 percent of the forest area) given the total forest complex (fifteen hundred square kilometers), a forest "endangered" by logging activities run by Polish State Forestry (Lasy Państwowe).

The Białowieża Forest spans Poland and Belarus (Belaveskaya Puszcza) and is divided politically by the border of the European Union. I am writing about the Polish side of this forest, where close to twenty-five thousand people live, scattered in villages and small towns. The village of Białowieża serves as a central communication center for forestry operations, tourism, and scientific research institutes. The arguments developed here are taken from ethnographic thesis research conducted between 2005 and 2007 and refer to events that transpired from the mid-1990s to 2005.[1]

There is a complex chain of alliances and technologies that prefigure the ontology of Poland's Białowieża Forest: whether the forest is primeval and threatened by logging or is a well-managed woodland and therefore foresters can speak about the forest authoritatively. Those alliances need to be understood and rooted in the environmental history of the forest, a history specific to nineteenth-century Russian and Polish relations, the period of state socialism in Poland, and now, when it serves as the boundary between the "democratic" European Union and the corrupt former Soviet space of Belarus.[2] Bringing the cultural boundaries of science into this particular environmental history (EH) challenges dominant notions in EH about what can possibly count as science when scientific practices are contested by various actors who believe they understand science as well as or better than the "scientists." Scientific practices in the Białowieża Forest refract "west" and "east," "local" and "universal," "popular culture" and "scientific culture," to produce a mediated landscape that can never be "wilderness" alone. The landscape produced is a contested forest in the remaking due to post-Socialist negotiations about expertise. Science and technology studies (STS), and its ongoing concern with the cultural boundaries of science, draws attention to what kind of knowledge counts in the production of science, a cultural activity produced by the interaction of objects, technologies, and discourses.[3] Using Thomas Gieryn's work on the cultural boundaries of science, science is best understood by following it downstream of its production, taking nonscientists' interpretations of science to see the effects of scientific interpretation.

We can see the Białowieża case as what Gieryn refers to as a cultural map of science. Cultural maps provide a better way to understand how scientific controversies play out than just looking at what scientists say and do. "People navigate not just streets and highways but the culturescape: we wend our way through or around entrenched institutions, decide which rules apply where, subvert expectations by exiting, discern the signs given off by material objects, locate events in some historical narrative—and we routinely do so quickly and effortlessly."⁴ In other words, we learn more about science by seeing it from afar, as if on a map, rather than near.

There are two mirroring complications in assessing the cultural boundaries of science in Białowieża. The first is the question of who can gain legitimacy as a scientist at the local level, and the second is the question of who is local. After almost two decades of democracy in post-Socialist Poland, Polish society continues to discuss and debate who can legitimately rule the country, including its institutions, its memory, and even its forests. In many cases "local" refers not to democracy, but to the cronyism and networks of the Socialist past against a notion of an attainable standard of "Western normalcy." Scholars of post-Socialism mark the era by its stubbornness to let go of the Socialist past in the context of a "transition" toward a "Western" future that is always just out of reach.⁵ In other words, the period following the collapse of Socialism generates material discursive frameworks for truth, in this case the truth about nature.

What constitutes good science in the Białowieża Forest is caught up in an episteme of science and democracy. What constitutes the local is deeply embedded in an environmental history of the forest where various ethnic elites have competed for the loyalties of a formerly peasant/working-class population. The locals are critical to understanding how the forest will be used. Unlike individuals at many sites of nature conservation, local people in Białowieża *can* determine if the park should be expanded or if logging will continue. Ultimately the decision to expand the national park and/or stop logging rests on the Polish central government. However, the Polish Sejm, under intense lobbying by foresters working for State Forestry, passed "The Polish Law on Nature Protection" in 2001, stating that all decisions to expand nature protection (such as parks and preserves) must be approved by local governing councils prior to the central government supplying any funds. Many biologists and nature activists objected to the ruling, pointing to corruption at local levels and arguing that places such as Białowieża belonged to the nation and not just the local community.⁶

Polish wildlife biologists repeatedly referred to the locals as "undemocratic," "Eastern," possessing a "post-Communist syndrome," or "easily swayed by ideology." Yet the long-term residents often referred to as "the locals" scrutinized biologists' activities in the forest as "irrelevant," "not as good as in the West," and "cruel" and asserted that biologists did not belong, or try to belong, to the local. Biologists tried to distance themselves from "undemocratic" local practices "rooted in the Socialist era" by objectifying the forest via technologies used in the West, such as radio telemetry. Local discourses about scientific legitimacy thus impinged upon biologists' political power to define the forest as primeval.

Locals were discursively understood as people who *belonged* to the forested settlements. This vague and wide-ranging definition enabled Polish foresters, who were historically at odds with the local Belarusian population, to meet new criteria of belonging, which I will explain in more detail shortly. Biologists never met these criteria. Locals described biologists as wanting to control the forest resource through studying obscure elements of forest ecology. Furthermore, locals accused biologists of improperly using a new (mid-1990s) technology, radio telemetry, in order to torture animals.

All three groups, locals, wildlife biologists, and foresters, used technologies to make the forest legible, that is, to make the forest knowable as an object that could be regulated for either strict nature protection or logging.[7] Wildlife biologists used objectifying technologies, such as radio telemetry. Foresters applied a grid form of management over the woodland. And locals applied the expertise of a Western media production.

The story that is unfolding here is one where "authentic" nature is being mediated and contested by foresters, biologists, and local people. Recognizing that the local and nature are socially produced in ways specific to post-Socialism requires a flexible rethinking about the authority that scientists are presumed to have in conservation politics.

## HISTORY, IDENTITY, AND MATERIALITY

"They [biologists] are like Jehovah's Witnesses who come here and try to get us to change our faith, whereas, the forester fits in," Alicja explained. Alicja is a local manufacturer of ceramic bison sold to tourists, created in the barn behind her house. "He [the forester] has a holistic knowledge of the forest and people."[8] Like many residents of the forest Alicja speaks Polish to me and "Pa Swojemu" (what many locals and outsiders refer to as Belarusian but what roughly translates into "our own language") with her husband and neighbors. Alicja's husband, Marek, an unemployed middle-aged man who formerly worked as a low-level technician for State Forestry (as most of his neighbors did during the Socialist era) adds that a forester's thinning and pruning role goes back to when wild animals served the ecosystem function that the forester now performs: "The forester is like the auroch [the ancient progenitor of the domestic cow that roamed the Białowieża Forest until its extinction in the seventeenth century]. He keeps the canopy open for new growth." The forest would not survive without the forester, they explained. Despite their objection to an expanded national park, Alicja and Marek earn their income from tourism and keep an extra room open to rent to guests.

How did a Belarusian-speaking couple, of Orthodox religion, come to identify so closely with Polish state foresters, even as they lost their income from forestry and now survive off of tourism? The answer is rooted in forest history and the twists that create anew the meaning of the local in relation to forest use and protection.

Formal forest protection began in the fifteenth century, under King Jagiello of the Polish-Lithuanian commonwealth. Since that time the forest has received protection from clearing under Lithuanian dukes, Polish kings, Russian czars, and even occupying Third Reich Germans, all of whom used the forest as an important, if infrequent, hunting ground.

In the nineteenth century, when the forest was part of the Russian Empire, scientific forestry was not only the hobby of Polish gentry landowners but also a nationalist organizing front. As the science of forestry spread from its Physiocratic and Cameralist origins in Prussia to outposts of colonial influence throughout the world, Polish gentry used the science as a position from which to rethink the role of stateless Poland. Forestry schools in Warsaw were used to organize rebellions against Russia.[9] After the 1863 insurrections, when Polish landowners battled Russians in the Białowieża Forest, Czar Alexander gave orders to shut down Polish forestry schools. Russian opposition defined forestry early on as an independent, militia-like force battling against occupying powers.

After the Russian Revolution in 1917, occupying Germans cut down over one-third of the total forest area of what today is Poland and Belarus, leaving forest restoration for the returning Polish foresters, newly reorganized as Polish State Forestry.[10] The ethnic Polish foresters who arrived in Białowieża in 1921 brought a newfound sense of entitlement and a mission to convert the peasants, almost all of whom belonged to the Russian Orthodox Church, to Catholicism.[11] This sense of historic destiny, of reclaiming the forest in the name of a newly established Poland, began a long chain of alliances among the forestry elite.

Foresters needed as much timber as they could acquire to produce revenue for the new state, drained from the Polish Soviet War (1919–21), in which Poland felt it had to move its borders as far east as possible to contain the Soviet Union's territorial ambitions. Instead of relying solely on the local Belarusian-speaking population, many under the influence of Bolshevik ideology, especially those working in the new sawmills, ethnic Polish foresters hired labor from Central Poland.[12] Worker colonies sprang up in many pockets of the forest, with makeshift housing for the newly arrived men and Polish Catholic churches to serve their spiritual and doctrinal needs.

At the same time that Polish foresters returned to Białowieża, Polish field biologists began studies that continue today. For example, Europe's last wild bison, which had gone extinct in the rest of Europe hundreds of years earlier, roamed the Białowieża Forest. World War I had set a path of destruction not only for the old trees but for the bison as well, with the last Białowieża bison shot by a poacher in 1919. Zoologists and botanists operating out of the Forestry Research Institute and the newly formed Białowieża National Park set out to reconstitute the bison population from seven genetically distinct individuals found in zoos throughout Europe.[13]

State socialism's privileging of science in the postwar years spawned two new research centers in the Białowieża Forest, the Warsaw University Geobotanical Station (founded in 1952) and the Polish Academy of Sciences' Mammal Research Institute (also founded in 1952). Unlike the Forestry Research Institute, which had been established at the same time as state forestry and the national park in 1921, these new research stations diverged along the line of basic research, without direct relevance for forestry.[14]

When the Solidarity movement challenged the political elite in the 1980s, many younger biologists and a small number of foresters took active roles in organizing demonstrations for free and open elections. Many of the biologists who had agitated in the 1980 Solidarity demonstrations possessed the vision and organizational skills to rethink their institutions' role in changing society and the approach to protection of the forest. Already in the late 1980s, before free elections in Poland, biologists brought in international Earthwatch volunteers to help assist their research and established themselves as a group ready to take advantage of exchanges and opportunities with international science. In 1991 twenty biologists, as well as a few schoolteachers, wrote a letter to Lech Wałęsa (then Poland's president), explaining the value of the Białowieża Forest and calling for an expanded national park. This move was followed with interest by dozens of Western conservation organizations that wanted to be involved, most notably the World Wildlife Fund in the 1990s and then Greenpeace after 2001.

Foresters, like biologists, have long sought to advance local people, but on much different terms, especially in reference to the Socialist past. Foresters spoke of their organization as a transhistorical power that had survived Socialism and kept its autonomy from the post–World War I era. Foresters did not identify with Communism, but rather with the endurance of their own organizational structure, which was top-down and provided jobs for the locals. During the Socialist period nearly everyone worked for State Forestry. At every subsequent attempt to expand the national park, foresters objected, arguing that a forest with rotting woody debris and standing dead timber was not a healthy forest but a breeding ground for bark beetles and disease. Biologists identified this suite of forest power and forest knowledge as "ideological." "They don't understand biodiversity functioning in the ecosystem," one biologist explained. "In many cases their knowledge is from the nineteenth century."

Linking democracy to ecology, biologists shared a view of rotting woody debris as essential to biodiversity functioning and modernity. A "biologically correct" view of the forest, however, was not part of every political activity. In the mid-1990s biologists went to study the forest by day and planned activities to build new social structures in their off-work hours. They supported new independent newspapers, boosted locals who wanted to open private food shops, and pressed the national park education center to offer workshops on the function of natural forests. In the mid-1990s much park education was linked with identifying trees and their usefulness for humans. Such activities had the effect of politicizing ethnic relations in the forest communities.

By the 1990s the demographic of Białowieża had change considerably. At this time there was a Polish "minority" of foresters, scientists, and forestry workers; a "majority" of Belarusian-speaking inhabitants who farmed or worked for State Forestry or possibly the national park; and many intermarriages between Catholic and Orthodox inhabitants. During the Socialist period Polish Communists had denied Belarusian inhabitants within the region minority status: all inhabitants were officially referred to as Poles. But now ethnic status would be brought into a new set of politics, including the politics of nature protection.

A handful of biologists pinpointed attachments that "local people" had to Socialist-era mentalities. Local people, biologists alleged, were still too "Russified," meaning that their religion and ethnicity left them vulnerable to ideology, especially that espoused by foresters—"ideology" about the forest needing the foresters. Biologists' activism to reform local society was short-lived and restricted to the early and mid-1990s, a time when a few of them also tried unsuccessfully to be appointed to local governing councils.

Unlike in the post–World War I era, foresters now promoted Belarusian identity in the post-Socialist era as if they belonged to it. Foresters told local people that only they could protect locals from the uncertainties of the new capitalist era, equating themselves with the protections of the former state, while condemning biologists as dictatorial for pushing national park expansion.

One senior forestry official explained how democracy and the forestry tradition worked in Białowieża by contrasting Poland with neighboring dictatorial Belarus: "In Belarus [on the other side of the transboundary Białowieża Forest] you can have this big national park because the residents of that side of the forest all work in the park, but here in Poland we have ordinary people. Those are people who have rights given to them to settle directly from the Polish kings and from the czars. On what principles could we take those rights away from them? We could change that situation, but that would be Communistic. We don't know how to do that, but we can't do this in a spirit of conflict [with the local people]. In Communist time you could have created this park, and no one would have disputed it. But now things are different, and you must speak up to get what you want." For this forester, known throughout the region as a high functionary in the Communist Party during the Socialist era, Belarus was an authentic site of Communist-style politics, while in Poland, local Belarusian people were an important element of the forests' long history.

Local residents (both ethnic Poles and Belarusians) reacted in populist form when biologists grew closer to their goal of expanding the national park. Angry protesters threw eggs at the environmental ministers' announcements of an expanded national park in 2001 (a plan that has yet to materialize) and held signs in Polish and Belarusian blaming biologists for their poverty, even as many enjoyed increased spending power from the tourist industry.[15] Their actions prompted the nature protection laws I mentioned earlier, in which the Polish Sejm transferred decisions about nature protection to local councils.

Biologists responded with discourses about meritocracy, blaming foresters for inciting locals and pinpointing ways that foresters' behaviors belonged to the previous era. Biologists accused foresters of bribing politicians at the highest levels to stop national park expansion; of controlling local opinion through control of a key resource, wood for heating; and of supporting locals in keeping Socialist-era monuments and street names.[16] Many biologists were convinced that I was too young and too American to understand what it was like living under a totalitarian regime that according to one biologist went on "the total offensive with any criticism, denying any mistakes, questioning the competence of professionals trained in the west, and ridiculing your opponents." He said that foresters today were acting similarly in defense of their institution's forest management.

Thus, local people's understanding about the facticity of Białowieża's nature (a primeval forest complex threatened by logging or a well-managed commercial forest) was embroiled in the politics of the local, which reaches far back to the environmental history of the forest. The environmental history of the forest shapes a story of ethnic relations and resource use. Actors are understood as constituencies with agendas (foresters, biologists, and local people). Science appears to be what biologists do. Although forestry is the applied science of foresters, foresters invoke science as a tool that is on their side, rather than calling themselves scientists.

When STS's concept of the cultural boundaries of science is incorporated in this history, then the popular media, the history of scientific forestry, and a diverse array of business interests (i.e., ecotourism and the Discovery Channel) complicate attempts to define universal science against local interest.[17] A contemporary making of the "local" and "science" is socially produced.

## TRACKING HOME RANGES

In one significant way biologists asserted their authoritative knowledge about the forest, which marked the new era. In the mid-1990s they acquired radio telemetry equipment, instruments that required biologists to capture and release animals in order to place transmitters onto wolves, bison, lynx, and other mammals. Radio telemetry was one small part of biologists' modernization of their equipment and facilities. These modernizing moves symbolized their distance from the past and their seriousness about producing topnotch science that would make significant contributions at the international level. Yet radio telemetry in particular sparked an enormous deal of controversy at the local level.

Use of radio telemetry transmitters and receivers provided much of the data that helped scientists participate in international mammal science and conservation biology, together with other techniques, such as animal censuses and DNA laboratory analysis.[18] Radio telemetry permitted scientists to track home ranges of mammals and to compare ranges with densities in both historical periods and the present.[19] Patterns of space utilization proved to be an effective way to see which animals could migrate from Białowieża to other forests, such as those within Poland, and, through genetic interchange, even to Siberia. Biologists took this data and used it to publish in international journals, such as *Ecology,* the *South African Journal of Wildlife Research,* and the *Biological Journal of the Linnean Society,* which led to invitations to international conferences and to work opportunities abroad in Scotland, Panama, and the United States. The Mammal Research Institute received several distinguished awards for its accomplishments in science, including the status of a European Union Centre of Excellence, known as "BIOTER."

Locals, however, told me that biologists tortured and unnecessarily killed animals in pursuit of their "international" careers. Foresters also used this logic, reinforcing their position as the best long-standing guardians of the forest. Foresters' careers were not international but rather focused on fitting in to the local community they joined. They were voted into positions on the local governing council (*gmina*), the same council that could decide if the park should be expanded. They talked about their long-standing respect in the local community. One forester told me that this respect originated in the nineteenth century, when the local people addressed the forester as "Pan Leśniczy" (Sir Forester).[20]

Biologists, however, never gained the same kind of respect, even after many of them spent their entire careers living and working in the forest. Locals not only challenged biologists' expertise about the forest, but they dismissed Polish biologists' abilities to properly use Western equipment. Biologists on the Discovery Channel, working in such "exotic" locations as the savannahs in Africa and the jungles of South America, knew what they were doing, according to locals. They knew how to capture, tranquilize, and treat animals without harming them. Polish wildlife biologists pursued scientific careers at the expense of both the animals they studied and their place in the local community.

Locals also referred to their internationalist understanding of science when they repeatedly brought my attention to one Discovery Channel episode in particular that circulated among residents on a home video cassette, a Yorkshire production, translated into Polish, titled *Tańcząc z wilkami* (*Dances with Wolves* [1999]). Many locals based their evaluation of biologists by referencing evidence in this videotape. In this episode a team of British volunteers (much like the Earthwatch biologists invited to Poland in the 1980s) helps Polish biologists with their wolf research. Biologists initially accommodate the volunteers, shown as training them in forest knowledge and telemetry reading. As the film wears on, the volunteers discover more and more about the biologists' methods and their cultural disrespect for the locals.

In one scene, elderly inhabitants shuffle in the winter cold on their way to an Orthodox Sunday service in the Białowieża village. While the choir chants back and forth with the priest in Old Slavonic, viewers are indulged with the rich interior of the historic structure. Then the film juxtaposes this scene with the face of a focused scientist climbing a church tower to check radio transmitters.

What the episode doesn't tell you is that the radio tower is located on top of the Catholic Church, not the Orthodox Church, and that scientists, most of them devout Roman Catholics, set their own parameters of respect for their religion. The scene makes it look as though insensitive scientists climb church towers during services with no regard for "minority" religious beliefs.

From there the international volunteers want to film the capture of wolves, but the Polish biologists block them, as they say they fear for the safety of wolves and the biologists. Without the biologists' permission, the film crew tracks the biologists to the capture site, where the biologists scuffle with the volunteers to keep the film crew away from the captured wolf. Yelling and hostility ensue, portraying a biological cover-up of animal cruelty. Viewers see a shaky zoomed-in closeup of a wolf, struggling in a pinned-down net.

The story line then takes an investigative journalist approach, bringing in a rival researcher who works for State Forestry. The researcher, who has long conducted counts of game mammals for the forestry districts, is shown from the porch of her cottage. As she gazes off contemplatively into dense forest, she condemns the cold nature of Polish biologists in Białowieża.[21]

In a shocking conclusion to the episode, she and the chief forester pull a dead lynx from a shed. They explain how a radio collar choked the animal to death. The overall message is that Polish biologists are hapless products of the wrong kind of search for modernity in Europe's last *primeval* forest, which is best left to its Eastern character.

The wildlife biologists who appeared in the film took great pains to explain to me how the lynx shown in the film had been killed by a dog, as evidenced in an autopsy conducted by a veterinarian. Polish courts vindicated the biologists after animal rights activists took them to court.

What does it mean that local people interpret the ability of biologists and foresters to make credible facts about the forest where they live? The most prominent landmark for visualizing and understanding the Białowieża controversy is the post-Socialist context in which the radio telemetry details can be set. Biologists pursue careers and forest protection simultaneously as a means to speak authoritatively to reform their society. Becoming a "real" scientist means participating in standardized practices shared by an international community of wildlife biologists. Unless biologists acquire and use certain technologies, they cannot legitimately speak for the forest at the level of the international scientific community. Technologies give them an "objective" picture, one that evokes scientific authority over the body of the forest.[22] The more entrenched biologists become in pursuing careers and using new technologies, the further they appear from belonging in the local community. Radio telemetry is not the device that will settle the question of what is the proper management of nature for Białowieża or even what is "good" science. Yet the "facts" of the technology's interpretation and application at the local level speak volumes about which truth counts.

But foresters are also using objectifying technologies to know the forest, and their method derives from a long tradition of the applied science known as scientific forestry. The forest is divided into a grid of even parts, and the forester monitors the growth of trees for the volume of timber in each of them, with prescribed cycles for cutting, thinning, and pruning. As James Scott has pointed out, scientific forestry first rationalized forests through viewing a forest only in board feet.[23]

One biologist differentiated Białowieża's scientists from foresters on cultural terms: "Biologists don't wear uniforms, address each other as 'sir,' or kiss women's hands." She was pointing out that biologists do not have the same relationship of patronage with the locals as the foresters.

Local people know the forest in many ways that do not involve "sophisticated" technology. Most inhabitants seasonally harvest mushrooms for their own consumption and for sale. Some collect antlers and herbs. But the era when local people went to the forest weekly for firewood, work, or socializing passed with the changing economic reality. Local people's use of a television program to contest scientific authority suggests the importance (for them) of a valid source of Western information about their "local" Polish scientists.

Gregg Mitman and Donna Haraway have shown how nature films have long been complicit in creating illusions of reality about nature and the scientists who study it.[24] Attracting large audiences and capturing the public's imagination are at the very core of nature shows for television. Nature films, as Mitman tells us, are supposed to give viewers direct, unmediated access to wildlife. Their authoritative appeal is their scientific veneer. In Białowieża, local people spend less time in the forest and, based on the ubiquitous presence of satellite dishes on houses, more time consuming television. While increasing consumption of television might be one of the clichés of the developing world, scholars cannot overlook the power and potential of this technology in creating new social relationships between people, but also between people and nature, especially nature in the place where they live.

Biologists took exasperated offense to the episode and local people's use of it, blaming local people's mentality on a few animal rights activists who had ties to one prominent researcher at State Forestry. Biologists wanted to conduct their science with reasonable barriers determined by professional communities. Given that biologists needed to access animal movements throughout the whole forest, including the commercial part, they spoke jokingly about hiring a public relations manager to improve their image with local people.

Given the complex history of lies and manipulation by the secret police and others during the Communist era, the Polish intelligentsia have been particularly eager to find social tools that produce "truth" in a democratic era. *Science,* understood as a detached form of inquiry for producing truths, had to belong to the side of democracy and social development. Biologists wanted to protect and even police the boundaries of what counted as good science and democracy.

For biologists, the attack on their radio telemetry activities was misdirected and frustrating. In my interviews, biologists never drew my attention to the "Western" nature of the television production that had generated such great misunderstanding. For locals, the production provided evidence of what biologists "really" do in the forest, that which locals can't see by going to the forest, confirming rumors spread between neighbors.

Locals viewing the Discovery Channel believed they were drawing upon international notions of good scientific practices. Local people used these discourses to reinforce an idea of belonging (foresters and themselves) rooted in the deep natural history of the forest, while also showing their new cosmopolitan identities. Biologists persisted in viewing locals as provincials misled by the resource logic of foresters and their gullible predispositions to believe television's "truths."

When locals used the Discovery Channel to contest biologists' use of radio telemetry, they opened negotiations on who gets to be a proper scientist. They applied scrutiny to "local" biologists' methods by using an international media production that they viewed as a legitimate source of information about what proper (Western) scientists should be doing.

Thomas Gieryn suggests that scientists compete for epistemic authority, while ordinary people take shortcuts in understanding which science to believe. If we want to learn something about the way science works, the answers will be found by looking not only at the labs and practices of scientists but at which facts are trusted by the people whose lives are affected. Thus, local people are particularly important in arbitrating facts about the forest that would lead to the forests' protection and/or use. But local people do not just interpret scientific facts, if we take science to be a transcultural activity to which many actors contribute. Local people contribute in the making, stabilization, and overturning of scientific facts. They are caught up in a distinctly post-Socialist set of relations with foresters and wildlife biologists, part of a vast network of institutions and infrastructures that link expertise to knowledge. They spread notions of what counts as good science and in doing so reinforce the status quo of a commercially logged forest surrounding a small national park.

## PROVING THE FOREST IS PRIMEVAL

Neither biologists nor foresters say that their evidence proves whether the forest is really primeval or whether logging destroys the forest. Rather, biologists and foresters send messages that say "listen to us," to our potential for knowing the forest, based on our commitment to knowing the truth in our respective ways. They point to their established legitimacy as a reason that people should trust their knowledge, rather than arguing about the particular biological details of their studies.

Local people embrace the idea that they can decide for themselves who is a legitimate scientist and what should be done with the forest. "Good science" is a judgment they apply to Polish biologists, but they do not use this measure to grant epistemic authority over the forest. Local people apply standards of belonging and Socialist-era notions about the state providing them with jobs and security. Their usage of the Discovery Channel episode marks them as actors who want to participate in international discourses about science, as well as participants in civil society development. They believe they are capable of understanding complex scientific practices.

Today's socioeconomic reality of market capitalism has changed and charged relations among biologists, foresters, and locals. All employ international forms of legitimacy and the Socialist past to produce a forest that is simultaneously past, present, and future. The present symbolizes a period of potential and uncertainty, marked by the ongoing campaigns by biologists to protect the area as a national park and the tourist economy. The tourist economy guarantees no one a job and symbolizes a present where the value of the forest and of people's lives lies in appearances. Foresters battle fears in the local imagination by extending the image of their authoritative power. Foresters behave as though the state can still take care of the local population through a managed, sustainable yield of timber, as if their technocratic order over the forest can extend to the social order. Yet in real economic terms, tourism supplies most residents with work. The state liquidated most forestry jobs, leaving a skeletal crew of foresters and forestry workers.

In Białowieża, biologists', foresters', and locals' aspirations about the future are tied to projections about modernity and the West. Biologists link themselves to the ideal of universal scientific knowledge, valid in all times and places, as a way of distancing themselves from the past and of working toward a future where Białowieża's local society can trust the right kind of experts. Foresters project a forest future that can provide people with living wages, even as their organization has downsized. And local people assert their ability to continue living in the place they always have lived with a newfound power to judge science and scientists, as well as to determine the future of the forest (based on the 2001 law giving them the power to decide its fate). Local people do not want to be artifacts of the past (peasants or low-wage workers with no real spending power). Engaging in debates using the Discovery Channel episode, they challenge biologists' authority to speak for the forest and cement their alliance with foresters.

Credibility contests about what nature is doing are rarely won by science alone because science is a cultural activity. Once we are able to see how science and scientists are embroiled in social worlds, we also see how their claims are subject to local social rules for belonging and acceptability. What is interesting about this case for STS and environmental history is the way credibility is caught up in an era, in this case the era of post-Socialism. In many ways the economic differences between the member countries of the European Union (developed and less developed) provide a constant reflection of the securities, as well as the corruptions, of the former and present eras, which fuels the forest conflict. Science operates in the shadow of the imagined West, and on the forested border of the European Union, because local actors refer to Western normalcy as an unobtained yet achievable goal. Such historical stickiness binds the history of the forest with credibility contests for what nature is in the forest and for the best way to protect and use that nature.

PART III

NETWORKS, MOBILITIES,
AND BOUNDARIES

# CHAPTER 9

# THE PRODUCTION AND CIRCULATION OF STANDARDIZED KARAKUL SHEEP AND FRONTIER SETTLEMENT IN THE EMPIRES OF HITLER, MUSSOLINI, AND SALAZAR

## TIAGO SARAIVA

IN MAY 1944 Heinrich Himmler wrote to the SS Brigadenführer Körner urging him to take charge of a karakul sheep flock recently arrived at the SS-Truppenübungsplatz Böhmen, a vast Waffen-SS training area located in the territory of the Protectorate of Bohemia and Moravia.[1] The flock had been brought in by agriculture officers of the Military Administration of Russland-Süd and arrived in Bohemia following rapid German retreat on the Eastern front. Himmler made clear his personal interest in the fate of the sheep herd and even inquired about the possibility of buying part of it. The ownership of the animals was contested territory and demanded careful negotiations among the SS, the Ministry of Agriculture, and the Ministry for the Occupied Eastern Territories. Here, I don't intend to explore the well-known rivalries of the different Nazi state agencies in the German colonial overtaking of Eastern Europe, namely the tensions between Himmler's SS and Rosenberg's ministry.[2] Instead of delving into the Nazi bureaucratic maze, I prefer to follow sheep in illuminating plans for German expansion into the East.[3]

Karakul sheep were highly valued animals whose pelts were, and still are, used to produce the famous Persian fur coats also known as Astrakhan.

The growing interest in these sheep since the second half of the nineteenth century is part of the wider story of the replacement of vanishing wild animals by domesticated ones for the supply of fur markets in the context of the expansion of the luxury demands of urban consumers.[4] In Canada and Siberia, for example, mink and fox farmers would progressively replace hunters as frontier folk producing animal skins for international markets.[5] This text follows the role of karakul sheep in the settlement of the frontier for the three fascist empires of Hitler, Mussolini, and Salazar. The ability of karakul to thrive under harsh environmental conditions and their high value on the fur market allegedly enabled the settlement of brave pioneer communities, making karakul a perfect companion species for the expansionist ambitions of these three regimes.[6] More than just building a comparative argument, I intend to offer a transimperial narrative weaved via the travels of karakul to start exploring the significance of the frontier experience for fascism.

In the very difficult military context of the spring of 1944, Himmler's detailed inquiry about the climate and soil conditions of the Böhmen military camp, very different from those preferred by karakul sheep, may come as a surprise. Such attention already indicates Himmler's high expectations regarding the role of karakul in the "would-be" German colonial rule in Eastern Europe. Sheep farms were to be part of the colonies of settlers of the General Plan East, the Nazi blueprint for the future Eastern Europe drafted by the geographers, demographers, rural sociologists, and landscape architects of Himmler's Commission for the Reinforcement of Germandom (Reichskommissar für die Festigung deutschen Volkstums).[7] The plan, built on the killing, expulsion, and enslavement of tens of millions, was to convert the region into a planner's Eden, with towns, villages, forests, settlements, and industrial areas carefully distributed in the landscape and connected by a network of railway lines and Autobahnen.[8]

In mid-July 1942, when Hitler approved the plans for the settlement of these open spaces of the East, Himmler confessed it was "the happiest day of his life" and boasted about the "greatest piece of colonization that the world will ever have seen," making the case for looking at the German invasion of the Soviet Union as the last great land grab in the long and bloody history of European colonialism.[9] In the outskirts of the empire, in its most eastern regions, the new villages were to be inhabited by Aryan armed peasants forming a defensive wall repelling the barbaric Asian hordes. The "Go East" push was to be driven by the establishment of these colonial outposts trusted with the double task of "defending the ultimate ownership of the land conquered by the sword" with SS militias and increasing "German blood" to guarantee demographic expansion. If, following Frederick Jackson Turner's thesis, the experience of the western frontier was the basis of American democracy, for Nazi ideologues the settlement of the eastern frontier was to become the source of German cultural rejuvenation and the materialization of Germany's manifest destiny.[10]

The 1944 arrival in Bohemia of the karakul flock from the Ukraine is a prosaic reminder of the short life span of such grandiose visions. At the moment Himmler wrote Körner, the Red Army had already expelled the Wehrmacht from much of the Ukraine, a key region for the establishment of the New Order. Archived records don't explicitly tell us where in the Ukraine the sheep came from, but it is reasonable to suppose they originated at the karakul experiment station, created at Kriwoj Rog in 1941 by Rosenberg's Ministry for the Occupied Eastern Territories.[11] This city, located in southeastern Ukraine, was signaled in the General Plan East as a "base of colonization" for the Gotengau, which means it would be settled by SS "warrior-farmers," German frontiersmen.[12] As Himmler's physician, Felix Kersten, recalled, all members of the SS dreamed of the grand estates in the East that had been promised to them as the first fruits of victory. But as the General Plan East also reminded them, "settlers have the obligation to be exemplary managers and pioneers in the agricultural, technical, and economic aspects of farming."[13] Settlers were to make intensive use of tractors, chemical fertilizers, and new seeds and animals bred to increase productivity. The karakul experiment station at Kriwoj Rog was thus part of a technoscientific infrastructure built to produce the proper environment for the flourishing of settlers' communities, in this case supplying "warrior-farmers" with industrialized organisms and offering expertise on "scientific animal husbandry," as suggested in the General Plan East.[14]

The head of the Kriwoj Rog experiment station was Hans Hornitschek, who had temporarily left his post at the Animal Breeding Institute of the University of Halle to join the Ministry for the Occupied Eastern Territories project in settling the frontier.[15] When Rosenberg's ministry tried to hire Hornitschek for a permanent position, the director of the Halle Agricultural College, Robert Gärtner, vehemently refused, arguing his importance in the handling of Halle's own karakul flock. Hornitschek had been considered the main German expert in karakul breeding since the premature death in 1940 of his mentor, Gustav Frölich, with whom he had coauthored the karakul bible: "Das Karakulschaf und seine Zucht" (Karakul sheep and their breeding).[16] Gärtner reminded Rosenberg that the Halle karakul sheep were the purest flock in the whole of Europe and had already contributed to the ongoing colonization efforts through the distribution of certified rams in the eastern territories. If Hornitscheck left Halle for Kriwoj Rog in the Ukraine, the quality of the Halle animals was doomed to decay, with severe consequences for the production of karakul pelts all over Europe.

## KARAKUL AT THE UNIVERSITY OF HALLE

Let us then follow more closely Hornitschek's research at Halle to understand the role of the Halle karakul flock. Or in other words, let us use the tools of the trade of science studies scholars by delving into experimental practices and connecting them with political and social projects. Most of Hornitschek's work dealt with the developmental genetics of curly hair in karakul sheep.[17] He explored the development of hair on karakul skins through histological, morphological, and physiological methods, having demonstrated specifically that the bending of hair follicles determines the patterns of curl formation. Hornitschek used the practices normally associated by historians of science with the *Phënogenetik* (phenogenetics) studies of Valentin Haecker, chair of zoology at Halle. Phenogenetics aimed at bridging the gap between hereditary factors and traits through following development backward, starting from mature phenotypes and tracing their causes into the germ cells, instead of starting with the fertilized egg onward, as in experimental embryology.[18] One took the normal and variant forms of a well-defined trait and then followed their developmental stages back until one could identify a branch point prior to which their patterns of development were identical.

Only by analyzing the origin and development processes of hair curl was it possible to establish a "true" biological classification of the sheep, by contrast with classifications based solely on the phenotype, that is, the final appearance of the sheep pelt. Traditionally many different types of curl had been identified, in a great mess of names, which revealed no true difference, for they were just different developmental stages of the same curl type. This research was expected to have significant results not only for the specific classification of karakul but also for that of many other domestic animals that had "similar" processes of "curl formation . . . such as in dogs, pigs, horses, chickens, goose, and also in humans."[19] Thus, karakul was used as a model organism to illuminate the processes of hair curl formation, allowing proper biological classification to replace unreliable crude phenotypic methods. As stated by Gustav Frölich, karakul deserved attention as a research object not only for its economic value but also as a model organism in the developmental genetics of domesticated animals.[20] Karakul sheep importantly conflated science and technology in their dual nature as laboratory model organism and as market industrialized organism.

This conflation was not automatic: it demanded the experimental work of Halle scientists. Breeders at the University of Halle were the ones who established the standards for the classification of types of curl in karakul pelts. For commercialization of the furs, there was no need for any knowledge of the ways the curl developed; their value would be the same if hair orientation was tail to skull or the opposite, as long as the fur presented a homogeneous pattern. But to make decisions about selections in a karakul flock, such information was crucial. Hornitschek stressed that several selection programs had been undertaken trying to fix pure lines of karakul for different types of curl with no knowledge of curl development processes, thus leading to different lines that were actually biologically identical. Their different furs represented only different stages of development of the same type. As he bluntly asserted, "selection of karakul following the character curl type can't be solved by sheep raisers in their herd, and private initiative has no role in this class of tasks. Only scientific agriculture institutes may and should undertake such research work."[21]

Experiments undertaken since the 1910s had tried to prove the Mendelian behavior of the main characters in karakul, namely brightness, color, curl, and tail shape. In 1917, the opening volume of the collection *Bibliotheca genetica*, edited by Erwin Baur, was dedicated specifically to Mendelian heredity in karakul sheep.[22] Importantly, the research confirmed that those traits that gave furs their value had the great advantage of being dominant, thus allowing for crossings of karakul rams with other sheep varieties. Sheep farmers, after acquiring a karakul ram, could start producing valuable furs very quickly by crossing the ram with their local varieties. The negative side of this was the inability to identify "pure-blood" karakul and the proliferation of rams sold as such when they were actually the result of recent crossings. To have a transparent market of karakul rams, it was necessary to have genealogical certificates asserting the origins of the animals in question.

The breeders at the University of Halle were the ones who established such normalized certificates.[23] The records first used to experiment with the university sheep became the standard for German commercial breeders. Each ram's file included information about genealogy (four generations), birth date, brightness score, curl score, mating register, and so on. As another crucial element of the ram's file, a photograph had to be included. And as expected, the photograph procedure was also standardized as to illumination, exposure time, distance, and so on. The curl type had of course to be recognizable, but the crucial thing was that the picture be taken twenty-four hours after birth. After that point a ram's hair starts to lose its characteristics, and the curls undo themselves. It should be mentioned that this is the reason why karakul farmers, whose business is to produce furs that are valued exactly for those curly patterns, sacrifice lambs immediately after birth. To produce the much-appreciated astrakhan furs, karakul are not supposed to live more than forty-eight hours. Karakul farming was as profitable as it was brutal.

Without the genealogical books developed at Halle, sheep farmers wouldn't know how to evaluate a ram sold by a commercial breeder. Further,

Halle breeders systematically crossed karakul with other sheep varieties, advising farmers which sheep to cross with their expensive karakul ram to still get the valuable furs. Only by crossing karakul with local varieties were karakul farmers able to establish themselves in very different geographic settings.[24]

## CIRCULATING KARAKULS I: UZBEKISTAN, GERMANY, SOUTH WEST AFRICA, EASTERN EUROPE

Indeed, this was the reason the first karakuls arrived in Halle in 1903. Julius Kühn, then director of the Agriculture Institute, who was responsible for introducing animal breeding as a main area of activity at the University of Halle, imported from Buchara in Uzbekistan the first lot of four rams and twenty-six ewes. The initiative was undertaken jointly with Paul Albert Thorer, head of a major pelts firm in Leipzig. The aim was to bring karakuls into Germany and explore the possibility of raising them in the sandy soil of the country yet maintaining their characteristics and thus dispensing with the importation of pelts from Central Asia.[25] Kühn's first typical acclimation experiments demonstrated that the curl patterns of karakul were kept constant with a diet of German fodder, against the common perception that Buchara's unique environment was responsible for the prized curl patterns. However, the major success story of the Halle sheep was not located in Europe but in South West Africa. The association between the academic Kühn and the pelt merchant Thorer was also responsible for launching karakul farming in the German colony of South West Africa (SWA), today's Namibia.[26] In a letter written in 1906 to the governor of the colony, Friedrich von Lindequist, Thorer informed the governor about his eagerness concerning karakul breeding in SWA.[27] For the promotion of pelt production in SWA, he would "donate three rams and 23 ewes from the stud at Halle" and also acquire animals directly in Buchara. In February 1909, 22 rams, 252 ewes, and 14 lambs were off-loaded at the port of Swakopmund. These animals were the basis of the Government Stud in SWA, placed at Research Station Fürstenwalde, fifteen kilometers west of Windhoek, which subsequently was to be the source of private pure flocks for karakul ram production as well as for karakul cross-bred flocks for pelt production.

Now, anyone slightly familiar with the history of German SWA would place this apparently innocent circulation of sheep in the violent context of Germany's colonial experience.[28] The projects for the establishment of karakul farms in SWA were a direct consequence of the wars against the Herero and Nama, held from 1904 to 1907, which resulted in the annihilation by German colonial rulers of around 80 percent of the Herero and 50 percent of the Nama. This has been repeatedly depicted in historiography as Germany's first encounter with genocide in the twentieth century, and it has been used to make comparisons between colonial practices in Africa and Nazi behavior in occupied Europe. This literature has nevertheless recently fallen under harsh criticism because of its difficulty in establishing concrete historical ties between the two imperial experiences.[29] Here, I want to illuminate such relations by connecting African and European experiences of the frontier through karakul. It is not just a question of formal comparison, but of material circulation of industrialized organisms.

Under the "Imperial Decree of 26 December 1905 Pertaining to the Sequestration of Property of Natives in the Protectorate of South West Africa," issued by von Lindequist, the whole territory of Hereroland (central SWA) and later the whole of Namaland (south SWA) passed to the possession of German colonial rulers.[30] General Lothar Von Trotha's brutal repression of the Herero and Nama rebellion paved the way for proper settlement of the territory, transforming it into a blank slate ready for the establishment of brave German pioneers. Analogous to the promise made to SS members of idyllic farms in the open spaces of Eastern Europe after the victory over the Bolsheviks, a huge portion of the new SWA settlers came from the military ranks after demobilization following the victory over indigenous rebellions. Military personnel were allotted units of five hectares a person, at the cost of thirty pfennigs a hectare.[31]

The size of each settling unit already gives a hint of the challenges faced by the white colonization project. It is difficult to imagine how the vast arid dusty plateau that makes up much of Namibia could support visions of wealthy communities of German settlers producing for world markets. And here enter karakul: if karakul farmers in Europe had to be very cautious of the dangers brought by excessive humidity, SWA had an environment very similar to that of the Buchara region, where karakul originated. The SWA's grasses and bushes were also perfect fodder for karakul, which couldn't feel more at home.

In the interwar years, under the administration of the Union of South Africa, karakul pelts would become the cornerstone of the territory's economy. By the 1940s karakul pelts were responsible for no less than four-fifths of overseas exports, overtaking diamonds as the basis of its economy, which justified the denomination of karakuls as black diamonds.[32] The "miracle" of making the desert produce valuable goods was achieved by dotting Hereroland and Namaland with extensive white settlers' karakul farms, which replaced the open spaces of the transhumance routes.

While in the first decades of the twentieth century the research undertaken at the Institute for Animal Breeding was tightly connected to German colonization of the arid areas of Southwestern Africa through sheep farms, in the Nazi period Halle scientists were praising and exploring karakul properties for the colonization of Eastern Europe. In 1937 three professors from the University of Halle, Gustav Frölich, Theodor Römer, and Emil Woermann, presented to the Kaiser Wilhelm Society their project for a new institute that would become the Kaiser Wilhelm Institute for Animal Breeding, inaugurated in 1939 with Frölich serving as director.[33] To confirm the importance of the subject of animal breeding for Nazi expansionist ambitions, suffice it to say that by 1944 this institute had the second-largest budget of the entire Kaiser Wilhelm society, second only to the plant-breeding institute.[34] And it is no surprise to learn that the Halle karakul flock traveled together with Gustav Frölich from Halle to Rostock, considered one of the major assets of the new institution. Nonetheless, after the premature death of Frölich in 1940, the flock would return to Halle.[35]

But let us stop for a moment at the Kaiser Wilhelm Institute facilities, for the buildings and location of the new flamboyant institution deserve mention.[36] The society had at its disposal a vast holding of some eight hundred hectares in Dummerstorf, near Rostock, in the province of Mecklenburg. The architects of the society adapted a previous farm into a scientific institution, preserving the original central building, from which radiated the many different departments and their laboratories of morphology, physiology, biological chemistry, genetics, and veterinary medicine. The institute, following Gustav Frölich, was to become a central research institution for the whole of German breeding; it would become the largest such facility in all of Europe, exploring the potential of artificial insemination to bring about rapid change in the animal breeding scene of the new colonized areas of Eastern Europe. The model institution also had residential buildings, a school, a swimming pool, a community building with social facilities, farm buildings, and of course many stables. All the facilities were surrounded by green areas, with every residential house granted its respective private garden. The streets and squares were also lined by dense rows of trees. As the annual report of the Kaiser Wilhelm Society proudly asserted, the architects were able to design a model settler community, like the ones Germans would establish on the Eastern frontier, following the General Plan East.

## CIRCULATING KARAKULS II: GERMANY, ITALY, LIBYA

The travels of karakul don't finish here. Let us stick to the fascist connection and now follow the circulation of sheep from Germany into Italy. In April 1931—or, following the official calendar of Mussolini's regime, the year nine of the Fascist era—Francesco Maiocco, the head of the National Institute of Rabbit Breeding (Istituto Nazionale de Coniglicoltura) in Alessandria (Piedmont), presented to the Ministry of Agriculture a detailed account of Italian pelt production, in which rabbits played a central role.[37] In the previous years the institute had been developing standards of body weight and pelt quality to put rabbit producers and pelt merchants in accordance. Maiocco made his best effort to offer dignity to the modest object of research of his institution, reminding the minister that the city of Milan alone consumed no fewer than twenty-five thousand rabbits every week and that Florence needed a supply of some one million rabbits per year. In a country launched on a battle for self-sufficiency that would only become harsher in the following years, rabbits, following Maiocco, could become an important resource, supplying the flourishing national fashion market and decreasing Italian imports of proteins.[38] Such reasoning was of course well tuned to Mussolini's vision of Italy as an autarkic economy, able to release itself from dependency on the "plutocratic states" that dominated the world economy, namely the British Empire and the United States.[39] The closing of the gap with industrialized nations and the building of a Great Italy were to be achieved by a nationalistic development policy promoting home industries producing for internal markets and making intensive use of the country's own resources.

By the end of the 1930s, the Alessandria Institute had established formal relations with some six thousand rabbit growers, forming numerous local breeding circles whose statutes were also designed by Maiocco himself.[40] He actually worked closely with Fascist mass organizations such as the Dopolavoro, or "After Work," which the Nazi Labor Front would later translate into the Kraft durch Freude, or "Strength through Joy," organization. The task of mobilizing the immense number of "breeders and friends of rabbits," in Maiocco's words, was not easy, for we are talking of a highly dispersed group of small growers located mainly in the suburbs of major Italian cities, feeding the animals the produce of their home vegetable gardens. Only through the mass organizations of the regime, namely ones run by women, was it possible to reach this diffuse population and standardize their breeding practices.[41]

Eight years later, in 1939, Maiocco welcomed the Ministry of Agriculture to Alessandria once more, but this time he was joined by no less a figure than Mussolini in person.[42] He boasted again of the great work of his institute in contributing to the yearly Italian production of fifty million rabbits, for which he received the praise of Il Duce, along with four hundred thousand extra lire for his research work. In exchange for this generous donation, the head of the regime demanded that production be doubled to one hundred million rabbits, in order to significantly contribute to the national autarky effort. Maiocco apparently didn't shy away from the dictator's request and in addition now made promises concerning a new pelt animal developed by the institute: the karakul sheep. The pledge now was that mass production of karakul would totally cover the needs of the Milan fashion industry, which consumed some two hundred to three hundred thousand astrakhan pelts of Afghan and South West Africa origin, purchased in the two big world markets in London and Leipzig, thus saving Italy an appreciable amount of foreign currency. Embodying his promise, Maiocco offered Mussolini a karakul lamb from his institute's herd.[43] The Alessandria National Institute of Rabbit Breeding, following the example of Halle, was fashioning itself as a center of karakul circulation, supplying rams to the brave Italian settlers in the Italian colonies in Northern (Libya) and Eastern Africa (Ethiopia, Eritrea, and Somalia).

But before we explore the circuit from Italy to Africa, we still have to understand how Maiocco was able to form his pure karakul flock in Alessandria. His interest in karakul had first been raised by his visit at the end of 1930 to a pelts fair in Leipzig, where he saw furs exhibited by the Thorer company. But more important, in March of the following year, the Halle Animal Breeding Institute brought to the twelfth Milan Pelts Fair a small group of karakul sheep, whose picture made the April front page of the journal *Coniglicoltura* (Rabbit breeding), edited by the Alessandria Institute.[44] Although a few private Italian farmers had already imported some karakul rams from several European countries, the naive nature of these first assays drove Maiocco to try to emulate the Halle example and seize the opportunity to launch karakul production in Italy through the standards and methods established by his own institute. More than just offering a traditional extension service to farmers, he designed a research program concerning acclimation and the crossing of karakul rams with local ewes. Following the German model, he also established a stud registry to be managed by the Ministry of Agriculture or, alternatively, by the Alessandria Institute.

To undertake such a program, the crucial thing was to be in possession of the expensive animals, which were acquired immediately from the flock the Halle institute exhibited at the Milan Pelts Fair. In addition, Maiocco traveled that same summer of 1931 to Germany, where he not only visited the major German karakul breeders but also stayed at Halle to become acquainted with Gustav Frölich's work in crossing karakuls with European breeds.[45] In the following years the Alessandria Institute's journal would offer detailed accounts of the growing of its karakul flock, celebrating the fine qualities of the newborn lambs. From the first small group formed from the imports from Halle, the herd climbed in 1940 to 130 pure-blood karakul sheep.[46] And the numbers were not the only thing that was growing: plans around karakuls were also becoming more and more ambitious.

The original aim of importing karakul to Italy was to cross them with local breeds such as leccese-moscia and to establish karakul farms in the impoverished areas of the Mezzogiorno, namely in Puglia, Calabria, and Campania, the internal frontier of Italy. But in the 1930s, to settle the empire with peasants from the overpopulated regions of Italy, diverting migration headed for the Americas to the Italian possessions in Africa would progressively become a cornerstone policy of the Fascist regime.[47] Fascist ideologues boasted again and again that Italian colonialism had a nature totally distinct from the imperial undertakings of the plutocratic powers, namely the British Empire, for the rationale of the enterprise was not capitalist greed, but the establishment of settler colonies to absorb the Italian population surplus.[48] Or, in other words, Italian exceptionalism was to be derived from the frontier experience. And when looking at the climatic conditions of many of those territories, with the vast desert and semidesert areas of Libya, Eritrea, and Somalia, karakul farms offered the hope of reproducing the German miracle in South West Africa: producing gold, in the form of furs, out of the desert. Of course, as previously with the German colonial experience, useless desert regions in the eyes of the colonizer present a more complex reality to those inhabiting them. The establishment of brave Italian settlers in the new Italian frontier was also to be preceded by a violent story of genocide.

Although Italian atrocities during the occupation of Abyssinia in 1935–36, with ruthless gas attacks and terrorist tactics, were the object of open disapproval in the international press of the time, it has not been easy to acknowledge the violent character of Italian colonial rule more generally. But scholars such as Giorgio Rochat, Angelo del Boca, Nicola Labanca, and Alberto Sbacchi have been responsible for producing a decisive change in our understanding of Italian colonial practices.[49] They have explored namely the grim realities of the sixteen concentration camps that operated between 1930 and 1933 in the Cyrenaica region, in the eastern half of present Libya. Following Nicola Labanca's account, almost one hundred thousand people of the nomadic and seminomadic populations—that is, roughly half the population of eastern Libya—were forced to settle in the camps.[50] Punishments, executions, and death by starvation were daily occurrences. Only sixty thousand came out alive. This was the dark balance of the pacification of Cyrenaica by Governor Pietro Badoglio, paving the way for Italian peasant colonization. By October 1938 Italo Balbo, one of the most popular figures among the Fascist leadership, who was appointed governor of Libya in 1933 after Badoglio, organized in grand style a fleet of seventeen ships that brought in some twenty thousand colonists to occupy the twenty-six new model villages designed by Italian planners.[51] The truth is that the scale of planning here was no less ambitious than the plan for Eastern Europe designed by Himmler's experts just one year later. The long-term aim was to establish 6.5 million Italians in the colonies. As described by E. J. Russell, the director of the Rothamsted Experimental Station, the British national agriculture experiment station, "As this Libyan colonization is by far the largest group settlement ever undertaken its progress will be watched with the deepest interest by all concerned with colonization, and certainly by many administrators in the British Empire."[52]

And as in the German case, karakul farms were to be an integral part of the new settlements, namely in the Cyrenaica. These farms were to be managed exclusively by white settlers, since the detailed recording procedures necessary to properly conduct a karakul herd immediately excluded the possibility of any native farmers.[53] Also, karakul farms were cautiously delimited by barbed wire, constituting a typical sedentary operation not deemed appropriate to the local nomadic and seminomadic customs.[54] Already in 1932 Maiocco was asked by the head of animal breeding at the Sidi Mesri Experiment Station in Libya to deliver two karakul rams and four ewes.[55] The animals arrived in the colony in May 1934, with certificates from the Zootechnic Institute of Prague, whose flock had been started, not surprisingly, with animals from Halle. Three months after the karakul arrival in Libya, one of these expensive rams died, forcing crossing experiments with the local fat-tail breed, the barbaresca, to be conducted with a single karakul ram. In the following years the Alessandria Institute would start shipping rams from its own flock to local experiment stations scattered through the Italian Empire, including Ethiopia from 1936 onward.

But Maiocco was the first to recognize that, independent of the growth of his Alessandria flock, it would always be too small to accomplish the grand plans aimed at karakul sustaining Italian communities in the semiarid areas of the empire.[56] The problem grew only more serious when taking into consideration the increasing difficulty of purchasing karakul rams all over the world. Russia and South West Africa, the two main world producers, forbade any commerce with Italy, and it was extremely onerous to buy rams in Germany, which sold each ram at the scandalous price of twelve hundred marks in 1939, in a situation of extreme scarcity of foreign currency in the Italian economy. All these difficulties supported Maiocco's contention that the development of karakul farms was totally dependent on his institute, which was not able to deliver more than one hundred pure rams to farmers by the end of 1939, half of them to the colonial territories.[57] In good conditions these rams would be able to fertilize in one year some four thousand ewes, which would be highly insufficient to cover the national demand for pelts.

The solution lay in making intensive use of artificial insemination techniques, which were being developed in the recently founded experimental Italian Institute Lazzaro Spallanzani for Artificial Insemination in Milan, inspired by German and Russian experiments.[58] Telesforo Bonadonna, the director of the institute, reported results from his experiments with karakul on the order of one thousand ewes inseminated by a single ram in a year. The fifty rams Maiocco promised to deliver to the African colonies would thus be able to cover fifty thousand ewes every year. The first experiment started in Libya in 1939 with only eighteen ewes and several sperm concentrations, to establish the conditions for the use of the technique in the climate of the

Cyrenaica. The four basic steps of artificial insemination—collecting, diluting, conserving, and inseminating properly—were to be undertaken under strictly standardized rules. After that, 1,592 local barbaresca ewes were inseminated, with encouraging results.

Several experiments were also made to investigate the possibility of shipping the ram's sperm via airplane from Italy to the colonies, namely between Milan and Addis Ababa, the Ethiopian capital. As Bonadonna emphatically made clear, this was to be considered the first shipment of sperm from Europe to African tropical countries.[59] And it was experiments like this one that made it possible to conceive of gigantic undertakings, such as the 1939 project of a karakul farm of some twenty thousand hectares, or forty-nine thousand acres, in Giggiga, not far from Harar in Eastern Ethiopia.[60] To inseminate the eight thousand local ewes, one apparently didn't need more than a few karakul rams, whose sperm could actually be shipped from Alessandria. The truth is that Italian imperial dreams would abruptly come to an end only two years later, with the emperor Haile Selassie reentering Addis Ababa on May 5, 1941.

## CIRCULATING KARAKULS III: SOUTH WEST AFRICA, GERMANY, ANGOLA

While the Italian fascist regime was desperately fighting to keep its empire, the Portuguese, profiting from their neutrality in World War II, were engaging that same year in Southwest Angola, in the Namibe desert, their last war of pacification in their African colonies. Southwest Angola, immediately north of Namibia, was properly identified as the last frontier of the Portuguese empire, inhabited mainly by the nomadic Kuvale—who are none other than the Herero from Angola.[61] As in the Italian and German cases, the imperial dimension was also crucial for Salazar's fascist regime, which promised to revive the Portuguese imperial tradition in a country that claimed to be naturally entitled to colonize and settle distant territories.[62] To keep with the brutal tone of this chapter, although now the scale was much smaller, in 1940–41 the Portuguese army mobilized one thousand soldiers, in addition to some one thousand native troops and two airplanes, to fight a local population of no more than five thousand Kuvale shepherds, accused, probably rightly, of stealing cattle from settler ranches in the highlands.[63] This was a typical frontier story that formed the basis of many westerns. During five months the Portuguese military embarked on a manhunt throughout this desert area, executing prisoners in barbaric ways; confiscating some twenty thousand head of cattle, roughly 90 percent of the total number of animals in possession of the Kuvale; and imprisoning more than thirty-five hundred people, deporting the vast majority to the cocoa plantations of the island of São Tomé or sending them to the diamond mines in the northern region of Angola.

With the example of German settlers in South West Africa just south of the border, one didn't have to be very imaginative to start envisaging ways of making this desert area contribute to the imperial economy. In 1944 the governor of the region demanded that M. Santos Pereira, responsible for local veterinary services, investigate the possibility of reproducing the German experiment with karakul in this desert fringe between the sea and the highlands. Pereira traveled around Namibia and South Africa and was quick to give a positive answer to the governor's request, although reminding him that acquiring purebred rams from the Namibian breeders would be impossible, thanks to their fear of competition from their northern neighbor. The first rams were thus bought only after the end of World War II, from the US Department of Agriculture flock in Washington. Nevertheless, it was only after a second import of karakul sheep from Germany, in 1952—ten rams and two ewes, descendants of the Halle flock—that more encouraging results started to appear.[64]

A Karakul Experiment Station was founded in 1948 in the middle of the desert, directed by Pereira, with the entire arid area of Southwest Angola— some ten million hectares, a bit larger than the size of Portugal—earmarked as Karakul Reserve. Farms were to spread around the sixteen thousand hectares of the experiment station, which constituted the nucleus of this gigantic frontier settlement. By the beginning of the 1950s there were thirty karakul farms occupying an area of three hundred thousand hectares, with the experiment station at its center. Following Pereira one needed at least a total of five thousand sheep to maintain a viable operation, with the station lending its pure rams to inseminate farmers' flocks. Artificial insemination was now a common practice, with a single ram able to fertilize some fifteen thousand ewes every year. This deep relation between the station and the farms was enough to cast Pereira in the role of the true father of this brave pioneer community, settled in an arid region that had previously been uncolonized. And Pereira didn't limit himself to distributing the genes of his precious karakul rams: he could also be found four days a week crossing the region in his Chevy truck, overseeing the different farms, checking barbed-wire fences, water stations, or the management of the dry grasslands. He boasted that he drove no fewer than nine thousand miles every month. Back at the post, his wife was responsible for maintaining a typical Portuguese atmosphere in the house, which had been built as a model settler home, demonstrating that one could, and should, keep one's civilized European manners in the middle of the desert. A few yards away, for the accommodation of the station's native workers, fifteen huts were built in cement, forming a circle around a fountain, an adaptation of local building practices of which Pereira was very proud.[65]

Fortunately enough work has been done by anthropologists on the Kuvale that we can now understand the profoundly disrupting effects karakul farms had on local populations.[66] For shepherds whose transhumance trajectories were based in a profound knowledge of local environments, the location of the farms exactly at those points where water was accessible also prevented access to richer pasturage areas. And even when new ditches were dug by colonial authorities to serve Kuvale cattle, the farms now occupied several ogandas, the most important of the settled places, which included the cemetery and ritual sites.[67] It is important to remember that the Kuvale, after returning from their exile to Southwest Angola, always refused to be employed by the karakul farms, preferring instead to keep their seminomadic practices.

In 1975, less than one year after the overthrow of the dictatorship in Portugal, karakul farmers organized their last karakul pelts auction in the city of Sá da Bandeira, the most important city of Southwest Angola, known as the white capital of the colony, in an effort to produce enough international currency for their escape through Namibia and South Africa amid the violent turmoil of the independence period.[68] Karakul farmers thus joined the massive exile that followed; half a million Portuguese arrived in Portugal after the independence of Portuguese African colonies. The karakul farms, not surprisingly, were ransacked, and the sheep that survived now mixed promiscuously with local varieties. Nowadays the area is again the territory of the Kuvale and their mobile herds.

This end implies a profound interdependence of karakul sheep and fascist regimes. The end of the story of the karakul in Angola coincides with the end of the dictatorship in Portugal. But it also demands careful exploration of the nature of such relations. In following the karakul in the frontier territories of the three regimes, this narrative never suggested a causal relation, never suggested that whenever you have a karakul farm you have a fascist regime. One can think of fascism without karakul and of karakul without fascism. The first world producer of karakul furs was the Soviet Union, thus eliminating any easy identification of karakul with fascism. This said, why then should one follow the sheep and care about the work of breeders dealing with curl patterns and artificial insemination techniques?

This chapter tries to make the case that karakul were not only interesting for scientists interested in developmental genetics but also interesting for historians dealing with the imperial dimensions of fascism. By exploring the trajectories of these organisms, one is able to understand how the expansionist ambitions of fascist regimes were to be materialized in the frontier landscapes of Southwest Angola, Cyrenaica, and the Ukraine. This claim for integrating nonhuman animals in our narratives is well in tune with traditional environmental history and its ability to build relevant historical accounts by paying attention to bison, pigs, or mosquitoes.[69] Here, making use of the science studies toolkit, I suggest in addition the interest of looking in detail at the experimental practices that standardized these nonhuman animals. Ignoring the work undertaken by animal geneticists at the University of Halle, we would not understand how karakul traveled to Italy and from there to North Africa. The circulation of karakul was enhanced by the experiments in hair development at Halle but also by the trials on crossing with local sheep breeds and the artificial insemination undertaken in the local experiment stations in Libya and Angola.

Science studies scholars and historians of science have shown the great importance of looking in detail at the standardization of life-forms for understanding the production of biological knowledge in the twentieth century. Robert Kohler's drosophila, Karen Rader's mice, and Angela Creager's tobacco mosaic virus are now common elements in narratives dealing with the biological sciences.[70] Widespread circulation of such standardized model organisms has been importantly identified with the expansion of the communities of researchers built around them. But it seems fair to recognize that these historians, focused on the material practices of laboratories, have not been as successful in dealing with vertical circulation—between laboratories and more vast social and economical spheres—as they have been in accounting for horizontal circulation—between scientific spheres.[71] On the other hand, environmental historians and historians of technology, who have followed the circulation of standardized forms of life between different spheres, have not paid much attention to the laborious processes involved in making industrial organisms, taking standardization for granted.[72] This text explores in a single narrative how genetics research produced standardized organisms and how they circulated between different spheres. It mixes the typical approaches of science studies and the history of science, looking in detail at scientists' practices, with environmental history and history of the technology, blurring the distinction between model organisms and industrialized organisms.

Karakul are not interesting because artificial insemination techniques were later applied to humans or because curl formation patterns also pointed to human classifications. Karakul sheep are interesting in themselves because one could organize settler life around them. To put it more bluntly, settler life in the colony could be performed by breeding karakuls. Anthropologists have no problem recognizing the role of animals in many of the seminomadic societies encountered, or exterminated, by the settlers considered in this text. We should probably follow them in acknowledging the need for a deep understanding of the place of model and industrialized organisms in colonial societies.

# CHAPTER 10

# TRADING SPACES

## TRANSFERRING ENERGY AND ORGANIZING POWER IN THE NINETEENTH-CENTURY ATLANTIC GRAIN TRADE

### THOMAS D. FINGER

A SHARED COMMITMENT to uncovering the complicated process of recursive feedback is the best place to look for overlap between science and technology studies (STS) and environmental history.[1] This chapter centers the process of nature-human feedback on the concepts of energy and power. As certain individuals harness and organize energy from nature, their ability to apply that energy to realize a modicum of power is increased. Often, that power is then redirected simultaneously toward other humans and the natural world. Energy is often harnessed to create power, and that power tends to accelerate the flows of energy on which it draws. This chapter seeks to discuss the relationship between energy and power by following the career of Liverpool's Rathbone family within the Anglo-American grain trade between 1830 and 1890. I will describe how energy is structured to achieve power, how human networks and technological systems arrange energy and power, and how energy and power are moved around human societies and the natural world.[2]

I begin my analysis from two central concepts of STS: actor-network theory (ANT) and technological systems thinking. I suggest that ANT and systems thinking describe two different ways of organizing energy. ANT is interested in how networks are formed, why they are held together, and the reasons why they break apart.[3] Technological systems thinking helps explain the evolution of complex bundles of human values, institutions, and technology that gain increased coherence over time.[4] Unlike actor networks, technological systems tend to become more stable over time as they develop internal dynamic and logic. This is because while actor networks account for the power of material objects to influence human decisions, their internal logic remains in place only as long as the actors accept their enrollment in the network. Technological systems, as Thomas Hughes notes, can achieve a high degree of stability from the "mass" of material goods, the "velocity" of relevant skills and institutions, and the "direction" shaped by generally agreed-upon cultural goals. The "mass" and "velocity" can stabilize technical systems even after a breakdown of "direction." This process Hughes terms *momentum*.[5] No such force exists to keep actor networks stable over long periods.

ANT and systems thinking are analytical categories that both may be applied to describe the same constellation of humans, artifacts, and nature. But there is some utility to establishing a difference based on the internal consistency of the network and their differing reliance on social mechanisms (networks) and technology (systems). Systems thinking assumes a modicum of human control over energy and therefore a greater amount of power over a particular arrangement. Actor networks also represent a way of organizing energy to achieve power, but this way of thinking describes an arrangement that is less stable over time.

Both actor networks and technological systems need a steady input of energy.[6] Power is the extent to which that energy is applied within particular arrangements by one individual or group to limit the action or decision-making ability of another.[7] It is no coincidence that both ANT and technical systems thinking employ concepts from physics—momentum in the case of technical systems, entropy in the case of ANT—to explain their evolution and internal distribution.[8] Actor networks break up because they are social arrangements subject to considerable entropy; power is held only when all participants agree to the terms of involvement. In technological systems, power tends to increase over time because human groups organize their activity around technologies that stabilize the flow of energy, thus reducing the energy lost to the system through entropy.

Environmental historians can apply the energy-power continuum to explain why certain individuals have greater power than others to initiate environmental change at particular places and times. In this chapter, I use ANT and systems thinking to help explain why environmental changes in the American prairies and grasslands (well documented by environmental historians) accelerated in the 1870s. Beyond explaining accelerated environmental change in the American Midwest, this chapter also outlines the ways in which energy and power can tie together nature and social organization.

In 1853, William Rathbone, scion of a Liverpool merchant dynasty, steered his family from the profitable cotton trade into the grain trade. In 1884, William's brother Samuel sent an exasperated letter to his partners, noting the family could no longer profit from the grain trade and would have to drastically scale back their operations. These moments bookend a story of energy and power—told through the Anglo-American grain trade—that connects the expansion of wheat agriculture in the American West with the expansion of the British industrial economy in the nineteenth century.

The nineteenth-century Anglo-American grain trade transported food energy stored in wheat from American farms to the stomachs of English industrial laborers. It filtered money from England to the United States, triggering the development of transportation and financial infrastructure in the United States that made the energy flow more stable and regular over time. Beginning as a mere trickle in the early 1800s, the Anglo-American grain trade had reached monumental proportions by 1900. Grains and flour were the most valuable import into Great Britain by the beginning of the twentieth century, surpassing even cotton.[9] The food energy provided by this trade was instrumental in supporting the vast urban populations of England's industrial centers, and capital sent to the United States from Great Britain was central to many of the economic developments there in the nineteenth century.

The Anglo-American grain trade can largely be characterized by two overlapping periods that I will trace through the career of the Rathbones using the concepts of ANT and technological systems. From circa 1800 to 1870, the grain trade was dominated by loosely connected family merchant firms that relied on a complex network of personal relationships. This actor network was held together by common understandings of trust and respectability. Following the 1860s, the grain trade evolved into something more closely resembling a technological system dominated by organized companies that controlled transportation and storage technologies according to the doctrines of integration and efficiency.[10] Both arrangements were, at their heart, methods of dealing with the natural variability of agroecosystems and the cellular composition of the grain itself.

The Rathbones achieved power within the merchant network but found themselves increasingly marginalized within the system. As other merchants increasingly turned to technology as a management strategy, the underlying structure of the grain trade transformed. Merchants who controlled the technologies increasingly dictated the terms of the trade, slowly removing the relationships formed around respectability and reforming relationships based on standardization and efficiency that dramatically increased

TABLE 10.1. The Grain Trade as Actor Network and Technological System

|  | Actor Network | Technological System |
|---|---|---|
| Dates (approximate) | 1820s–1870s | 1870s–1920s |
| Underlying values | Trust and respectability | Increasing volume and efficiency |
| Predominant business strategies | Diversification, correspondent networks, family firms | Integration, speculation, companies |
| Characteristic technologies | Ocean packets, canals | Steamships, railroads, elevators |
| Trade policies | Protectionism | Free trade |
| Flow of energy | Intermittent | Regular |
| Volume of energy | Small | Large |
| Location of power | Controllers of information and arbiters of respectability | Controllers of transportation and storage |

the volume of trade and corresponding environmental changes throughout the American Midwest and West (see table 10.1).

## NETWORKING ENERGY TO ACHIEVE POWER

ANT helps us describe the many problems the Rathbones faced within the emerging grain trade, how they attempted to deal with those problems through the maintenance of complex social relationships, and why in the end they chose to adopt a technological solution to these problems. Beginning in the 1820s and 1830s, the Rathbone family moved increasingly from the cotton to the wheat trade. In doing so, they isolated a series of problems at the same time as they became increasingly interested in the trade's potential for profit. Finally, the firm jumped headlong into the grain trade in the 1850s and formed a transatlantic merchant network formed around standards of respectability and trust. The Rathbones employed these social relationships and the information they garnered to exert a modicum of power over the Anglo-American grain trade between the 1850s and the 1870s. These episodes map well onto the periods of problematization, *interessement*, enrollment, and mobilization described by ANT.[11]

The Rathbones were slow to move into the grain trade because it presented a series of entwined political, environmental, and technological problems. In 1812, the British government passed a prohibitive tax on imported grain to ensure domestic production.[12] Critics maintained that these Corn Laws made the price unnaturally high and increased the likelihood of bread shortages in years of dearth.

If the Corn Laws made it largely unprofitable to import foreign grain to feed Britain's hungry, the natural volatility of the trade itself made it even more problematic. Grain prices rose and fell constantly, rendering the grain trade a risky way to turn a profit. The trade required perfect timing and a knowledge of up-to-date weather conditions in both the United States and Great Britain. Drought, excessive rain, rust, rot, and the dreaded Hessian fly all made grain production highly variable from year to year in the United States and Great Britain. Grain merchants turned a profit when they could time demand in one area with abundance in another. Grain merchants, in short, needed to maintain a constant stream of knowledge about the climatic and soil conditions of grain-producing regions the world over.

In addition to problems presented by the politics of international tariffs and the variability of agricultural ecosystems, the grain itself gave cause for trepidation. While cotton breaks down very slowly over time, wheat and flour lose their nutritional (and hence economic) value quickly. Bales of cotton can sit in warehouses for months, even years, waiting for favorable prices. Wheat had to be transported quickly to reduce the likelihood of damage from heat and moisture.[13]

Merchants also had to contend with a transport system predicated on canals and ocean packets that (1) was similarly dictated by the seasons, (2) required grain to change hands multiple times, and (3) took over a month to transport from American farm to British market.[14] The timing of shipments was of paramount importance, for, as the Rathbones found out, even a short rainstorm could send prices on the Liverpool market spiraling upward in a matter of hours. The trick for merchants was to time lumbering transportation with the quick pulses of market price. Rathbone reported on this complex jumble of price, seasonality, and transport that grain merchants prior to the 1850s were forced to negotiate:

> Consignments of flour . . . go forward in masses at two seasons when the supplies come in too large for the receivers to manage with ease. . . . After the canals open first day during May and June is the first period and Oct and part of Nov when the flour from the new which courses in is the second. These also are the best times to buy particularly when money is dear as this season. This is because they know that all the consignments go forward at once that the Flour-men are so anxious to have theirs held till the rush is over.[15]

A recurring problem throughout the 1840s and 1850s—a period when wheat production in the Ohio and Upper Mississippi valleys exploded—was lack of adequate transport to the seaboard. Even as late as 1868, William could report that "supplies are accumulating and the movement in the West is only restricted by the capacity of the means of conveyance."[16]

Despite these problems, the family entered into a period of *interessement* in the grain trade during the 1840s. While this was in part a generational shift (William Rathbone VI replaced his more conservative father in the early 1840s), it was also due to the new underlying energy needs of Britain's industrial economy. William VI knew that England's population was growing, and thus the market for wheat was growing as well. He began to write his father with increasing fervor in the 1840s, maintaining that the grain trade represented potential profits. Understood in the parlance of our times, William VI knew that Britain's economic and demographic surge necessitated an expanded energetic base. William wrote his father, "I prefer to have at least some part of our business in grain for the Mills which consume it are the last put on short allowances and death is the only stoppage they know."[17] Essentially, William argued that grain was perfect for the family business because people will always need to eat, and even in times of economic depression, caloric requirements remain the same even as demand for other commodities may drop.

William Rathbone traveled to New York City twice during the late 1830s and 1840s to enroll business connections that would help the Rathbones find access to grain. They did this by inserting themselves into a network that had coalesced in New York, Liverpool, and London in the 1830s around the powerful merchant firm Baring Brothers & Company. William VI had apprenticed in Baring Brothers' London office in the early 1830s and had crossed the Atlantic in 1841 with Joshua Bates, the influential Barings partner who coordinated much of that firm's American business.[18] While in New York, William met and established correspondence with a number of merchants specializing in the grain trade flowing out of New York and Ohio via the Erie Canal.

By the 1850s the Rathbones were convinced of the future prospects of the grain trade and had mobilized their organization to begin shipping grain. Prompted by their participation within the network, the Rathbones jumped headlong into the northern wheat trade, consistently importing wheat from markets in Baltimore and New York City by 1861.[19] These shipments usually amounted to approximately ten thousand bushels apiece, a number dictated by the size of farms, transport, and storage. Compared with the following decades, the comparatively low volumes of the grain trade in the 1850s and 1860s were the result of conditions within the network. The Rathbones remained diversified, engaging in wheat transactions when the price timing indicated a potential profit. The risk of loss was as great as the risk of profit due to the wide fluctuations in price and the slow pace and high cost of transportation. Grain changed hands multiple times as it worked its way through the network, and at each stage profits reduced as middlemen took their cut.[20] For these reasons, imports of American grain into Great Britain before the 1870s remained comparatively small.[21]

Until the 1850s and 1860s, the Rathbones obtained their wheat primarily from Upstate New York. During this period the grain-producing regions of the United States were located near the Atlantic Coast, in New England, the Shenandoah Valley of Virginia, the eastern valleys of Pennsylvania, and the Hudson and Genesee valleys of New York.[22] As environmental historians have argued, eastern farmers had over generations developed systems of agriculture that mixed Native American practices with traditional European styles.[23] Though these regions were undoubtedly feeling pressure by the 1840s and 1850s to modify their agricultural practices to suit market demands, farms in all regions were small family affairs. Agricultural practices on these farms more closely resembled methods adopted by the farmers' grandfathers than the mechanized and intensive farming their sons would adopt on the prairie a generation later. Girdling trees in order to preserve soil fertility and reduce erosion—a practice developed in colonial New England—was documented in New York's wheat districts well into the 1830s.[24] Mixed agricultural systems were also the order of the day. In New York and elsewhere farmers kept their soils fertilized with cow manure.[25] During this period, grain merchants had to move with dexterity from region to region. Crops failed in one place and abounded in another, and merchants could not look to one region alone to satisfy their orders. The origin of the next shipment was never taken for granted.

The essential method for dealing with the complex problems of the grain trade prior to the 1850s and 1860s was the management of social relationships to ensure trusted information about the location and quality of prospective shipments. If merchants could not be sure about the actual condition of crops, they contented themselves with information about the merchants who forwarded them. Rathbone spent as much time researching the status of merchants as he did the prospects of the trade itself. William sent a steady stream of information back to the Liverpool office, passing uncompromising judgments of coded merchants: "No. 33—Ruined by gambling," "No. 65—Entirely ruined," "No. 80—Prudent, understands his business well, acts as agent for Browns."[26] In an economic climate of uncertainty, in which the natural world and transport technology both represented potential problems, the Rathbones attempted to manage their network through the arbitration of social standing. Even this was an unsure method. As late as 1855, one merchant wrote to the Rathbones, "I will endeavor to get you some information but it is extremely difficult to get any that can be relied upon."[27]

As a direct result of the measures they adopted to deal with the problem of reliable information, the Rathbones saw their influence grow within the grain trade. The Rathbones had facilitated a flow of information that made them privy to knowledge that others did not have. Letters poured into Liverpool from Ohio, New York, Baltimore, and England's agricultural districts, all speaking to the potential of particular transactions. This network, over time, allowed the Rathbones a greater ability to dictate the price of transport and the price of their commission, thus removing risk from their organization and placing it at the foot of others. As a clearinghouse of information —however incomplete—the Rathbones became essential members of merchant organizations in Liverpool that structured the flow of energy in and out of that port. In 1851, William became chairman of Liverpool's American Chamber of Commerce and thus helped entice American produce to Liverpool by setting produce standards and merchant regulations. In 1856, he became a member of Liverpool's Dock Committee of the Corporation and its successor the Mersey Docks and Harbour Board.[28] With his participation in these merchant organizations, the Rathbones further structured energy flows to achieve power.

## TECHNOLOGIES, POWER, AND MARGINALITY

Beginning in the 1850s, in an attempt to further expand their power over the grain trade, the Rathbones began to invest in technology. In doing so, they inadvertently set the terms of their eventual marginalization. They could not have realized that by investing in technology, the family moved from a social network that they had created to a technological system in which they were only a minor participant. While the Rathbones enabled such a transition with their investments, they did not anticipate the considerable organizational changes such investments would provoke. The vast flow of grain these investments produced temporarily blinded them to these facts. Systems thinking provides a way of thinking through the various stages of this evolution. From a period of invention and development in which they did not actively participate, investment by the Rathbones helped circulate the means through which the system could grow, leading ultimately to momentum.[29]

The implications of this shift from network to system are best illustrated by describing the Rathbones' investment in land and transportation in Iowa during the 1860s and 1870s. The Rathbones saw integration toward the production of food energy as a risk-management strategy. While they had previously invested their money in actual shipments and devoted their efforts to researching individual merchants, the Rathbones began to more actively filter their money toward the sources of their supply, thus opening several sources of potential profit from the same system. Throughout the early 1860s, the Rathbones invested in over fifteen hundred acres of land in Hamilton and Grundy counties, Iowa. These were "prairie lands of prime quality."[30] An agent reported optimistically that "the great hope for this property is that someday a railroad may pass within a reasonable distance," in which case the lands "must become a source of value at no very distant day."[31] Not content to wait around for this eventuality, the Rathbones took action. They increased their investments in the Dubuque and Iowa City Railroad, which connected their lands to the Mississippi port of Dubuque, Iowa. They also invested in the Dubuque Harbor Company. These investments served to connect their lands in Iowa to the growing orbit of the Illinois Central Railroad, also included in the Rathbones' portfolio. By the 1870s the Rathbones had invested in every point of transshipment from their lands in Iowa, to American grain ports, across the Atlantic Ocean, to Liverpool docks, and thence on the English railway system to its eventual point of consumption.[32]

The Rathbones increasingly adopted a technological solution to information as well. The Rathbones were one of the earliest business organizations to make use of the 1866 Atlantic Cable to discuss investment opportunities and harvest prospects with agents and brokers in New York.[33] Their use of the cable foreshadowed problems they would continue to face as they ceded control of information to those who controlled technology. Atlantic Cable telegrams were frequently mixed up, delayed, and lost. In addition, the company worried that employees of the cable company were stealing information and selling it to their competitors. In July 1867, a mere twelve months after the opening of the line, one company agent wrote, "It seems almost impossible to get the cable honestly managed," and later, in reference to grain prices, "There is not the slightest doubt that the news was falsified in the interest of speculators who had combined to keep up the price of the article here. From the way the thing is worked here the cable is getting at to be a great nuisance."[34] It was beginning to dawn on the Rathbones that their investments in technology had forfeited some of their competitive advantage in the grain trade.

Success temporarily blinded them to these new realities. By the 1870s, the Rathbones, aided by railroads and the Atlantic Cable, were shipping immense quantities of grain. In October 1873, an agent reported, "We are shipping wheat as freely as possible and have already forwarded . . . 125,000 Bus. . . . Next week we hope to ship from 100,000 to 120,000 Bus more and as much more in the following week."[35] In the span of ten years, their capacity to ship wheat had increased by an order of ten.

Environmental historians have paid particular attention to the changes wrought by the spread of market demand into the Midwest and West.[36] Notice that in the span of ten years, the sources of the Rathbones' supply had moved from within four hundred miles of the eastern seaboard to the vast interior of the North American continent. As American farmers and British capital moved into the West, their ability to impose market demand on the environment accelerated. The measures taken by the Rathbones to improve their position within the grain trade can help describe why those changes accelerated in the 1860s and 1870s. As the volume sent through the Anglo-American grain trade increased, so too did the plowing of the prairies' grasses and the mining of their soils. The energy and nutrients extracted from the American interior powered industrial labor on the eastern seaboard and in Great Britain. The capital and credit provided by British investors flooded midwestern agricultural communities, bolstering the construction of railroads and providing farmers with the means to purchase new types of farm machinery. By way of example, one chronicler of Hamilton County, Iowa (in which the Rathbones had invested), noted that the expansion of wheat growing and the use of elevators coincided with the arrival of the Illinois Central and the increased flow of outside money in the 1860s.[37] Farmers across the prairies of Illinois, Wisconsin, Iowa, and Minnesota quickly altered their practices to suit the new market realities. Farmers in these regions increasingly adopted wheat as a single crop, mining the soils of the rich grasslands without added nutrient inputs, rather than spending time tending cows in a mixed system.[38] The epitome of this new style of extensive monoculture farming were the vast bonanza wheat farms of North Dakota's Red River Valley.[39] Environmental changes in the American Midwest were thus bound up with the momentum of the transportation and financial system that had originally sought to contend with environmental variability within the Anglo-American grain trade.

By the late 1870s and early 1880s, Rathbone Brothers and Company slowly came to realize the implications of these changes. The technological system in which they operated had slowly shifted the location of decision-making power from family firms like the Rathbones toward companies that controlled transportation and storage. The Rathbones had spent two decades building an organization that sought to deal with the dramatic variability of the grain trade. When the transportation system began to stamp that variability out, the Rathbones found their organization ill suited to the new reality.

In 1884, Samuel Rathbone—who had taken over management of the company from his brother William—wrote his partners to suggest a dramatic reorientation and downsizing of the company's business away from commodities forwarding to focus solely on shipping and finance. In that letter, he outlined five reasons why the nature of business necessitated such a move. First, "our business has continued to be carried on too much on the lines of the old fashioned merchant, and that in several trades, her services are no longer required." Second, "the telegraph and steamers have made it far more difficult than formerly to make money out of anticipated fluctuations in prices." Third, "the telegraph has enormously increased the power of Banks to stimulate operations in produce, and has encouraged the use of credit and capital." Fourth, "the effect of the electric telegraph has further been to introduce a permanent, speculative gambling element into many trades which I fear will be fatal to the profits of a legitimate kind." And fifth, "the general tendency of prices of many articles must for sometime be downward as every day new fields of production, new methods of agriculture, new methods of manufacture are being brought into play, and the most vigorous and enterprizing [sic] races are emigrating from countries where their energies had little scope to new lands, where with the aid of steam their productive powers are wonderful."[40] By 1890, the firm had virtually ceased shipping grain.

Samuel had isolated a fundamental difference between the two periods of the grain trade, between the inner workings of actor networks versus technological systems. He described a structural shift in the nature of long-distance trade that favored some business organizations over others.[41] Energy was now structured in new ways, changing the locus of power. He noted that long-distance transport, up-to-date information on harvests, and increases in grain production flattened the variability in prices that the old merchant network was designed to exploit. The logic of the system favored the organizations that had grown up within it: integrated and hierarchical firms that employed large numbers of individuals stretched out over long distances.[42] These companies kept transportation and distribution within one organization, thus eliminating the difficulties between multiple organizations that likely had subtly different goals.

Rathbone Brothers' involvement in the grain trade following the 1840s traces an arc followed by many other grain merchants in the nineteenth century. By 1900, the family outfit had virtually disappeared from a grain trade dominated by railroads and ocean liners. These new companies were predicated upon a vast shipment of grain at the lowest price possible, and their control ensured that grain traveling from the American interior to Manchester, England, would only have to change hands two or three times, as opposed to the five or six it would have under the guidance of firms like Rathbone Brothers. Just as William had foreseen the connections that would bind Britain and the United States together through the grain trade, so too did Samuel now isolate the conditions of the trade that would eventually leave the firm marginalized.

## ORGANIZING NATURE, STRUCTURING POWER

In 1800, Great Britain had a population of just over nine million. It imported 1,265,000 quarters of grain, only a small fraction of which came from the United States.[43] In 1900, Britain had a population of 32,249,000. It imported 68,669,000 centals of grain, 32,588,000 of which came from the United States.[44] That means that over the course of the nineteenth century, grain imports into Great Britain from the United States grew 217 times.[45] From 1878 to 1900, the United States exported an average of 30 percent of its grain crop.[46] During that period, these US exports represented on average 47 percent of all wheat imports into Britain.[47] While these numbers reflect the growing scope and scale of the North Atlantic grain trade, they are also indicative of a vast increase in the abilities of humans to initiate environmental change in the nineteenth century.[48]

The Rathbones' participation in the grain trade provides environmental historians with an example of how STS can help environmental historians analyze the relationship between environmental change and social organization. Business in the grain trade necessitated a constant stream of information and knowledge of ecosystems that was constantly reproducing itself in market logic. The Rathbones and other grain merchants utilized this information, along with investments and personal relationships, to structure the flow of energy from the American interior to British industrial cities. In an attempt to collect this energy, grain merchants connected many ecosystems and social groups into shifting but increasingly stable relationships between 1830 and 1880. The marginalization of the Rathbones was tied to this stability. As extraction of energy from agricultural ecosystems in the American West accelerated and increased in regularity, the network forged by the Rathbones to contend with variability lost its relevance. Power derived from one method of extracting energy from the environment does not transfer to another. In this way, environmental historians can apply ANT and systems thinking to analyze complex arrangements of humans, nature, and technology in ways that can (1) provide a framework for describing the systemic origins of accelerated environmental change and (2) highlight the ways in which networks and systems connect that environmental change to the rearrangement of social relationships.

Again, let me emphasize that I could have told this entire story using only ANT or systems thinking. I could have noted that the increasing reliance on technology to deal with variability only altered the terms of enrollment with the actor network, which kept evolving as the natural, human, and technological actors imposed new realities. I could have also emphasized how the canals and ocean packets centering on Liverpool and New York in the

1830s were a technological system largely controlled through the whims of a few British banks. Both types of analysis could have told the story very well. But they would have told different stories than the one I wanted to tell. Highlighting a difference between ANT and systems thinking allows STS and environmental history a way to more closely analyze the relationships among sociotechnical arrangements and the scope and scale of environmental change.

ANT and systems thinking describe two different ways of structuring energy and power. Actor networks organize energy primarily through social interaction, while technological systems stabilize energy flows through the very materiality of artifacts themselves. This difference is crucial for coming to grips with not only how changes in American agriculture were connected to the British grain market in the nineteenth century but, on a more general level, how humans are able to increase the spatial extent of anthropomorphic environmental change at any given time and place. Networks and systems describe different relationships among humans, their technologies, and the natural world by emphasizing how control over underlying energy regimes dictates the speed, strength, and structure of relationships among places.[49] The internal consistency of systems channels energy and power with greater structure and purpose than networks.

The natural world sits at all points within the energy-power equation. Nature provides the energy that all humans require for action, sets the terms of extraction, and serves as collateral damage in human ambitions for power. STS can help environmental historians fully describe this relationship by highlighting how power dynamics within society are bound up in the strategies humans employ to extract resources from the natural world. As humans rearrange the natural world and direct its energy, so too must they organize and restructure power in society. This the Rathbones found out the hard way.

# CHAPTER 11

# SITUATED YET MOBILE

## EXAMINING THE ENVIRONMENTAL HISTORY
## OF ARCTIC ECOLOGICAL SCIENCE

### STEPHEN BOCKING

FOR CENTURIES THE Arctic has captured scientists' attention. Features found nowhere else—polar bears, tundra, permafrost, months-long days and nights—underscore its distinctive nature. These phenomena have long attracted and challenged scientists, as have the region's extreme conditions, remoteness, and reputation as pristine wilderness.[1] There is also a rich history of cultural meanings associated with the Arctic, expressed through art and literature.[2] As a region with special intellectual significance, yet situated on the margins of human activity, the Arctic presents an opportunity to explore how scientific practices and knowledge relate to the environment: how, in other words, one might write an environmental history of science.

However, an environmental history of arctic science must also acknowledge that for scientists the region's distinctive environment was not their only concern. Scientists have moved readily into and out of the Arctic, applying knowledge and techniques from elsewhere, while arguing that arctic knowledge could be relevant to other regions of the world. How did they manage this apparent tension between studying a distinctive environment and moving knowledge, and themselves, in and out of the region? Certainly, scientists themselves often asked this question, as they considered how they should situate themselves and their work. And environmental historians must as well, when they examine place and movement in the history of scientific knowledge.

Place and movement are also essential themes in science and technology studies. Two concepts from this field are particularly relevant to the environmental history of science. "Situated science" acknowledges that scientific knowledge and practices should be understood in terms of particular places, such as laboratories or field sites. "Mobile science" refers to the circulation of scientists and knowledge and, in particular, to the capacity of knowledge to "travel"—that is, to be considered reliable beyond the sites of its production. Both concepts can help us understand how arctic science was both situated in this region and made mobile beyond its boundaries.

This paper examines ecological research in the Canadian Arctic during the 1960s and 1970s. This was a time of active interest in arctic ecology, particularly in three specialties: animal population ecology, boreal ecology, and the ecology of fire. Each specialty had a complex relationship with the Arctic. While some ecologists viewed themselves as arctic scientists, asking questions and applying methods distinctive to this region, others saw themselves as members of scientific communities based elsewhere. Concepts of place and movement—of situated and mobile science—were therefore central to arctic ecological research. The potential tension these present, between locality and circulation, is the focus of this chapter.

## SITUATED AND MOBILE SCIENCE

In recent years historians and other scholars have examined the relations among place, space, and science, exploring how social, political, and cultural circumstances, institutions, and research sites present scientists with constraints and opportunities, thereby shaping scientific agendas, methods, and knowledge. Knowledge production may be situated anywhere: in laboratories, field sites, or museums; or in cities, regions, and the Earth itself. Scientists have also contributed to constructing the identities of spaces and places. Certain places, characterized as "truth spots," have emerged as privileged sites for research, because of their role in lending credibility to scientific claims.[3]

Laboratories have long been considered the paradigmatic location for science. For scientists, laboratories have distinctive advantages, of control, exclusion, and standardization, that enable them to claim that knowledge produced there is objective—that is, uninfluenced by local circumstances. Essential to this claim is the assertion that all laboratories are equivalent and so, in effect, are located nowhere.[4] Historians have also described how places outside the laboratory—anywhere from ships to schools, as well as field stations, urban and workplace environments, and landscapes associated with health or illness—relate to scientific practice and knowledge. Certain field sites have played special roles in forming knowledge, presenting opportunities for distinctive "practices of place" through which knowledge can be produced.[5] Episodes in the history of ecology—Charles Elton on Bear Island in the early 1920s, Raymond Lindeman at Cedar Creek Bog in the 1940s, Herb Bormann and Gene Likens at Hubbard Brook since the 1960s—testify to the interdependence of knowledge and place, with research at each site contributing to formation of ecological concepts of wider significance.[6] Making a place for science in the field involves justifying choice of research site, by demonstrating that it presents phenomena of interest to scientists elsewhere and is conducive to producing credible knowledge. Technology—cameras, remote sensing, airplanes—contributes to defining these relations between science and place: distancing observer from object, reinforcing the authority of scientific knowledge over that gained by direct sensory experience.[7]

Science in the lab and science in the field represent distinct attitudes toward place: denying it or incorporating it within the practice of research. In practice, however, this distinction is often blurred: field study intertwines with laboratory work, with data and specimens moving from field to lab for analysis, field sites described as "outdoor laboratories," and laboratories placed in the field. Scientists may choose only certain aspects of a field site as relevant, downplaying its "messiness" by omitting whatever is considered contrary to the laboratory ideal of control and standardization. Alternatively, they may form hybrid research sites that combine the advantages of both places. As agricultural scientists and others who have placed themselves between standardized lab and messy field can attest, each site has epistemic virtues, and all can serve as complementary sources of credibility.[8]

Science is also situated on larger scales than lab or field. Cities, regions, nations, and the Earth itself have served as both places and objects of study. Science has been enrolled in surveying national or colonial territories, rendering them legible, serving as an instrument of national or imperial authority.[9] Community and indigenous knowledge are commonly considered inseparable from their site of origin, with this feature sometimes invoked to assert a boundary between "local" and scientific knowledge—but also to assert the legitimacy of local knowledge against the simplifying and universalizing forces of global science. Just as global political imperatives, such as climate change, have been expressed in terms of science, so too has local knowledge become a tool for asserting a voice in global politics.[10]

One consequence of situated science is that the history of science and environmental history necessarily intertwine. Places for science are not fixed and stable but dynamic and open, distributed unevenly across landscapes, exemplifying the interaction between knowledge and place.[11] Knowledge

has motivated human actions, which have then had environmental consequences; conversely, environmental change—land clearing by settlers, the
arrival of exotic species, shifting climate regimes—has often compelled scientific responses. Through science rivers have become known as sites for
producing salmon, a coastal estuary became famous for its oyster production, and cities were redefined as unhealthy environments.[12] The relations
between knowledge and place have been especially evident in contrasting
views of health: as the product of a reciprocal relationship between bodies
and surroundings or as a characteristic limited to individual bodies.[13]

Science is not just situated: it moves. Mobile science has been the focus
of extensive theoretical and empirical work, examining, among other issues,
the movement of scientists and knowledge; the construction of scientific networks; the assertion of knowledge as relevant and credible at places other
than the site of its formation; and how some knowledge becomes classified
as universal, while other knowledge, in a more disparaging key, is viewed as
merely local.

Knowledge does not move spontaneously—diffusing from center to periphery, driven by the imperative of authoritative knowledge. Instead, given
the situated nature of science, its movement has become something that
must itself be explained. Scholars have accordingly described how knowledge moves in all directions, interacting at multiple sites. A variety of categories and concepts have been applied to understanding the place and
movement of knowledge: centers of knowledge, contact zones, standardized
packages, boundary objects, immutable mobiles, and hybrid knowledge.[14]
Scientific networks—associations, journals, conferences—play an essential role in moving science, as do economic and political power structures.
Movement was essential to the roles of science in empires, as embodied in
scientific expeditions and reports and transfers of plant and animal specimens. Imperial scientists circulated knowledge obtained at particular sites,
such as tropical islands (regarding conservation of fragile environments),
or through professional and disciplinary networks (as was the case for imperial forestry practice).[15] Analysis of mobile knowledge has been extended
to the postcolonial world and to institutions of governance and assessment
that impose globally authoritative approaches to analysis and management.[16]
Making knowledge mobile by redefining nature has become essential to contemporary disciplines and professional practices. For example, the transformation of industrial hygiene into toxicology involved defining as irrelevant
the complex relations between workers and their environment. Instead, only
the factor under study was to be considered, thereby making knowledge of
its toxicity applicable to any situation.[17] In opposition to this imperative of
objective, standardized knowledge, environmental justice, and the assertion

of local and indigenous forms of knowledge demonstrate how relations between knowledge and place have political implications. A postcolonial perspective demands attention to how power relations between places can be embodied in knowledge.

While the notions of situated and mobile science are apparently contradictory, in fact they complement each other. Making science mobile is a local accomplishment. The practices by which scientists make knowledge mobile—standardization of research conditions, training, techniques, specimens, and measurements, as well as tests of rigor and reliability and relationships of trust—reflect the places where science is done.

## SITUATED AND MOBILE SCIENCE IN THE ARCTIC

These diverse relations among scientific knowledge, methods, and agendas, and the circumstances in which science is situated and mobilized, have been of continuing importance during the recent history of science in the Canadian Arctic. Scientific activity in this region expanded after World War II, driven by scientific, strategic, political, and economic considerations. Over several decades increased funding and new facilities, including research centers, laboratories, and field stations, as well as new technologies for survey and research, have reshaped how scientists travel to the North and what they do there. Arctic science has also served a variety of national and strategic goals: imposing legibility, asserting political and economic control, and defining the region as a strategic asset. Aerial surveys have been especially important in making the Arctic legible for both scientists and government.[18] A more recent feature of arctic science and politics has been the assertion by northern indigenous peoples of the value of local knowledge, as part of their claim to a role in governing the Arctic.

Arctic science has been in part a product of place. Local features have been defined as scientifically interesting: ice-free refugia in northern Yukon that are relevant to plant evolution and biogeography, sea ice that is essential to climate regulation, and habitats like the Mackenzie Delta and Lancaster Sound that are highly productive and diverse. Other features have attracted scientific attention because they have presented opportunities to study phenomena of more general interest. One is the Smoking Hills in the Northwest Territories, where bituminous shales have burned for centuries, providing a rare opportunity to examine the long-term impact of acid rain on an otherwise undisturbed environment.[19] In the early 1970s the long winter night and absence of photosynthesis provided limnologists at Char Lake on Cornwallis Island with an opportunity to study lake metabolism.[20] These are examples of what Robert Kohler has described as "nature's experiments," in which local circumstances mimic experimental manipulation of environmental variables, reproducing in the field the advantages of the controlled laboratory, while allowing ecologists to invoke the authenticity of field-based evidence.[21]

Other efforts to construct the Arctic as a place for research have emphasized its remote location and undisturbed state. Distant from industry, the region provided something that ecologists and other scientists had often sought: pure nature, where natural processes could be observed directly, without the confounding influence of human impacts. As such, the Arctic presented some of the same epistemic advantages as a laboratory. As Canadian writer James Woodford noted in 1972, "the Arctic might best be seen as a vast outdoor laboratory where the basic secrets of the complex natural systems of the globe may at last be discovered and understood."[22] Similarly, Max Dunbar, a marine biologist at McGill University, noted in 1968 that the polar regions can be "a special laboratory from which results of general significance can be elicited." To emphasize its potential to provide results of general interest, Dunbar downplayed the region's distinctive climate, arguing that "the Arctic has problems for life which extend considerably beyond ice, snow, and cold water."[23]

Such efforts to assert the wider relevance of arctic knowledge have a considerable history, demonstrated by attempts to incorporate the Arctic within broader scientific concepts, such as Darwinism, Humboldtian science, or Euclidean geometry.[24] Scientific networks such as the International Polar Years and the International Biological Program, as well as various global and circumpolar agencies, have linked arctic science to the rest of the scientific world.[25]

The situated nature of arctic science has also been a matter of negotiation, with scientific perspectives drawing on observations and theory from both within and outside the region. For example, in the 1950s research in "survival science" sought ways of insulating bodies from this environment, to enable personnel to work efficiently. In contrast, since the 1980s much arctic research has focused on pesticides, and researchers have had to acknowledge that the boundary between bodies and the environment is highly porous and that they must draw on local environmental knowledge, as well as general principles of environmental chemistry and toxicology.[26] A second example of arctic scientific ideas drawing on both local and more distant knowledge is the view, once widely accepted among ecologists, that the Arctic is a distinctively fragile environment. This was believed on the basis of both observations specific to the Arctic and more general theory regarding the relation between the diversity and stability of ecosystems. When in the early 1970s ecologists became more skeptical of this view, it was similarly because of evidence from the region itself, an evolving political context (in the Arctic and elsewhere), and shifting theoretical perspectives within the discipline of ecology.[27]

Throughout the postwar era the identity of scientists working in the Arctic has evolved in ways that implicate situated and mobile science. The distinctive identity of arctic science has declined, as modernist assumptions of homogeneous research spaces and scientific practices have taken hold. Originally basing their claims to credibility on personal characteristics associated with the demands of fieldwork in this region (such as endurance, resourcefulness, and ties to indigenous peoples), in the postwar era arctic scientists have instead relied increasingly on the authority provided by membership in disciplinary research communities based in the South.[28] This has been accelerated by new technology, especially aviation, that has made the Arctic more accessible and by the creation of laboratory-like facilities for arctic research. New scientific networks have also enabled disciplinary communities such as wildlife biology and climate science to impose their practices and forms of explanation in this region. Arctic scientists may still do fieldwork in the north, but they then typically interpret their results in southern labs, for global audiences. This evolution reminds us that scientific disciplines, like knowledge itself, have geographies as well as histories—that they are situated in space as well as time.

Given the evolving identity of arctic science—situated in a distinctive environment yet highly mobile—the place of arctic research has remained ambiguous. This ambiguity has often been a concern to arctic scientists, encouraging efforts to define the region and its boundaries and to explain how research there relates to the rest of the world. One approach has been to define the "essential" features of the region or its boundaries, so as to make sense of the Arctic's many identities—whether these are defined in terms of natural features (such as permafrost or the tree line), geography (such as the Arctic Circle), political boundaries, or scientific perspectives. For example, in 1951 the botanist Nicholas Polunin specified precise criteria by which to define vegetation as "arctic." A decade later, entomologist J. A. Downes did the same for insects. Efforts to define the Arctic reached their apogee in Louis-Edward Hamelin's effort to develop a quantitative measure of "nordicity."[29] Each of these schemes constituted a different way of relating knowledge and place, by constructing scientific cartographies of the Arctic that define the boundaries of their study site. Arctic scientists' preoccupation with science and place also makes this region especially suitable for a study of situated and mobile science. Accordingly, this chapter seeks to identify how at least some arctic scientists—ecologists, to be precise—thought about place and movement. It will do so by examining their work not only in terms of empirical results and theoretical conclusions but with an eye to how they constructed their places of study and how they related these places to the rest of the world.

## NORTHERN ECOLOGY

In October 1969 scientists, scholars, and officials gathered in Edmonton, Alberta, for a conference titled "Productivity and Conservation in Northern Circumpolar Lands."[30] As the first conference focused on circumpolar ecology, it was a milestone in the history of arctic science. It illustrated how scientists were identifying new research opportunities in the region and also highlighted the evolving political and economic circumstances of the Arctic, including concerns regarding the environmental impacts of northern oil and gas development. This was reflected in the involvement of new institutions in arctic research. Scientists in northern Canada had previously worked mainly for federal agencies, such as the Canadian Wildlife Service and the Fisheries Research Board of Canada, or at universities, such as McGill, that had close ties with the federal government. However, by the time of this conference Canadian arctic research had expanded to include scientists from a wider range of universities, government agencies, and industry.

Not only new scientists but new research areas were becoming evident in arctic ecological science. Three will be examined here: animal population ecology, boreal ecology, and the ecology of fire. Each had distinct institutional identities, research problems, and theoretical orientations. Ecologists working within each also asserted distinctive perspectives regarding how science related to the arctic landscape, how knowledge from elsewhere could contribute to understanding this landscape, and how arctic knowledge could be applied elsewhere in the world.

### POPULATION ECOLOGY

Northern animal populations have long fascinated observers. Surges and crashes in lemming populations (and their mythical proclivity toward mass suicide), unexplained cycles in abundance of other species, and massive migrations of caribou have intrigued generations of explorers. In the early 1920s British ecologist Charles Elton conducted some of the earliest and most influential studies of animal population ecology on Bear Island and elsewhere in the Arctic. He, like many of his Oxford colleagues, was attracted to study in a remote site, but he also viewed the island's relatively simple ecological community as an opportunity to derive conclusions of general relevance. In subsequent research he and his colleagues at Oxford's Bureau of Animal Population tracked changes in snowshoe hare and other populations in northern Canada, seeking to identify the factors responsible for cycles in their abundance. Some ecologists saw these cycles as a distinctive feature of northern populations. Elton did not, but even so he was drawn to study Canadian animal populations because of the availability of population records and by the opportunity to study ecological communities composed of relatively few species that extended over a large land area and so were suitable for studying the effects of factors, like climate, that were not merely local.[31]

By the 1950s Elton had turned away from northern population research, choosing, instead, to focus on intensive study of animal populations at specific sites in England. More generally, ecologists were seeking explanations for changes in population that could be framed in terms of regulatory mechanisms relevant to animal populations everywhere. This was evident in the approach adopted by Charles Krebs, who in the late 1950s set up a field study focusing on lemming populations at Baker Lake in the Northwest Territories.[32] Just as Elton had several decades earlier, Krebs (like Frank Pitelka, who was doing analogous population research in Alaska) saw northern animal populations as potentially a key to understanding general mechanisms of population regulation. Krebs chose to study in the North not because he was interested in factors specific to that region, but because it was a suitable site for the study of factors found in all animal populations. His approach involved focusing on selected features of population cycles. A starting point was a more specific definition of a lemming population cycle: "typically 3- to 4-year fluctuation in numbers in microtine rodents characterized by high body weights of adults in the peak summer."[33] In Krebs's view, this definition clarified the confused state of understanding of cycles, clearing away accumulated notions and myths specific to the North, such as legends of mass migrations of lemmings. Constructing this definition was essential to making knowledge of these cycles mobile, because it provided a basis for relating northern observations to what was known elsewhere about populations. By doing so, he could derive conclusions of interest to his own intellectual community (the Bureau of Animal Population at Oxford and the Department of Zoology at the University of British Columbia) and to the wider international community of animal ecologists.

The history of studies of animal populations in the Arctic, as represented in the transition from Elton to Krebs, therefore exhibits a shift in perspectives on how this research should be situated. Initially focused on understanding factors specific to the North, it shifted to become a search for regulatory mechanisms of relevance to populations everywhere, with the phenomenon of population fluctuations redefined to be portable. At the same time, as Krebs also noted, this research remained situated in the North, because he considered this region suited to generating knowledge that could be applied elsewhere.

## Fire Ecology

Until fairly recently, fire in nature was widely considered a destructive force.[34] Northerners shared this view, with arctic and subarctic ecosystems considered especially vulnerable, particularly because of its impact on caribou—an iconic northern species. In the 1950s and 1960s Canadian Wildlife Service biologists based their argument on the observation that caribou depend on the climax boreal forest, where they find their preferred winter food of lichen. Once burned, lichen requires a long period for recovery. Fire could therefore be a limiting factor for caribou populations. Some viewed the increased incidence of fire that was said to have accompanied northern settlement as a factor in declining caribou populations.[35]

However, by the early 1970s scientists from other federal research agencies and universities were becoming involved in northern fire research. Support for their work came from a new source: the Arctic Land Use Research Program, put in place by the Canadian government to support studies of the impacts of development on terrestrial ecosystems. These ecologists instigated a reconsideration of fire in northern ecosystems. On the basis of fieldwork and statistical analysis of fire occurrences, they redefined fire as a natural event essential to the ecology of the boreal forest. As two ecologists argued, "fire is an integral part of the Subarctic. . . . Fire is neither unorderable, infrequent nor external to the Subarctic ecosystem. To suggest that fire in the Subarctic is not part of the natural course of things is to suggest that the weather and land are not also."[36]

This view can be traced, in part, to wider shifts in ideas and knowledge, beyond the Arctic. These included a new perspective on forested ecosystems as habitats that are not timelessly stable, as in a climax state, but are marked by change, disturbance, and heterogeneity. Fire fit well within this view, with its role in creating mosaics of different successional stages and thus a more varied habitat. These ideas were also seen as relevant to northern forests. So were ideas regarding other roles of fire in forest ecosystems, including nutrient cycling (an awareness tied to the attention devoted by ecosystem ecologists to this process) and habitat renewal. Finally, ecologists became skeptical of generalizations about the impacts of fire: recovery was seen instead to depend on local circumstances.[37]

Factors specific to the North also played a role in shaping attitudes toward fire. These included the impacts of fire on tundra, on the relation between tundra and forest (i.e., the tree line), and on permafrost. With new participants in northern ecological research came new topics, methods, and evidence that together encouraged new attitudes. Local events also played a role: in 1968 a fire near Inuvik presented an opportunity to examine how fire affects the tundra and frozen soil. A new awareness of the history of northern fire was also important. Paleoecological evidence (sediment cores, tree rings, and soil traces), together with old forest inventories and field journals, was used to reconstruct this history. This work demonstrated that fire was not a recent phenomenon, originating with settlement; the prevalence of lightning- over human-caused fire also indicated that fire was a natural process in the boreal forest.[38] In addition, Henry Lewis, an anthropologist, demonstrated an extensive history of Aboriginal burning, which assumptions of increased fire with settlement had failed to consider.[39] And finally, there was a reassessment of the impact of fire on caribou populations. By the mid-1970s the assumption that they were dependent on lichen was being questioned. This was motivated by, among other factors, a new view of caribou diet. Caribou were now seen as generalists, able to adapt to fire and periodic loss of lichen. Overall, therefore, shifting attitudes toward fire in the Arctic exhibited the influence both of ideas generally held by ecologists and of factors that were more specific to the region.

Fire ecologists acknowledged the issues involved in the movement of knowledge into and out of the North. While they drew from more general ideas regarding the ecology of fire, they also criticized the too-quick application of ideas from temperate regions—what they referred to as "Smoky the Bear thinking." They especially emphasized the need to rely on local empirical evidence and to be aware of the region's distinctive conditions (such as lichen and tundra) and its environmental history.

## Boreal Ecology

Population ecologists and fire ecologists drew on knowledge from elsewhere in developing their views of arctic ecology; some also argued that insights from arctic research could be applied elsewhere. However, other ecologists resisted these assertions of the mobility of arctic knowledge. Instead, they argued that the distinctive arctic environment, and especially snow and ice, should be at the center of research in this region. They had accumulated evidence of the pervasive impact of snow on the lives of arctic animals, including access to food (with caribou particularly sensitive to its thickness, hardness, and density). Many animals were adapted to life in snow: it regulated their distribution in winter, and small mammals relied on behavioral adaptations to snow. Study of the ecological implications of snow required careful attention to its characteristics, which could be best described using terminology from arctic languages, such as Lappish or Kovakmuit (Inuit): thus, for example, *qali* referred to snow on trees, and *siqoq* was drifting or blowing snow.[40] Wind, in association with snow, was also a significant ecological factor, as was sea ice, which determined the distribution of polar bears, seals, and other marine mammals. Russian ecologists had done much of the most detailed work on snow, and some ecologists urged study of and even translated Soviet ecological literature.[41]

These ecologists also argued that snow and other environmental circumstances specific to the Arctic were not only an appropriate focus for research but justified a distinct discipline—boreal ecology. William Pruitt of the University of Manitoba was among its most active advocates. As he explained in 1978, "boreal" meant northern, and thus boreal ecology was the study of regions in which distinctive northern conditions, and especially snow cover, affect plants and animals. Boreal ecology could even be characterized as the study of the ecology of snow: a specialty that could not be reduced to general ecological theory.[42] As Pruitt and other boreal ecologists argued, this had implications for the mobility of knowledge. Citing their colleagues in other northern nations, and particularly the Soviet Union, they argued that only knowledge gained in similar physical conditions—that is, snow and cold— was truly relevant to understanding northern ecology.

## PLACING AND MOBILIZING ARCTIC ECOLOGY

The expansion of ecological research in the Canadian Arctic during the 1960s and 1970s brought new people, institutions, and ideas to the region, encouraging new areas of research and fresh perspectives in areas of long-standing scientific interest. Ecologists focused on a variety of biological and ecological phenomena—from lemmings to fire to snow—and applied diverse approaches. Writing the history of this science requires attention to the essential dimensions of contemporary science: the theories and methods applied within scientists' disciplines, the political and managerial contexts of research, and the institutions in which scientists worked. Thus, study of the history of arctic population ecology, fire ecology, and boreal ecology must examine, among other issues, how Krebs's work related to ecologists' ideas about the practice of population research, how ecologists studying fire responded to demands for knowledge relevant to managing northern resource developments, and how all ecologists pursued institutional support for their research.

But if this history of science is to be an environmental history, then it must also consider ecologists' working environment: how they situated their work in the field and how they related their research sites to the rest of the world. In understanding the historical relations between scientists and their environment, both interpretative extremes must be avoided: the assumption that environmental features or problems determine what research will be done or the assumption that scientists' agendas are simply imposed on an otherwise passive environment. Just as in environmental history generally, more subtle and complex interactions between human activities and environments must be imagined.

The concepts of situated and mobile science can guide this imagining: suggesting lines of inquiry and hinting at patterns in how arctic scientists talked about place and movement. Among the relevant issues are the relations among knowledge, practice, and place; the formation of the identities of places; the spatial distribution of scientists and their audiences; and the circulation of knowledge. Arctic science also provides an opportunity to examine the social and environmental circumstances in which place becomes a significant factor in asserting the credibility and mobility of knowledge. Attention to such issues acknowledges that ecologists produced not only knowledge but ideas about the Arctic as a place distinct from yet similar to other parts of the world.

Place has obviously mattered to arctic ecological research. The natural environment and the social and political contexts of arctic science have influenced the design of research practices, identification of relevant theory, and construction of credible knowledge. As arctic ecologists often noted, the landscape was essential to their research. Small mammal populations exhibiting unusual fluctuations, forest and tundra fires that had distinctive impacts and patterns of occurrence, the inescapable influence of snow and ice on plants and animals: these features exhibited the importance of place to arctic scientific practice. They also implied, at least for some ecologists, a requirement that this region be understood on its own terms. This became evident in studies of fire ecology, which were justified by concerns regarding the arctic environment, and by the Canadian government's need to demonstrate effective management. In contrast, Krebs justified study in the Arctic on the basis of it providing conditions for producing knowledge that would be of interest to population ecologists everywhere, not just in this region. Boreal ecologists like Pruitt also justified study in specific sites because their results would be relevant elsewhere—but only within the Arctic. These diverse approaches to justifying the place of research demonstrate how ecologists' presence in the Arctic was contingent on specific historical circumstances, including the priorities of their own research communities.

A similar diversity in relation to place was evident when ecologists reported their results. Fire ecologists emphasized features specific to their research site, such as the local history of fire. Boreal ecologists' reports stressed local snow and ice conditions. In contrast, Krebs's conclusions generally omitted local details, emphasizing instead aspects of the dynamics of mammal populations that were of general relevance—in effect, detaching these aspects from their surroundings. These diverse perspectives reflected how ecologists tended to view the Arctic as either an object or a place of study. Their choices reflected both their audiences (e.g., environmental managers or scientists elsewhere) and their own identity (as arctic specialists or as scientists with a disciplinary affiliation not specifically associated with the Arctic). One strategy—evident, for example, in research on the ecology of fire—was to emphasize details that could help scientists understand the Arctic on its own terms, as an object of research. In this situation, as in other arenas in which science has been applied to practical environmental issues, credibility depended on relevance to local conditions. Boreal ecologists pursued a similar approach, albeit with a different motivation: they sought to associate their discipline with this specific region, pursuing an approach analogous to that followed by scientists who have constructed other situated disciplines, such as tropical ecology. In contrast, Krebs emphasized details that would be considered relevant and credible elsewhere. Thus, he viewed the Arctic as a place, not an object of study.

In asserting the credibility of their results, arctic ecologists pursued in diverse and historically contingent ways the epistemic advantages of both field and lab. Fire ecologists acknowledged the complexity of their field site and sought to identify and account for all factors that could affect the incidence and effects of fire. Boreal ecologists also situated their research in the field, asserting that their results had the authenticity of fieldwork, reinforcing their argument regarding the distinctive nature of this environment. In contrast, Krebs asserted the lablike nature of his work: populations, he suggested, existed in an environment whose local particularities were not relevant to their results. Together, these cases demonstrate how scientific credibility, and truth spots, could be constructed in diverse ways from local materials, as ecologists responded to the challenges and opportunities presented by specific places. Across this diversity, however, ecologists also adhered to standards of scientific practice that were in general use elsewhere: extending laboratory practices into the field and using standardized field and analytical techniques—a practice conveyed through the rhetoric of "outdoor laboratories." These standards displaced the demonstration of local skills and capacity for endurance that had once been obligatory for arctic scientists and explorers, thereby illustrating how these ecologists shared in the decline of the once-distinctive regional identity of arctic science.

While arctic science has been situated in diverse and specific ways, it has also been on the move. The Arctic has always required lengthy travel by scientists to their research sites. Once there, they have often interpreted the Arctic in terms of their experience elsewhere and have reported their results to distant audiences. This raises several questions relating to the mobility of ecological practices and knowledge. One relates to the extent to which knowledge from elsewhere could be applied to understanding the Arctic. Ecologists had several views on this issue. They drew selectively on perspectives emerging elsewhere regarding the role of fire in ecosystems. Population ecologists were highly receptive to knowledge from elsewhere. Boreal ecologists considered knowledge gained in any snow- and ice-dominated ecosystem to be relevant. In each case, ecologists' explanations combined local evidence with knowledge and theories from elsewhere: constructing arctic knowledge in a way similar to that seen, as noted above, with respect to ideas about fragile arctic ecosystems. These combinations of knowledge from the research site and from elsewhere were distinctive to each site and helped impart to each a distinct identity, thus illustrating the historical contingency of scientific descriptions of the Arctic.

Arctic ecological knowledge circulated not only into but out of the region. Although often viewed as a marginal space, the Arctic does not fit the simple model of a peripheral region receiving knowledge from the center. As we have seen, asserting credibility was an important part of how ecologists related their practice to place. It was similarly essential to making knowledge mobile. But mobilizing knowledge involved more than credibility. Just as ecologists had diverse views regarding the movement of knowledge into the region, they differed on where knowledge could be circulated and to whom. Fire ecologists considered their results relevant only to their study site. This reflected their chief audience: local land-use managers. In contrast, boreal ecologists sought to mobilize their results to wherever snow and ice were important, as they considered conditions in arctic Canada to be relevant to arctic regions elsewhere, but not to ecosystems outside the circumpolar region. Krebs, as a population ecologist, circulated his results to others in his discipline, speaking to a narrow but widely distributed audience located at centers such as Oxford's Bureau of Animal Population and at other universities that formed nodes on the international network of population ecologists. Overall, therefore, mobilization of knowledge out of the Arctic was selective, reflecting ecologists' purposes and audiences and their managerial or disciplinary imperatives.

The history of arctic ecology illustrates how science can be both situated and mobile. Far from contradicting each other, these concepts provide for environmental historians complementary ways of understanding the interaction among scientific practices, knowledge, and the environment. Different views of the Arctic as a place, as expressed through various forms of ecological knowledge—about animal populations, fire, or snow—reflected different conceptions of the audiences for this knowledge. Each of these audiences had its own geography: the international community of population ecologists, local land-use managers, or ecologists located in circumpolar countries with an interest in boreal ecology. Strategies for mobilizing science were place based, and places were defined in terms of what aspects could be mobilized: in effect, constituted amid the circulation of scientists, knowledge, and practice. These diverse ways in which arctic ecology could be both situated and mobile exemplify the historical contingencies of science in a region that has itself been located on both the geographic margin and at the center of human awareness.

# CHAPTER 12

# WHITE MOUNTAIN APACHE BOUNDARY-WORK AS AN INSTRUMENT OF ECOPOLITICAL LIBERATION AND LANDSCAPE CHANGE

## DAVID TOMBLIN

IN 1960 D'ARCY MCNICKLE—the famous Salish scholar, activist, and writer—predicted that American Indians will "probably use the white man's technical skills for Indian purposes" and that "Indians are going to remain Indian . . . a way of looking at things and a way of acting which will be original, which will be a compound of these different influences."[1] This prediction came true. Perhaps one of the best-kept secrets in American history is the political resurgence of some Native American nations in the latter half of the twentieth century.[2] Some nations have demonstrated more success than others, but one key aspect of many of the relatively successful cases is the assertion of control over natural resource management science and technologies on tribal lands, a phenomenon that remains relatively unexamined.[3]

A historical case study of the White Mountain Apache tribe's institutional struggle to control and restore their ecocultural resources affords an excellent opportunity to explore the significance of this appropriation process as a mode of resistance to federal cultural assimilation and land dispossession policies.[4] The main argument of this chapter is that the Apache institutional appropriation of Western restoration and land-management techniques provided a political platform for a liberatory form of boundary-work that proved

necessary to maintain cultural identity, protect tribal resources, resist paternalistic knowledge production practices, and reassert tribal sovereignty. At
the same time, because of the tribe's imposed dependence on Euro-American society, boundary-work supplied tools necessary to integrate Apache
traditions with Western traditions as a strategy for persisting in the context
of a twentieth-century capitalist society. This institutional appropriation process played a powerful role in reasserting a mediating influence over knowledge production practices and control over ecocultural resources on the Fort
Apache Indian Reservation, redirecting them toward the benefit of the tribal
community rather than Euro-American interests. Furthermore, this shift
in knowledge-production practices toward developing partnerships based
on politically equivalent terms with non-Apache management organizations
had some positive consequences for the ecological productivity and health of
the Apachean landscape.

## BOUNDARY-WORK AND SOCIAL JUSTICE

In the second half of the twentieth century White Mountain Apaches
incrementally combined elements from both Apache and Euro-American
traditions to reimagine not only their ecocultural landscapes but also their
political and economic systems. In this quest, Apache political and environmental organizations regulated, traversed, defended, and established four
types of boundaries: political, cultural, epistemological, and geographic.
Ever since the Fort Apache Indian Reservation's political boundary was dictated to them in 1870, they have attempted to police it from the intrusion of
outside interests that wished to exploit Apache resources.[5] With the inception
of the Indian New Deal in 1933, Apache ecological restoration and land-management efforts began to figure into the maintenance of all four boundary
types. Because political, cultural, epistemological, and geographic boundaries are inherently fluid and porous, these technosciences became mediation
devices between Apaches and a whole host of non-Indian interests (federal
and state land-management agencies, environmentalists, academics, NGOs,
and industry). Thomas Gieryn's boundary-work concept helps frame this
mediation process.

Gieryn originally applied the boundary-work concept as a model to explain the self-preservation strategies of scientists. As science grew in stat-
ure, scientists developed "an ideological style" that demarcated science as
superior to other forms of knowledge production. He identified three basic
areas of political discourse that scientists employed to achieve that end: expansion of authority and expertise into other professional domains, monop-
olization of resources, and autonomy from responsibility for the unintended
consequences of research.[6] Subsequently, the boundary-work concept has
become a very useful analytical tool for investigating the interactions be-

tween science and society (e.g., who gets to do science); power relations that inform the disciplinary formation process (e.g., restoration ecology, wildlife management, forestry); and defining what is "normal" (e.g., baselines for restoration, sustained yield, healthy wildlife populations, the existence of a species, ecological integrity, etc.).[7] However, this conceptual tool has yet to be employed to understand socially unjust distributions, applications, and consequences of science and technology. Just as boundary-work is useful for explaining the political maneuvers of scientists defending disciplinary turf, acquiring resources, or claiming autonomy, it can also be seen as a liberatory tool implemented by disenfranchised, marginalized, and non-Western groups seeking to redistribute the benefits of Western technosciences, redefine technoscientific practices, and legitimize local knowledge.

While Euro-American scientists generally enjoy the advantage of employing boundary-work from a privileged position, other groups seeking to enroll the benefits of science typically don't—the very reason being that scientists practice boundary-work to protect their domain. For instance, health activists, environmental justice groups, and grassroots restorationists all work to procure credibility for their knowledge-production practices.[8] Native Americans also struggle to legitimize the use of Western science in conjunction with their knowledge-production systems.[9] Non-Western cultures in particular have difficulty succeeding in this goal, especially since Western society, in a broad act of boundary-work, unjustly discredited indigenous knowledge and grossly underestimated the intellectual capabilities of these cultures. In this study, I reframe the boundary-work concept as a liberatory tool for marginalized cultures. As White Mountain Apaches worked to regain control over their ecocultural resources, they also strived to influence, mediate, and adapt knowledge production and ecological restoration techniques through boundary-work.

Ultimately, Apache boundary-work forged a pathway toward scientific and technological empowerment, which led to freedom from federally mandated technocratic oppression and Euro-American industrial exploitation. This process evolved over four historical periods significant to Native Americans: the Indian New Deal (1933–45), termination (1946–60), self-determination (1961–89), and self-governance (1990–present). Each of these periods represents an important step in a long-term appropriation process that resulted in the cultural reconfiguration of Euro-American restoration techniques.[10] Of course, in highlighting the appropriation process, in no way is it my intention to suggest the primacy of Western science and technology over Apache epistemology. Apaches didn't choose to employ these technologies because they found their own technologies and knowledges inferior but, on the contrary, complementary. They weren't seeking to replace their traditions but, as all cultures do, to adapt to changing circumstances. What I aim to demonstrate is that Apache values and knowledge systems were selective filters that determined which Western traditions suited tribal needs. The Apache institution that evolved through this process—the system of organizations, along with the values and rules used to govern reservation resources—built political leverage through boundary-work, while incrementally incorporating and redefining increasingly complex layers of social and technical resources from Apache and Western traditions. Early in this process the focus was on the appropriation of economic, institutional, and technological resources to garner more control over reservation lands. Later, boundary-work served to delineate Apache epistemological practices from Euro-American practices, which led to the legitimization of a new knowledge-production system— a local system that hybridized Apache and Euro-American traditions.[11]

Apache institutional boundary-work occurred on multiple fronts. Filtering the outside world through restoration and management techniques, Apaches had to police the political, cultural, epistemological, and geographic boundaries between themselves and other cultures of science and technology (e.g., federal experts, state experts, industry, Salt Valley water users, environmentalists, hunters/fishers, and tourists). Additionally, boundary-work occurred within the Apache community, as different Apache factions (e.g., elders, ranchers, and the Apache government) had diverse positions on how the tribe should address encounters with Euro-American technosciences. However, for purposes of conceptual clarity, this chapter focuses on Apache intuitional struggles with outside interests.

## THE SLOW MARCH OF LIBERATORY BOUNDARY-WORK
### The Indian New Deal and the Groundwork for Ecopolitical Liberation

During this period, Apaches experienced an intense encounter with Euro-American restoration and land-management techniques. John Collier, commissioner of Indian affairs from 1933 to 1945, and his allies conceived the Indian New Deal and the Civilian Conservation Corp–Indian Division (CCC-ID) as a remedy against the cultural and ecological degradation that modernization imposed on Native Americans and their lands. Unfortunately, the CCC-ID represented a double-edged sword, maintaining a paternalistic knowledge-production structure while introducing new technologies. Despite Collier's good intentions, his policies relegated American Indians to a marginal role in management decisions. Because CCC-ID training programs contained relatively little theoretical and detailed technical content, American Indians remained unable to develop comprehensive management plans within the context of twentieth-century development projects. Furthermore, since they couldn't create their own plans, they had no choice but to "trust" Bureau of Indian Affairs (BIA) experts to manage tribal resources. Although Collier wanted to instill self-sufficiency and political autonomy, his program left Native Americans dependent on the federal government and unintentionally facilitated the further exploitation and ecological simplification of Indian country.[12]

Despite these unintended consequences, Indian New Deal programs opened a gateway for some Native nations to reassert influence over knowledge production on reservations. For Apaches, these programs provided enough economic infrastructure and technical education to lay the groundwork for later appropriation of Euro-American technosciences. For instance, perhaps one of the most important CCC-ID projects involved the construction of the Williams Creek National Fish Hatchery in 1939.[13] Williams Creek became an instrument of boundary-work that aided the Apaches' journey toward restoring control over ecocultural resources, providing a focal point for the first successful Apache business, the White Mountain Recreational Enterprise (WMRE).[14]

## Termination Politics and the Foundation of Institutional Boundary-Work

The White Mountain Apache Tribal Council officially established the WMRE in 1954, amid the post–World War II conservative backlash against Indian New Deal programs, when Congress threatened the withdrawal of federal services, the dissolution of tribal governments, and once again land dispossession.[15] Unlike with Indian New Deal programs, the tribe received obvious, lasting economic and technical benefits from the venture. By 1964, the enterprise boasted sixty permanent employees and raked in $1.2 million. And these numbers increased over time. However, its primary mission remained the protection and conservation of "wildlife, recreational, and natural resources . . . for the members of the Tribe and the general public."[16] Of equal significance, the WMRE instituted a foundation for boundary-work with Euro-American society, marking an intensification of Apache cultural appropriation of Western natural resource management techniques. For the first time, Apaches had formed an independent institutional boundary between themselves and the dominant society.

Predictably, the growth of the WMRE instigated resource battles with Euro-Americans. Fortunately, its establishment provided Apaches with the organizational force to take on institutions that had historically exploited the tribe's resources without consequence. For example, the Hawley Lake controversy, which took place roughly between 1954 and 1959, represented the Apache government's largest show of successful resistance to date against outsider attempts to control reservation resources. When the WMRE laid plans to build a dam to create a recreational lake, the Salt River Valley Water Users' Association objected to the construction of the reservoir, claiming it had prior rights to the impounded water for irrigating farms downstream of the reservation.[17] Through this controversy, tribal leaders such as Clin-

ton Kessay asserted the legal and cultural boundaries of Euro-American resource exploitation on the reservation: "We . . . know that the right to utilize
our water will determine whether or not we are to exist and we therefore
firmly desire to protect the water that was set aside for our use. . . . Our water
and the right to use our water to develop our resources is vitally important to
us as a tribe, and we are therefore very anxious to know just what the Bureau
of Indian Affairs position is . . . if in the future the White Mountain Apache
Tribe's water development program is harassed by outside agents."[18] Kessay's boundary-work conveyed distrust in the BIA's ability to protect Apache
resources. In the end, instead of relying on the BIA, the tribal government
went on the offensive, hiring its own lawyer to guide the tribe through the
jungle of legal and bureaucratic obstacles preventing the construction of
Hawley Lake.[19] Besides Hawley Lake, in continuing to expand authority over
reservation resources, the WMRE had built another six reservoirs by 1960.[20]

Nelson Lupe, a cofounder of the WMRE, saw such projects as critical
to asserting tribal sovereignty over reservation resources, ensuring that the
Apache people would actually benefit from resource development: "Apaches
have always lived around here. . . . We always will. . . . If we fail to develop our
reservation, sooner or later outsiders will do it . . . for their benefit, not ours.
We know that the more tourists that come, the more secure our country will
be for our kids."[21] Here, Lupe's blatant boundary-work declared that Apaches
welcomed Euro-American science, technology, and economic strategies, but
on Apache terms. WMRE boundary-work, therefore, built political and economic leverage to manage resources on the reservation for the benefit of the
Apache people.

The more White Mountain Apaches pushed back, the more respect they
garnered. Even with the Hawley Lake controversy looming, the non-Indian
public and BIA personnel began to notice positive changes in the Fort
Apache landscape as the WMRE emphasized restoring natural resources,
"not only for proper and wise use . . . but also for their orderly development
and the expansion of their productivity."[22] Restoration projects increasingly
served as sites of boundary-work. Even before conserving endangered species became *en vogue* in the 1960s, concerned Apaches took action to save
the endangered Apache trout. Despite recurring interest from the Tribal
Council since the late 1930s in closing Apache trout streams to fishing, the
Fish and Wildlife Service (FWS) and BIA had taken no action. A representative from the FWS claimed that native trout "were abundant" and believed
that it was unnecessary "to close these streams to fishing."[23] As a result of
this neglect, Apache trout populations continued to decline. Fortunately,
some Apaches continued to believe this species was "in danger of extinction"
until they gained more control over their homeland. In 1952 the Tribal Council ordered the closure of six streams to fishing and camping in the head-

Figure 12.1. An Apache Youth Camp group working on an erosion-control project in 1956.
Courtesy of the United States National Archives and Record Administration.

waters of Mount Baldy that contained pure populations of Apache trout.[24]
This decision essentially gave the species a greater chance of surviving until
more interest emerged in saving nongame species in the 1960s, earning the
tribe essential political capital to persevere through future natural resource
conflicts.

### SELF-DETERMINATION AND OPENING SPACES FOR NEW KNOWLEDGE-PRODUCTION PRACTICES

The WMRE represented the beginnings of an Apache-mediated cultural
hybridization of Western science and technology with the Apache knowledge
system, a reformulation of Apache identity that maintained elements of past
traditions and developed new ones. On the surface, the WMRE embraced
the ideals of Western land-management and business techniques, becoming
an economic cornerstone for rebuilding Apache society through resource
development, revenue generation, and job creation. Upon a closer look, the
WMRE became a platform from which the Apache could begin asserting
influence over knowledge production on the reservation, providing technical
training programs that expanded the pool of Apache expertise.

However, appropriation remained culturally conservative until the late
1960s, with most Apache restoration and management efforts reflecting contemporary Euro-American practices. In contrast, by the late 1960s, Apaches
pursued more overt attempts to revitalize culture through integrating

Apache traditions with Western technosciences, such as the establishment of the Mount Baldy Wilderness Area, performing double duty as a sacred site and an ecological preserve. More indicative of this period, however, was the intensification of boundary-work delineating the differences between how Apache land managers would employ science and technology relative to the past practices of Western land managers, with this political rhetoric leading to a major tribal institutional expansion of influence over reservation lands.

For instance, in the 1970s, building on past restoration and management successes, Apache organizations took on the Arizona Game and Fish Department (AGFD). After witnessing a decline in reservation elk populations, tribal officials claimed they were better equipped to manage this species than the state. Explicit in this rhetoric were claims that Apache local knowledge was superior to knowledge generated by the AGFD. Phil Stago Jr., the first Apache director of the WMRE, contended: "They have to consider what's out there in the woods. You can be a biologist down in a Phoenix Game and Fish Office, with a Ph.D. in biology, and make all kinds of decisions . . . but you also have to know what is going on in the mountains. You don't learn that from a book, believe me. That is where we [AGFD and the tribe] . . . differ. We see it and we live it every day by being out in the field."[25]

Stago's act of boundary-work asserted that the source of AGFD's management problems arose from a fundamental belief in a standardized, top-down mode of knowledge production. While drawing a distinct philosophical line between the management ethos of tribal organizations and the AGFD, he maintained, "We have better and more wildlife on this reservation than any place in the southwest, because we do not over-kill, we do not have excessive hunters and everything is controlled." His essential message pronounced that the formula of mixing Apache local knowledge with training in the Western sciences made for a superior management philosophy, implying that the AGFD had something to learn from the "Apache way."[26]

Reestablishing sovereignty over fish and game meant that non-Indians had to contend with the existence of a newly legitimized form of local knowledge. The power of local knowledge, as Stago articulated it, was that it incorporated the ecocultural concerns of the Apache community. This was something that bureaucratic organizations such as the AGFD could never accomplish from their impersonal perspective. As an alternative to Euro-American management regimes and standardized knowledge, the Apache began developing a new brand of local knowledge, a hybridization of Apache and Euro-American epistemologies constructed for the benefit of the Apache people. Even though the "Apache way" of managing ecocultural resources remained publicly vague at this time, articulating the rebirth of this knowledge in political dialogues with non-Indians furnished it with a substrate from which its relegitimization could grow.[27] Its evolving constitution would slowly be revealed and negotiated as Apaches utilized intimate understandings of the reservation's natural history, animal behavior, plant distributions, and environmental idiosyncrasies to manage and restore eco-cultural resources. By the 1990s, the content of the Apache hybridized knowledge system became more obvious, as land managers integrated Western practices with past Apache traditions, including philosophical dispositions toward land management, land-management techniques, Apache histories of the landscape, cultural relationships with nonhuman nature, and moral frameworks that tell people how to live on the land.

In the meantime, buttressed by past management successes and favorable court decisions, the tribe won the right to manage elk on the reservation, allowing an already successful recreational program to flourish. Elk hunting in particular became world renowned among hunters seeking trophy specimens. With the elk hunt in demand, by 1983, the WMRE charged $7,000 for a seven-day permit and by the 2000s fetched over $20,000 per hunting license.[28] Beyond economic gains, this coup had significant ecological implications. The tribe invested these funds back into the WMRE, which helped build expertise and reshape the Apachean landscape into ecologically more productive homeland. In terms of building expertise, the trophy elk-hunt program helped staff the Tribal Game and Fish Department (TGFD), a newly formed division of the WMRE, with college-educated, non-Indian wildlife biologists. Under the guidance of Stago, these biologists trained Apaches in increasingly sophisticated wildlife management and restoration techniques.

For instance, these wildlife biologists helped train tribal game wardens in radio telemetry to trace the movement of elk. Knowing more about their movements gave specific information about habitat requirements, individual home ranges, and migration patterns. They in turn used this information to assess potential habitat problems, targeting areas for restoration. The tags helped determine the survival rate, age distribution, and sex ratio of the reservation population, which they applied to make inferences about future population trends.[29] For example, in the late 1980s, surveys revealed that cow elk numbers had decreased on the West End, indicating this population might be in decline. To rectify the situation, game wardens trapped thirty young cows in the northeastern portion of the reservation, where they remained abundant, and released them in the West End.[30]

The TGFD, with the aid of the BIA Forestry Department, also conducted large-scale habitat restoration projects, employing controlled burns to restore the grazing capacity of elk habitat.[31] Prior to federal fire-suppression policies of the early twentieth century, natural and Indian-induced fire shaped the reservation's Ponderosa pine ecosystem, creating prime ungulate habitat of lush meadows interspersed throughout the forest.[32] Reestablishing control

over an ancestral Apache land-management practice, they restored forest
and range damaged by past BIA fire-suppression and grazing policies.[33] As a
result of these restoration efforts, by 1989, the reservation elk herd became
so large that the TGFD had to institute a cow elk culling program.[34]

This restoration work contributed to the growth of one of the most pres-
tigious trophy elk hunts in the United States. Over the years the hunt has
drawn celebrities such as Jack Nicklaus, George Strait, Dale Earnhardt, and
Kurt Russell. Other clientele consisted of wealthy white male corporate exec-
utives, while some came from European royalty.[35] In short, the participants
were privileged, white male hunters, a breed not so different from those who
started the Boone and Crockett Club in the late nineteenth century. To aug-
ment the experience, the TGFD maintained elk populations that steadily
yielded trophy-caliber bull elks. In most years, hunters experienced a 70–80
percent success rate, with many of them passing up sure kills because they
would rather wait for a chance at a record-size elk.[36] Currently, the TGFD
claims that over a hundred kills from Fort Apache have landed in the record
book.[37]

In a sense, the tribe offered, and perhaps even restored, a past way of
life—a time when Progressive Era adventurer sportsmen such as Teddy
Roosevelt ventured into the "wilderness" to test their character and bring
home legendary kills. From re-creating elk herds with record-setting bulls,
to maintaining a wilderness setting, to providing Apache guides, cooks, and
skinners, the TGFD sparked the romantic imagination of the rugged indi-
vidualist.[38] They had restored a hunter's paradise.

## Self-Governance and Ecocultural Restoration

As restoration lent itself to increasing the economic and political stability
of White Mountain Apache organizations during the self-determination era,
these organizations acquired flexibility to be more creative with restoration
work. During the self-governance era, which emerged in the 1990s, tribal
organizations solidified their foundation for self-sufficiency by increasingly
incorporating tribal cultural concerns with ecological and economic goals
when designing restoration projects. To that end, the 1990s saw an explosion
of restoration projects that integrated elements of Apache culture, which often meant rediscovering Apache knowledge and applying it to these projects.
The emphasis on Apache cultural elements also played a major role in shaping the ecosystems of the reservation. For example, the tribe acted to restore
culturally important sites, culturally important species (cattails, medicinal plants, Mexican wolves, pronghorn antelope, elk, and Mexican spotted
owls), incorporate culturally sensitive materials in restoration work, use the
knowledge of elders, reconstruct places based on Apache legends, adopt an
ecosystem management philosophy, and protect sacred sites such as Mount

Graham and Mount Baldy. The Tribal Council even established a permanent Land Restoration Fund to support the recovery of "tribal ecosystems to a condition that better reflects their condition prior to suffering damage from the mismanagement and to fund the education of Tribal members in the disciplines related to natural resource management." In addition, the council gave priority to projects that developed comprehensive watershed planning activities; incorporated community-based efforts; and included "activities that promote traditional cultural practices, the Apache language, and the education of tribal members."[39]

The work of the Tribal Watershed Program to rectify past mismanagement of reservation lands illustrates this shift toward ecocultural restoration in the 1990s. The BIA had long promoted policies that encouraged timber overexploitation, overgrazing, and increased water runoff to the Salt River Valley. As a consequence, the reservation's watersheds were ecoculturally degraded by the 1980s. Two of the most controversial BIA programs were the Cibecue and Corduroy Watershed projects of the 1950s and 1960s. The BIA, in conjunction with the Arizona Water Resources Committee and the US Forest Service, designed this multipurpose program to slow erosion resulting from overgrazing, improve timber production, and increase water yield to reservation ranchers and Euro-Americans in the Salt River Valley. This combined restoration and water management experiment ended in controversy in the early 1970s. Instead of repairing degraded land, the experiment exacerbated erosion problems and negatively impacted plant communities along targeted waterways, including the eradication of 843 acres of locally coveted cottonwoods.[40]

Cottonwood trees provided landmarks for place names. When invoked, these places relayed moral stories of how one should act in the world.[41] These stories also stored knowledge about changes in the landscape, natural history, and lessons on how to treat the land. One such place, "cottonwoods joining," constituted an origin site for a White Mountain Apache clan. In some cases these place names signified local group and clan identities, such as the "cottonwoods extending to the water people" or "at the standing cottonwood people."[42] In essence, the eradication of cottonwoods literally threatened the eradication of Apache identity, moral and social structure, and knowledge about the landscape. Therefore, watershed restoration in the 1990s went beyond an ecological imperative, imbuing cultural significance while implicitly critiquing the local knowledge gaps inherent in standardized knowledge-production systems. In addition to the cultural significance of these projects, demonstrating effective restoration and management of wetlands, streams, and rivers would endow Apache organizations with more political capital in water rights disputes.

Exchange with Euro-American organizations helped guide restoration

Figure 12.2. An Apache heavy-equipment operator working on a juniper-eradication project associated with the Corduroy Watershed Project in the late 1950s. Courtesy of the United States National Archives and Record Administration.

efforts. A prerequisite of these arrangements, however, was that partnerships had to benefit Apaches. Tribal land-management organizations still lacked adequate expertise, so they continued to hire outsiders to provide knowledge and training for the restoration of tribal ecocultural resources. For example, Jonathan Long, a Euro-American hydrologist, helped train tribal members in wetland and stream restoration techniques, while working with a new generation of college-educated tribal ecologists: Delbin Endfield, Mae Burnette, and Candy Lupe. Along with many other members of the tribal community, they worked to restore ecologically damaged wetlands and streams.[43]

For instance, a number of Watershed Program efforts combined Apache and Euro-American knowledge in considering the local needs of Apache people, including transportation safety, livestock grazing, traditional use of plants, stable stream banks, and unimpeded stream flow. To build community support, these projects taught Apache youth land-management techniques. Perhaps more important, the youth gained a sense of heritage, as they learned from Apache elders, who served as ecocultural advisors, about stream ecology and the medicinal and ceremonial use of plants such as cattails, which are important elements of Apache curing ceremonies and puberty ceremonies.[44]

Work on springs symbolized the restoration of Apache knowledge and

WHITE MOUNTAIN APACHE BOUNDARY-WORK 191

culture. All reservation springs remain culturally important to Apaches.[45] Along with their cultural significance, these sites represent collaborative intersections between Apache knowledge and Euro-American knowledge. At Swamp Spring in 2002, where postfire erosion threatened wetlands, a federal land manager not familiar with tribal customs suggested the use of metal erosion-abatement structures to stabilize stream banks. In response, Mae Burnette advised against such measures. Along with observing frequent failures of these structures, she noted that metal compromises the spiritual integrity of ceremonial springs. As a remedy, Burnette and Long applied an adaptively designed culturally sensitive erosion-control system that was "remarkably similar to traditional erosion control practices." Instead of using metal, they installed check dams and riffle structures constructed of local rocks and plants.[46]

White Mountain Apache restorationists express considerable pride in their restoration accomplishments, which manifests itself in tribal land managers employing boundary-work that actively delineates Apache land from adjacent federal and private lands. For instance, in addition to the Watershed Program's restorative duties, its projects became an opportunity to promote the health of the Apache landscape to outsiders. In 1996, the tribe invited University of Arizona ecologists to monitor the reservation's "pristine" highland springs for comparison to streams in the adjacent Apache-Sitgreaves National Forest. In exchange for training tribal members in cutting-edge stream-analysis technology, the tribe permitted these ecologists to conduct this comparative study. Mae Burnette, a benefactor of the training program, proudly claimed, "Our . . . visitors said that they liked our stream better than the ones they had been to on the Forest Service [National Forest], which they said had smelled bad."[47]

Apache pride in their restoration work remained quite evident during my visit with Apache restorationists in 2007. Tribal restorationists revealed a number of their successful wetland and stream restoration projects. On our stop at Soldier Spring, a culturally important spring that flows east off the reservation into the Apache-Sitgreaves National Forest, they highlighted the contrast between the stream bank within the reservation boundary and the off-reservation stream bank. Indeed, evidence of erosion and down-cutting was negligible on the reservation, while signs of erosion were considerable as soon we entered the adjacent national forest. Demonstrating the relative health of the Fort Apache ecosystem and their stewardship capacity remains extremely important to Apaches for political, cultural, and social reasons. Such evidential comparisons bolster their growing reputation as land managers and provide political leverage in their ongoing battles to control reservation ecocultural resources.

Tribal restorationists such as Mae Burnette play an integral mediating role in these projects. Employing boundary-work, her expertise with both Apache knowledge and Euro-American land-management techniques allows her to negotiate the epistemological boundaries between Apache and Euro-American cultures. Her presence requires non-Indian experts to address local cultural and environmental concerns. Although Burnette stood to learn from non-Indian restorationists, her epistemological breadth allowed her to determine whether the interests of non-Indian restorationists were in line with Apache interests. As a consequence, she could absorb new knowledge, while simultaneously protecting Apache assets. She and other tribal restorationists now embody the process of boundary-work that cautiously erects selective boundaries that delineate between knowledge that benefits Apaches and knowledge that lacks application to Apache circumstances. Because of such homegrown expertise, the integration of elder knowledge about the reservation with Euro-American knowledge, the encouragement of community participation, and the inclusion of Apache cultural, political, economic, and social concerns into projects, Apache restoration and management programs mark a significant departure from earlier natural resource management initiatives under the auspices of the BIA that encouraged dependency rather than autonomy.[48]

## OF NEW KNOWLEDGE-PRODUCTION SYSTEMS AND LANDSCAPES

Apache restoration projects evolved to perform the tripartite function of healing the negative impacts of Euro-American overexploitation of reservation lands, establishing a degree of economic freedom from the federal government, and restoring and maintaining cultural traditions. Ultimately, though, cultural appropriation of Euro-American restoration and land-management techniques had enduring effects on the Fort Apache landscape. Apache institutional construal of technoscience shaped the reservation ecology differently from adjacent lands.

While Apaches worked to restore recreational and ecocultural resources, towns on the northern edge of Fort Apache—Pinetop, Showlow, and Springerville—experienced extensive economic growth as a consequence of both increased tourism within Apache-Sitgreaves National Forest and Fort Apache. Motels, restaurants, gas stations, and retail services swarmed to this region to capitalize on the increased tourist activity. With the intrusion of these businesses came population growth and suburban and exurban sprawl. While Euro-American communities didn't limit the influx of retail and restaurant chains, the Apache government did. As a consequence, these countervailing land-use decisions made Fort Apache an attractive destination for nature seekers. Actions to protect sacred sites, rehabilitate culturally important springs, reintroduce culturally important plants and animals, limit allowable timber cuts, increase elk populations, restore the Apache trout and other endangered species, restore watersheds, build reservoirs, establish a wilderness area, conduct prescribed burns, and limit the intrusion of national business chains resulted in a noticeable contrast between the reservation and adjacent off-reservation landscapes.[49]

However, one should be cautious in assessing the capacity of Apache organizations to efficaciously manage resources. The documented successes are generally tenuous and subject to the whims of federal funding. And although the tribal economy has made great strides since the 1930s, it doesn't adequately provide for the needs of the Apache people. Benefits from economic growth have not been experienced equally, with many tribal members still jobless or in poverty. As the Apache institution has become more self-sufficient, this progress has been countered with the withdrawal of federal services, many would argue prematurely.[50] Consequently, the institutional infrastructure of the reservation still doesn't adequately support the health and safety needs of the Apache people. And even the fairly robust ecosystem management capacity of the tribe is vulnerable to large-scale disasters outside of its control, as the Rodeo-Chediski Fire of 2002 demonstrated. Despite temporary federal aid, this conflagration, which destroyed 276,000 acres of forest resources and effectively shut down the tribal sawmill, drained most financial resources from tribal management organizations. The primary effect has been that many of the ecocultural successes of the 1990s have either been put on hold or discontinued due to insufficient funds.[51] Even the highly successful trophy elk program has drawn criticism from both reservation and off-reservation farmers and ranchers who claim the large herds produced by the program have resulted in degradation of agricultural and grazing lands.[52] Nonetheless, the continuing institutional struggles evident on Fort Apache should not discredit the political, legal, and epistemological successes gained. Even partial success provides an instructive model for understanding strategies of empowerment for historically marginalized cultures.

The ecocultural and political transformations evident on Fort Apache are directly linked to an eighty-year pattern of Apache institutional boundary-work. To reiterate, initially federal land-management agencies during the Indian New Deal held a paternalistic relationship with White Mountain Apaches and thus held control over knowledge production in restoration projects. At first, Apaches had no "legitimate" recourse to challenge federal plans. Despite this predicament, tribal organizations slowly transformed ecological restoration into a vehicle for boundary-work. Slowly the tribe acquired the technical expertise to understand the environmental and cultural consequences of BIA restoration work. In addition, they developed the skills to conduct restoration projects that benefited the Apache community. This long and arduous journey eventually led to the Apache people regaining almost complete control of reservation resources and laying the groundwork for the relegitimization of long-lost, forgotten, or marginalized Apache knowledge and traditions.

As Apache organizations incrementally appropriated Western technology and learned the language of Western science, through boundary-work, they created a new knowledge system, which changed a historically exploitative pattern into a more inclusive, ecoculturally sensitive form of land management. Furthermore, combining locally generated Apache knowledge with Western knowledge eventually helped avoid the inherent blind spots created by top-down knowledge-production systems devoid of local knowledge.[53] The epistemological restoration of Apache knowledge involved creating an assemblage of cultural traditions, integrating Apache stories, legends, moral frameworks, natural histories, philosophical dispositions toward the land, and local ecological knowledge with a wide variety of methods, standards, tacit knowledge, techniques, and instruments from Western science.[54] This hybrid knowledge-production system has helped fill gaps found in both Apache and Euro-American systems. Perhaps it could be said, within the context of the twentieth and twenty-first centuries, that neither knowledge system alone could adequately address the environmental and social issues evident in the Apache homeland.

The White Mountain Apaches' context-specific, historical appropriation of ecological restoration and management techniques is an instructive example of how a local community can positively react and adapt to the introduction of environmental technologies. Capitalizing on the interpretive flexibility of ecological restoration, Apaches molded this technoscience to address the needs of the Apache people.[55] This intervention broadened the beneficiaries of restorative technologies to include not only non-Indians but also the Apache people and nonhuman nature. Ecological restoration became a site of boundary-work that created connections and equalized relations between the Apache and Western land management agencies. The Apache instigated this process to maintain a multidirectional flow of information across boundaries to sustain their emerging, but vulnerable, hybrid knowledge system. This emerging system remains in flux as Apache organizations and tribal members continue to struggle with reconciling tribal ecocultural practices with Western science and technology. Even though this struggle began over eighty years ago, the Apache people have just begun to explore the depths of lost and marginalized knowledge and how it can be applied to ecological restoration work. As this knowledge-production system matures, people will continue to reinvent and further elaborate the essence of Apache knowledge.

# CHAPTER 13

# NEOecology

## THE SOLAR SYSTEM'S EMERGING ENVIRONMENTAL HISTORY AND POLITICS

### VALERIE A. OLSON

FOR TWO DECADES, scholars concerned with perceptions of the global environment have examined how remote sensing technologies serve as social tools. They call attention to the scientific and political uses of satellite and astronautical views of Earth from orbit, asking how these downward views open up new ways to spatialize forms of governance,[1] legitimize forms of environmental knowledge of the Earth,[2] and, as anthropologist Tim Ingold asserts, promote a flat topology of global surfaces that obscures experiences of living *within* a three-dimensional environment.[3] These analyses overlook an equally interesting question: How do space technologies make it possible to perceive outer space as connected to the Earthly human environment—and to what ends? For over a century, astronomers, biologists, geologists, and human spaceflight proponents have been acting as if the human environment that matters, both scientifically and politically, is greater than the terrestrial globe. By extending senses, technologies, and even people *outward* from planet Earth, they redefine the features and boundaries of a broadly spherical human environment.

This chapter traces how solar system objects, namely asteroids and comets, have become environmental as well as astronomical objects, in both technical and political terms. In doing so, it addresses what environmental history gains by investigating this shift. I argue that environmental history obtains a broader view of the multiple and contingent understandings of environmental wholes that are emerging across disciplinary and social boundaries over time. To make this argument I demonstrate how space science, technology, and policy activism generate new ways to perceive the Earthly environment and the human ecosphere, despite the general exclusion of these topics from environmental history. These changes in perception result from redefinitions of scientific disciplinary as well as political boundaries; they are precipitated, in part, by intensifying interdisciplinary attention to what it means to live on a planet.

In what follows, I trace scientific, technical, and political articulations of a life-and-death environmental relationship between Earth and a group of comets and asteroids with Earth-crossing orbits now known as Near Earth Objects (NEOs). While the history of a human relationship with outer space has been largely the province of geography, technology history, policy, and ethics scholars, the contemporary human relationship to NEOs reveals a distinctly environmental history. This history is an example of what Steven Pyne calls an "extreme history" of uninhabited and out-of-bounds spaces that people nonetheless actively perceive and engage, like the deep sea, Antarctica, and the solar system.[4] Such an extreme history is recognizable as environmental history when, as in this case, we can follow how asteroids and comets were first categorized as erratic space debris then recast as defining elements of Earth's history and then how they have become the subjects of national policy making and policy activism. The space that matters in this extreme environmental history and activism is not a sedate solar system but instead, as scientists informed me during my ethnographic fieldwork at the National Aeronautics and Space Administration (NASA), a dynamic "heliosphere" of interactive matter and energy. This heliosphere is starkly natural but now also contains social topologies made by decades of remote sensing scans and spacecraft missions to its far edges. According to these scientists and policy activists, the heliosphere is the macroenvironmental context for terrestrial, and by extension human, history.

This extreme environmental history begins with the detection and eventual classification of asteroids and comets as environmental objects, a process that brings into view the underinvestigated astronomical dimension of environmental history. Two Western master narratives involved in making the solar system a social and political space have shaped this history. The first is a continuing narrative about outer space as discovered territory; the second and less-investigated narrative that I attend to here is an emerging tale of cosmos as ecology. I outline how telescopic technologies, like the microscopic technologies that science and technology studies (STS) scholars have examined more thoroughly, play a role in what Ingold terms the "topology of environmentalism" by remaking what count as the scales and boundaries of human political ecologies.[5] I make this case by deploying STS conceptual frameworks that consider the role of technically defined objects in the remaking of disciplinary and institutional boundaries. As Susan Leigh Star and James Griesemer's well-known theorization of how techno-scientific "boundary objects" consolidate new "institutional ecologies" would suggest,[6] social activities to categorize NEOs as environmental threats and resources indicate how diverse technical practices and political agendas have aligned over time and across social spaces. Analysts' typical focus on the function of boundary objects qua objects, however,[7] tends to background questions about what kind of new spatial perceptions emerge as a result of collaborative work with those objects.[8] As a result, I am concerned here with the shift in what kinds of objects NEOs are understood to be now, but also with how NEO-focused activities create new understandings of a whole and politicized human ecology. I trace how coordinated efforts to sense, track, and respond to NEOs as threats or resources dramatically scale up political commitments to managing risk and environmental security beyond the Earth's high atmosphere. The research for this chapter comes from archival and historical work I undertook during the course of an ethnographic study of environmental science and technology at NASA, where NEO activists coordinate a national and international NEO policy campaign.

NEO experts and activists want their audiences to recognize NEOs as universally threatening, accessible, and linked to Earth's history and ecology but also, through expert acts of foresight and preparedness, to environmentally unlimited and socially hopeful futures. By detecting, tracking, and making national and international plans to send spaceships to NEOs to move them away from Earth or mine them for materials, astronomical practices get attached to other disciplines, from biology and geology to rocketry, that are concerned with the problems of life in a dynamic cosmos. NEOs, in this way, make socially relevant the interrelatedness of solar system objects, sites, and processes, which become "matters of concern"[9] for ongoing reconfigurations of what Jasanoff and Martello call "Earthly politics."[10] As material and disciplinary boundary crossers, NEOs trace, like no other planetlike objects, an extended ecological heliosphere of the Earth. They evince a human/space relationship that spans the atomic and macrosystemic, the prehistoric and futuristic, the technical and prophetic, the catastrophic and opportunistic. In conceptions of this deep space ecology, human evolutionary progress is contingent upon moving from passive to active engagement with heliospherically scaled space.

To document these heliosphere-bounded political ecological calculations and characterizations, I begin by discussing how NEOs became part of a solar system that required regimes of detection and tracking rather than simply observation. Then, as I follow astronomical and geological data collection processes that ended up linking NEOs with evolution and catastrophe, I detail how NEOs became policy targets for interdisciplinary risk management as well as astronautical mission programs. I end by describing an American-led international movement to petition the United Nations to begin an asteroid deflection and mitigation project. This activism promotes the value of detecting NEOs and speculations about their use as resources and, in doing so, articulates a political and ecological reimagination of a total human environment.

## "NOT YOUR FATHER'S SOLAR SYSTEM": NEW OBJECTS, DETECTION AND TRACKING REGIMES, AND THE REMAKING OF ECOSPHERIC BOUNDARIES

By the early twentieth century, accepted notions of what constitutes the "solar system," in dynamical and geological terms, had come undone. What unraveled was the orderly Copernican fabric of solar systemic space ruled by planets and within which the Earth abides as a separate and unique kind of place. What exactly outer space was made up of had been a matter of debate between vacuists and plenists, but the celestial mechanicists of the nineteenth century had begun to abandon the Cartesian proposition of a plenum full of small and large matter committed to separate vortices set in motion by a deity. The mechanicists focused instead on the Newtonian notion of a largely vacuum-filled space in which all planetary orbits reflect calculable laws of force, motion, and order. They were sure that calculative anomalies in the Copernican solar system model would be reconcilable as missing planets were discovered. Over time, however, the increasingly powerful astronomical observing technologies and techniques used to resolve these problems turned up new problems in the form of disorderly nonplanetary objects. What emerged in the late nineteenth century was a post-Copernican solar system full of matter in more or less orderly motion that required the understanding that the Earth exists in a celestial environment that is neither serene nor separate. This is a crowded and unpredictable environment that demands vigilant surveillance as much as observation and contemplation.

The appearance of small, strangely behaving nonplanetary bodies during the nineteenth and mid-twentieth centuries heralded a regime of detection and tracking derived from practices of astronomical observation, calculation, and classification. Science studies scholar and biologist Hans-Jörg Rheinberger has described laboratory "experimental systems" as the "smallest integral working units of research," their purpose materializing new questions and "epistemic things" that can remake ways to know research objects and situate them contextually.[11] Astronomical observational systems also do this kind of work, elaborating new sighting regimes and cross-disciplinary data sharing that bring to light previously hidden objects and interactive cosmological processes to be accounted for and characterized. As a result, the ongoing remote sensing detection regime inaugurated by searches for the weird orbits of "near Earth" asteroids goes beyond the knowledge-production processes of standard astronomical observation. Detection merges modes of making things visible with modes of accounting for invisibility, anomalies, and problems. The detection of out-of-order nonplanetary objects and eventually a ninth unruly planetlike object (Pluto) shifted astronomers' attention to new kinds of boundary crossings, both spatial and disciplinary, that coupled observation practices with detection and tracking regimes. This eventually put asteroids and comets into the category of outer space things not just to be watched but also to be watched out for.

## From Orderly Spheres to Wild Space

Between 1800 and 1930, European astronomers with telescopes worked to correct two disturbing problems of solar system spatial harmony. To detect the planet that must occupy the geometrically disorderly gap discovered between Mars and Jupiter, the late-eighteenth-century Vereinigte Astronomische Gesellschaft (United Astronomical Society) set up a group of "celestial police" that included William Herschel, Charles Messier, and Johanne Bode, whose astronomical rule (ruled coincidental by today's astronomers) required a planet to be there. In 1801 an associate of the celestial police, Italian astronomer Giuseppe Piazzi, sifted through evidence to track down a "minor planet" called "Ceres" that seemed to satisfy the problem, but the discovery of orbital anomalies in the outer planets, Uranus and Neptune, raised new questions. In 1930, American amateur astronomer Clyde Tombaugh discovered the so-called Planet X, the mysterious and hidden entity that must be perturbing Uranus's orbit. Stronger telescopes and new observing regimens were turning up other ambiguous-looking and -behaving solar system objects.

In the late nineteenth century, comprehensive observational star charting and the introduction of long-exposure photography for capturing light from the distantly strange and beautiful "Messier" objects, later identified as galaxies and nebulas, had revealed tiny, moving objects of indeterminate origin and orbit. In photographs they appeared as unlovely smears. These were the overexposed motion trails of objects soon to be known colloquially, because of their numbers and out-of-place appearance, as the "vermin of the sky."[12] German astronomer Maximillian Wolf's "blink stereomicroscope," which juxtaposed two photographs of a section of the sky so that the eye could flicker between them to discern differences of location and brightness that indicated motion, allowed astronomers to detect rather than directly observe these moving "asteroid" ("starlike") objects and to then systematically catalog them, although such traveling objects were not immediately assumed to be related to meteorites.[13] Clyde Tombaugh used this blink technique to image far-off Pluto, which he assumed to be a large planet with formidable gravitational power. From the mid-twentieth century to the present, the neatly defined properties and parameters of a sedate solar system were being recalculated and recharacterized to include problematic objects.

Astronomers of the mid-twentieth century faced a solar system with unstable outer objects; an unknown boundary; and new classes of objects, both visible and theoretical, that vastly outnumbered planets. From the mid-nineteenth century onward, astronomers cataloged asteroids they referred to as occupying, with the planetlike object Ceres, a "belt" between Mars and Jupiter. In 1950, the number of cataloged asteroids topped two thousand, with fifty-four identified as outside the main belt region, and new searches were on for a "Planet X" that must exist beyond Pluto. In trying to account for the newly documented speed and behavior of comets, Dutch astronomer Jan Oort argued for the existence of a "cloud" of comets with long-term planet-crossing orbits at a distance of twenty thousand astronomical units (AU) in 1950 (one AU is the distance from the Earth to the sun). In 1951 Dutch American astronomer Gerard Kuiper theorized that primordial material from the early solar system should exist past Pluto, as far as fifty to one hundred AUs, consisting of an estimated ten small objects that never coalesced into one planet. Searches for planetoid objects beyond Pluto yielded the definitive discovery of a "Trans-Neptunian Object" (TNO) space and object class, into which Pluto was installed, amid controversy, as a dwarf planet in 2006. The path from Pluto's discovery to its recent "demotion" from planet to dwarf planet reflected a shift, as STS scholar Lisa Messeri has recently argued, from commonly held "cosmological" agreements about the nature of the solar system and the definition of a "planet."[14] "Planets" were redefined as rare, orderly, and geologically complex, and other kinds of planetoid objects like comets and asteroids as old, prolific, disorderly, and simple. When "planet" was redefined and Pluto was reclassified, the International Astronomical Union (IAU) also divided the broad category of nonplanetary "small bodies" into two groups: dwarf planets (the big TNOs Pluto, Haumea, Makemake, Eris, and the large asteroid Ceres); and the remaining smaller TNOs, asteroids, and all comets.

At the time of this writing, the number of registered small bodies totals over four hundred thousand, and they continue to be detected and subclassified according to locations, morphologies, and space-traveling behaviors that put them all over the place. Astronomers considered and dismissed theories that the asteroids of the asteroid belt were simply the remains of a broken-up planet, determining instead that *all* the small bodies of the solar system are captured bits of early, undifferentiated "primitive" matter moving and interacting. All these small bodies are being pushed and pulled by each other and the planets and set into motion through collisions and, as a result, continue to become dislodged and create new collisions. This characteristic informed the IAU's 2006 declaration that "small bodies," as opposed to planets, have not "cleared" their orbital neighborhoods of most threats. Indeed, the spatial boundary of a small body over the duration of the solar system's existence is, theoretically, the whole heliosphere. To muddy the heliospheric waters even more, the contemporary IAU definition of "small bodies" technically includes objects of vanishing smallness, encouraging perceptions of a heliosphere criss-crossed by lots of material in motion and headed for interaction. Boundary-crossing "small-body" objects are thus not only fossil remnants of solar system formation and collisions; they also provide objective evidence of a still-active solar system that matters to astronomers, as well as geologists, biologists, and—as we'll see—policy makers. They cross boundaries to inscribe, with their orbits and actions, the heliosphere's fully dynamic—and dangerous—material space.

All told, the astronomical imaginary of a serene solar system gave way to views of a space that is less than orderly and more filled up and diverse than previously imagined. This reimagined solar system precipitated the twentieth century's attention to long-term processes of planetary and small-body relationships. In addition, the Earth/space boundary gained a new permeability. That extreme boundary is calculable as an artifact of planetary atmospheric attenuation and is also imaginable as acutely relevant to human life, opening the question of what such a permeable boundary means in contemporary calculations of risk and models of planetary ecology. Also attenuating since the late nineteenth century were strict disciplinary boundaries between astronomy and other sciences, which came together exactly over efforts to define and know ambiguous boundary spaces like atmospheres and mysterious cross-planetary features like craters.

## SMALL BODIES AND AN EXTENDED TERRESTRIAL ECOSPHERE

With twentieth-century astronomical efforts to catalog, classify, and track the prolific nonplanetary objects now known as "small bodies" came parallel, and eventually interconnected, astronomical and geological theorizations about the existence and role of cosmic collisions in planetary and biological history. New kinds of comparative reasoning emerged about the macroecological boundaries of terrestrial life's beginnings, ends, and extensibilities. From the midcentury onward, asteroids became objects of astronautical theorization about how to reimagine the scale of Earth's environment.

Sightings of meteors and comets and the collection of meteoric material on Earth have a long history, but until the late nineteenth century, astronomers and geologists didn't have or seek evidence to associate meteors with significant terrestrial features or events. This was in part due to conditions of evidence and its interpretation. As a geological and biological active planet, most of Earth's large impact craters are hidden by water, erosion, or growth. Any Earthly crater formations, and those subsequently located on other planets through telescopes, were interpreted as originating in a planet's own (or endogenic) geology and usually as volcanic. In 1876, British astronomer Richard Proctor put forth an unpopular "impact model" of lunar crater creation. This idea was reanimated by American astronomer Grove Carl Gilbert in 1892 and later by mining geologist Daniel Barringer in the early twentieth century, both of whom were captivated by Arizona's Coon Mountain—later renamed Meteor Crater. The puzzling Tunguska Siberian impact "event" of 1908 and its devastation of forest lands also called broader attention, despite lack of a crater, to the prospect that planetary atmosphere-crossing cosmic objects could be large and environmentally destructive. Astronomical evidence for a small-body-filled and -disrupted solar system was emerging alongside new geological arguments that the Earth's and moon's craters were largely the result of impacts and that the course of Earthly life might have been shaped by impacts. Significant nonendogenic geological events were not just historical but could happen now.

By the mid-twentieth century, attempts to understand cosmic collision evidence brought together disciplines that eventually constituted formal collaborative forums and spaces for what became "planetary science." In 1953, the journal *Meteoritics,* focused on the boundary-crossing exploits of meteorites, encouraged interdisciplinary opportunities to investigate nonterrestrial material, thus inaugurating a new cross-disciplinary domain (the journal later became *Meteoritics and Planetary Science*). Cross-disciplinary collaborations were steps toward breaking barriers to the production of comparative and generalizable knowledge about the Earth as a planet, signaling the appearance of what STS scholar Peter Galison calls "trading zones" for astronomy and other sciences.[15] Along with concepts like "planet," this zone shared the terminology and data of "detection," "collision," "impact," and "ecosphere" to analyze and argue with.

In 1981, a watershed planetary publication in the journal *Science* sketched for a broader public the outlines of a heliospheric ecology by offering a definitive solution to an Earthly mystery that bridged the boundaries of astronomy, geology, *and* biology. What was the relationship between a thin band of sediment dated at 65.5 million years ago and the mass species extinctions that marked the boundary between the Cenozoic and Mesozic eras? The answer presented not just a case for the existence of a grand cosmic impact on Earth, but the beginning of scientifically acceptable theorizations about the historical ecological roles of those impacts. Physicist and Nobel laureate Luis Alvarez, partnered with his geologist son Walter and two scientists in the Energy and Environment Division of the Lawrence Berkeley Laboratory, found that the boundary sediment in question contained an unusual abundance of iridium, 6.3 parts per billion, which indicated an extraterrestrial origin.[16] Using data from the Krakatowa volcano eruption and astronomical estimations of size range among the detectible "earth-crossing asteroid" population, the colleagues concluded that a 10 ± 4–kilometer asteroid must have impacted Earth at that time, creating, besides local catastrophic damage, atmospheric disruptions that stopped photosynthesis, disrupted food chains, and caused widespread extinctions. They suggested also that other passages of extraterrestrial material through the Earth's atmosphere, such as cometary ice, might have caused other extinction events. Alvarez et al.'s work stands as a generally accepted explanation for "the end of the dinosaurs" and seeded other attempts to investigate impacts as catalysts for Earthly processes, including the "panspermia" theory that interstellar traveling space rocks bought prebiotic molecules to Earth.

As asteroids became known as the solar system's unruly, hostile, and generative "small bodies," their material primitivism and nearness invited efforts to detect and calculate their prospective economic value. Soon after Alvarez et al.'s article, NASA's 1983 IRAS (Infrared Astronomical Satellite) mission began to return systematic spectral data on 1,811 asteroids, confirming ways to classify them according to their composition. These classifications are determined by the analysis of an asteroid's "albedo" (reflection): metallic, silicate, and carbonaceous. Using such data, University of Arizona planetary geologist John S. Lewis speculated about asteroid material trade value. His claims followed on imaginative 1960s asteroid resourcing speculations boosted by futurist NASA engineer Dandridge Cole's dramatic claim that it is possible to extract "$50,000,000,000,000 from the Asteroids" and solve the space program's budget woes forever.[17] Lewis became in the 1990s a spokesperson for commercial asteroid mining, arguing that evidence of pure undifferentiated metals promised "untold riches" that could finally justify human space exploration.[18] As with enthusiasm for harnessing cosmic energy from the moon or space, asteroid-mining proponents perceive the terrestrial incorporation of cosmic wealth as a kind of heavenly providential resource that would end the extraction-based degradation of Earth and could be spread across social boundaries like the perpetual shower of cosmic dust that bombards Earth (after all, iridium is a form of platinum).

Despite late-twentieth and early-twenty-first-century calculations of small-body resource benefit, impact threat remains the topic that garners the most attention across scientific and social boundaries. The evidence of prolific solar system bodies with disorderly orbits, interstellar rock and dust accretions on Earth, and craters on other planets and moons signals the contemporary existence of serious heliospheric small-body threats. With

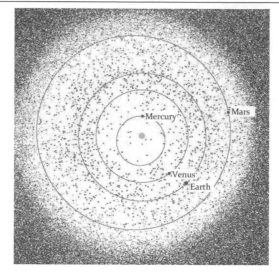

Figure 13.1. Location plot of known asteroids in the inner solar system (2007). Courtesy of
Scott Manley, Armagh Observatory.

Cold War policies for military-industrial technology expansion came accepted assumptions that national security had to be ever extensible,[19] and
cosmic-impact threat mitigation was identified as a prospective investment
site for multidisciplinary and multi-institutional collaboration. With the
cataloging of asteroid and cometary orbital behaviors and theories about
the history of an impacted Earth came projects to calculate the risk variables and develop the technologies necessary to manage a future impact-
threatened Earth. Starting in the 1990s, popular scientific articles about
asteroid and cometary threats contained generation-boundary observations
such as "it's not your father's solar system."[20] Another frequent opener,
echoed in science papers and in most of my interviews with fifteen small-
body experts, became: "Earth exists within a cosmic shooting gallery." The
boundary between heliospheric environmental perception and planetary
protection was being bridged.

## "IN A COSMIC SHOOTING GALLERY": NEOS AND A SOLAR SYSTEM RISK ECOLOGY

With interdisciplinary data sharing and theorizing, astronomical and
geological characterizations of Earthly asteroid impact risk became an urgent
project in the late twentieth century. What has emerged  in the twenty-first
century is a growing class of asteroids and comets known to a variety of
experts and publics as Near Earth Objects, some now even known by name,
such as 99942Apophis, which is due to make a "close" pass in 2029. NEOs
gained boundary-crossing characteristics that were not just astronomical,

biological, and geological but useful for making headlines or becoming the subject of predictions and policies, setting the parameters for a trading zone for heliospheric risk and social preparedness. This zone includes academic astronomers, geologists, and biologists, as well as the military, scientific and security arms of government institutions like NASA, and now impact-deflection activist groups.

In 1991, NASA Jet Propulsion Laboratory (JPL) astronomer Donald Yeomans wrote an article for the international Planetary Society's journal *Planetary Report* called "Killer Rocks and the Celestial Police" that transformed the historical celestial police detection role into a global threat surveillance service.[21] The contemporary celestial police, represented by an animated searching eyeball cartoon graphic on JPL's NEO Watch website, not only observe but also monitor the cosmos for threats to Earthly environmental and social order. This is an Earthly order with broadly spherical ecological boundaries that matter to nations and international relations. In general, these boundaries include zones of human-made orbital debris and incoming "space weather" (solar and cosmic ray) events but also extend in theory to the Oort cloud and beyond, where life-threatening interstellar neutral hydrogen atoms push into the heliosphere bubble in varying intensities. Such emerging discourse about existing "in a cosmic shooting gallery" exemplifies the spherical environmental cosmology concept Ingold describes, but with a decidedly contemporary cast. In this section, I describe the emergence of NEO surveillance and policy regimes that are formalizing the terms of Earthly planetary risk and the extreme spatial and temporal boundaries of a manageable human ecosphere. These are the terms with which NEO activist groups engage problems of destructive impact, as well space exploration potentialities.

## NEOs Cross the Policy Boundary

During the 1950s, astronomical and geological scientists interested in small-body behavior and impacts began routine data sharing and interaction with Cold War–era nuclear test scientists and human space program lunar reconnaissance surveyors. The long career of "astrogeologist" Gene Shoemaker illustrates these productive boundary intersections. In the manner of memorializing scientific knowledge and practice as a patriline, Shoemaker is the acknowledged "father of planetary geology," and asteroid policy activists I observed and interviewed refer to him as the "godfather" of what would become contemporary impact theories and deflection planning. When he started work for the US Geological Survey in the late 1950s, Shoemaker compared the structure and mechanics of craters caused by meteorites with nuclear explosions. Dedicated early on to the idea that impact was a fundamental force in solar system evolution, Shoemaker founded the 1973

Planet-Crossing Asteroid Survey at Palomar Observatory, which would become the model for tracking NEOs. During the early 1990s, the definition of "NEO" expanded to include long-period comets from the heliospheric edges and even the increasing orbital jumble of human-made objects stretching from low Earth orbit to geosynchronous satellite space.

In 1992, Shoemaker sat with Edward Teller, the father of the H-Bomb, at a round-table discussion about the use of nuclear weapons to mitigate near Earth asteroid threats.[22] This meeting signaled the dual concern of asteroids and comets to scientists and the military and the crisscrossing of military dual use and national catastrophe-preparedness technology proposals.[23] At the first International Academy of Astronautics Planetary Defense conference in 2007, it had become commonplace among military planners to call NEO mitigation "planetary defense" and not only to characterize Earth's location in space as within a "shooting gallery" full of "speeding bullet[s]" but also to counter astrobiological notions that Earth's orbit is simply life sustaining by calling it "hazardous."[24] While the quantification of the NEO risk in such military environmental terms legitimated discussions about using Strategic Defense Initiative weaponry that might be politically opposed if the enemy was another nation, there is later, in nongovernmental NEO mitigation activism, an alternative narrative that argues that appropriate anti-NEO technology cannot be weaponizable. Scientific evidence that weapons used against incoming NEOs would increase their threat (an exploded NEO hurtling toward Earth is not necessarily a neutralized NEO) has led to designs for nonweaponized NEO mitigation. As one engineer told me, the appropriate selection of anti-asteroid technologies should be an "intelligence test" for how humans will choose to survive and thrive in a solar system. The development of such technologies is at the heart of the pro-spaceflight and universalized technocratic notions of humanitarianism that motivate NEO deflection activism, as I detail further on.

The time between the 1980s and the end of the 2000s was a turning point for NEOs as they moved beyond the domains of science, science fiction, and futurism and into national policy circles. In meetings like NASA's 1981 "Collision of Asteroids and Comets with the Earth: Physical and Human Consequences" workshop, new characterizations of an Earth/space relationship began to suggest a hierarchy of heliospheric elements ordered by their relevance to human life, putting NEOs along with the sun and moon as objects of immediate consequence. The shocking discovery and close Earthly pass of a "Potentially Hazardous Object" logged as 1989FC in 1989 led to two congressional study mandates: one to develop a systematic NEO detection program, the other to determine new asteroid-moving or -destruction technologies. Unlike the moon, which waxed and waned as a cosmic policy object, NEOs were portrayed as an uncertain but ever-present threat to the Earthly environment. The NASA Colorado workshop report specifically emphasized the importance of Earthly impacts to the "ecosphere," as well as the importance of using technological ingenuity to avoid them, calling for a "gestalt shift" in how humans should think about their relationship with the NEO "population" and the solar system.[25]

That detection study became the congressionally mandated Spaceguard Survey, which orders NASA to detect NEOs and characterize what it calls a "threat environment."[26] The survey is named for an asteroid-watch program imagined by science fiction author Arthur C. Clarke.[27] Spaceguard's charge is to detect "asteroids larger than 1–2 km" that would cause "global scale events," understood to be the consequences of impact and structural and infrastructural collapse.[28] The Spaceguard survey attaches NEOs to national and Earthly futures by using orbit and mass data to calculate a "quantitative estimate of the impact hazard as a function of impactor size (or energy)." In addition, the survey was created to "advocat[e] a strategy to deal with such a threat."[29] These early 1990s activities and the dramatic and shocking 1994 impact of Jupiter by the Shoemaker-Levy comet (named after Gene Shoemaker) moved NEOs closer to being understood as environmental objects. That first eyewitnessed large planetary impact in modern astronomical history precipitated the relabeling of NEOs as "natural hazards," underscoring the socially meaningful contiguousness of Earthly and non-Earthly nature.

In the early 2000s, that meaning was elaborated in other ways, objectifying even further the features of an Earth/solar system boundary made collapsible by NEO collision. In 2004, Swedish astronomer Hans Rickman and Canadian geologist and environmental catastrophe expert Charles Bobrowsky received a request and grant to assemble a multidisciplinary panel to address on an "open platform" the "potential psycho-social and physical consequences of a catastrophic comet or asteroid impact on Earth,"[30] producing the science policy volume *Comet/Asteroid Impacts and Human Society*.[31] In keeping with Star and Griesemer's description of how boundary objects function to strengthen institutional ecologies, this work institutionalizes new disciplinary and conceptual associations that make visible, in scientific terms translatable to policy domains, Earth's nearly invisible but epic existence in a zone of extraterrestrial bombardment. Sociologist Ulrich Beck's description of a "world risk society," in which established ideas about the spatial and temporal bounds of risk and of what counts as an enemy and as defense became unsettled; the NEO threat creates a demand for mastery of a "*worlds* risk society" in which societies are called to account for not just the globally figurative but the extraterrestrially literal interconnectedness of heliospheric objects.[32] The Rickman and Bobrowsky volume's characterization of impact risk in universal terms has become the concern not just of scientific organizations and government institutions but of nongovernmental

NEO activism. This activism labels the NEO threat as more than natural. It maintains that in an astronautical age, to allow a devastating environment- and evolution-altering impact or to fail to use asteroids as environment-enhancing resources is ultimately a problem of global social policy.

## "WE INVITE YOU TO JOIN THIS ULTIMATE ENVIRONMENTAL PROJECT": NEO MITIGATION AS ENVIRONMENTAL POLICY

In 2008 American Apollo-era astronaut Rusty Schweickart mobilized the NEO committee arm of the Association of Space Explorers (ASE) to petition the United Nations to create an international asteroid-deflection and impact-mitigation program. He runs the B612 Foundation (named for the asteroid inhabited by Saint Exupery's Little Prince), whose website invites visitors to "join" in the "ultimate environmental project" that Schweickart with others launched because of their dissatisfaction with the "current lack of action" to protect the Earth from Near Earth Objects.[33] The ASE, made up of "350 people from 35 nations" who have flown in space, has observer status in the United Nations Committee for the Peaceful Uses of Outer Space. While the ASE petition reflects what social scientists Andrew Lakoff and Stephen Collier describe as "a profusion of plans, schemas, techniques, and organizational initiatives that respond to new kinds of perceived threats to collective security," their activism also promotes astronautical theorizations of human/environment interaction and evolution.[34] The environment they are concerned with is not just global but heliospheric, and they reframe the terms of human survival as tied to astronautically enabled success or failure. The ASE's UN petition presents a kind of universal moral imperative for developing authoritative technocratic schemes of rational planning and political preagreements about mechanisms of response, mitigation, and trust.

The ASE petition for the United Nations to take responsibility for NEOs, which is at the time of this writing still in committee, sets out the legal and policy basis for doing so by reinterpreting a key article of the 1967 Outer Space Treaty. The Outer Space Treaty, on which the United States is a signatory, represents space as a human commons of scientific and economic potential. Article 9 of the treaty states that space exploration must follow planetary protection protocols "so as to avoid . . . adverse changes in the environment of the Earth resulting from the introduction of extraterrestrial matter and . . . shall adopt necessary measures for this purpose."[35] Imaginatively, the ASE coalition rescopes this planetary-protection requirement by linking this article to a companion treaty resolution on remote sensing, "which calls upon states to promote by means of their remote sensing activities the protection of Earth and mankind, and share relevant information, whether it concerns a threat to the Earth's natural environment or resulting from natural disasters."[36] Although the coalition admits that this particular resolution was "not drafted with a view to asteroid threats, [it] should be interpreted *a fortiori* to entitle measures to be taken to avoid serious and adverse changes to the environment of Earth stemming from an asteroid threat."[37] This reinterpretation reframes space as a specifically environmental extension of Earth that requires surveillance and globally coordinated management. The coalition recommends specific actions that stem from the B612 Foundation's early work to design integrated detection and deflection systems, such as increasing NEO surveillance and building a nonweaponized robotic "gravity tractor" that could prod the asteroid away or, if possible, alter the location of ground zero, which is a worst-case scenario fraught with controversy.

In the coalition's petition, an impacted or unimpacted Earth is a determiner of both natural and social evolution. While effacing national differences and highlighting the interests of a universal category of humans as a planetary species, the ASE document writers admit that an NEO deflection project such as the gravity tractor may only turn, in Schweickart's words, an "act of God" impact into an "act of humankind" pathway of controlled impact points.[38] Here, the general term *asteroid impact* becomes a numeric and political calculation of impact consequences across national boundaries. A projected impact point has the potential to be deflected toward or away from, according to technical and political decisions. Projects to shift impact points from one site to another can create what the petition admits is an escalation of the potential for NEO mitigation projects to become, in Schweickert's words, "more political and difficult."[39] A *Wall Street Journal* article on Schweickart's description of the UN petition sums up his preference for internationally prenegotiated calculations to allay fears that the United States will use NEO deflection as a basis for weaponizing space:

> Now suppose the impact line for an asteroid begins over Country A, extends through Country B and ends at Country C. To nudge the asteroid so that it misses Earth completely, you first have to push it in one direction or another—in effect, toward either A or C. That means that residents of either A or C will bear a slightly greater risk if the rescue effort doesn't push the asteroid quite hard enough. Naturally, the citizens of A and C, and their political leaders, will be screaming for the asteroid to be pushed in the other country's direction and out of their backyard. Mr. Schweickart says the only fair way to proceed is to have a decision-making formula drawn up well in advance, thus unaffected by the political heat of an actual crisis.[40]

In this plan, negotiations for NEO impact are expected to mirror negotiations about pollution, dumping, forest management, climate change, and other cross-boundary environmental problems. In the UN petition, the authors nervously skirt the fact that impact-mitigation investors will have an advantage over noninvestors, going on to recommend that an appropriate plan would calculate in advance "the basis of the value of human life and property, independent of national political power or influence," and enforce that agreement as necessary.

Following established theorizations of impact risk in universalistic terms that elide inequalities, the ASE petition describes preserving the Earthly biosphere as the responsibility of a "complex and interconnected human society" with the obligation and unprecedented capacity to avoid becoming "victims" of this kind of environmental catastrophe.[41] In this plan, NEO defense is a form of responsible environmental policy and engineering, with "species" and "biosphere" salvation named as ultimate imperatives. Haunting the ASE document are the geographics of sacrifice and the fate of an evolutionarily unelect, made not to survive by dint of powerful negotiations that assess (in the document's terms) "human life and property" in ways that appear objectively "independent of national political power or influence." Like the heroic plan of a Hollywood asteroid-impact action movie, the ASE's goal is to unite humanity against a shared solar system threat, dramatically expanding the high-stakes arena in which the "Earthly politics" of global environmental governance and resistance is negotiated. As Collier and Lakoff argue, catastrophe preparedness logics, especially those centered on rare risks, can reveal extreme and powerful reorderings of value that are not based on humanistic units of preservation such as persons, communities, societies, or nations, thereby leading to schemas that value the saving of "vital" infrastructures over humans per se.[42] If undertaken, a UN NEO impact-mitigation program based on negotiating impact paths has to evaluate places in terms of their contribution to preventing what activists term "civilization" disruption and extinction, which may necessitate different extended site-units of risk calculation than "nations." The petition itself points to the need to remap the Earth in terms of protection-worthy spaces and those that, in the words of my activist interlocutors, must "absorb risk." This brings the Earth and NEOs together within a modern narrative about human fitness and destiny. In this way not only do NEOs serve as boundary objects that "inhabit intersecting social worlds and satisfy the informational requirements of each,"[43] according to Star and Griesemer's formulation of the concept, but they also make it possible to translate into social and political terms reasons for managing the intersection of terrestrial and extraterrestrial ecologies.

In sum, NEOs are boundary objects that enable environmental interpretations of cosmic history, Earth/space relationships, and human futures. Detected and tracked as threatening, disorderly, but accessible elements of a formerly detached outer space, NEOs make it possible to formalize the features and stakes of a near Earth and even far Earth environmental sphere in need of technical management. If, as Tim Ingold notes, environmental perspectives are "caught up in the dialectical interplay between engagement and detachment, between human beings' involvement in the world and their separation from it," then NEOs have moved from being only astronomical objects to being wholly environmental objects that mark a deeply socially engaged heliosphere.[44] In terms of the "thick/thin" dichotomy set up by environmental rhetoric scholars George Myerson and Yvonne Rydin to distinguish kinds of environmentalist discourses, the solar system that activists speak of as being shared by Earth and NEOs is a thick heliospheric ecology made up of dynamic material relationships, not a thinly diagramed system of mathematically ordered objects.[45] As new perceptions of what it means to live on a planet amid NEOs appear in national and international environmental imaginaries and policy regimes, they evince intersections of astronomical and astronautical history with environmental history. Joining STS and environmental history helps to illuminate these intersections, making it possible to trace their origins and emerging effects.

# PRESERVATION IN THE
# AGE OF ENTANGLEMENT

## STS AND THE HISTORY OF
## FUTURE URBAN NATURE

### SVERKER SÖRLIN

PRESERVATION AND CONSERVATION are standard tropes of environmental history, nowadays often cited as icons of the field's backward past rather than its bright future. In this chapter I argue that, on the contrary, there is currently a major transformation regarding how we understand the social-ecological processes of protecting nature and how previous dichotomies between nature and culture can be transcended in order both to better understand preservation as a social phenomenon and to inform policy. Environmental history can play an important role in this transformation. This chapter attempts to demonstrate that concept and theory from the field of science and technology studies (STS) can assist environmental historians very productively in that work. The focus is on preservation of urban nature, partly because urban nature is already an interesting field of preservation and partly, and perhaps most importantly, because cities are rapidly growing to become the hegemonic life-form of the large majority of the world's population and thus will likely shape, on an unprecedented scale, what we will come to understand in the future as nature, both inside and outside urban regions.

The general frame of thinking that I will apply is actor-network theory (ANT), here taken in its broad sense, starting with the sociology of *transla-*

*tion,* developed by Michel Callon and others in the 1980s.[1] This line of work has been refined and further developed in the work of Bruno Latour. It is particularly Latour's work on mobilization of resources and "actants" and "weights" for the shaping of new narratives that I will apply.[2] That has not been common. In fact, it is rather striking that protection of nature has not hitherto been an area with many applications of ANT. To preserve nature is to attempt something that goes against the grain of common capitalist uses of nature and real estate. It requires extraordinary efforts of mobilization of a kind that the sociology of translation can provide and where convincing counternarratives are essential. A possible explanation for the lack of interest is that early STS did not include a wider interpretive framework that could include nature. STS was chiefly about individual interaction in social microspaces, not about society and its interaction with natural macrospaces.

Work inspired by STS in a range of disciplines, including the social sciences, ecology, and design, has started to address these shortcomings over the last decade. From this work I would like to single out Bruno Latour's concept of *nature politics* as key, since it provides links between preservation and other applications of social thought to contemporary environmental concerns.[3] One concept that is particularly important in this regard is *entanglement.* Darwinian in origin, this is a concept that Latour uses as an image of the blurring and permeated boundary that exists between the human and nonhuman worlds.[4] Whereas previous narratives of modernity focused on the separation of culture from nature and emphasized the superiority of reason and science in relation to the natural world, entanglement signifies the rising currency of a rather different narrative. We become ever more enmeshed through science and technology and through the way we lead our lives and engage with nature everywhere. If *entanglement* is thus the general direction of modernity derived from STS's engagement with ecology, and *nature politics* the emerging arena for dealing with it, *translation* and *narrative* can be used as its analytical applications.

This is thus the interpretive framework, inspired by STS, that I would like to set in motion in this paper on preservation of urban nature. I will do so by drawing on empirical work done in many cities around the world but especially in Stockholm, where ecologists, architects, sociologists, and historians have been able to reformulate the very idea of what is going on when urban nature is defined, acquires value, and is protected.

## THE RECEIVED ENVIRONMENTAL VIEW OF PRESERVATION

The standard narrative of preservation is based on rich historical documentation and broad sources. Work on this history has been conducted by scholars of geography, the biological field sciences such as zoology, ecology, botany, and many others. In recent decades, however, much of the work has been conducted by environmental historians. Indeed, work on preservation, and its related concept of conservation, was part of the main thrust of the early development of the discipline, not least in the United States, with singularly important books such as Samuel P. Hays's *Conservation and the Gospel of Efficiency* (1959).[5]

Historiography in the early years was rather limited to Anglo-Saxon countries, notably former British overseas colonies such as the United States, and the Scandinavian countries.[6] In recent decades it has broadened to include other parts of Europe and countries and regions elsewhere.[7] The pattern that emerges is increasingly rich, and it demonstrates clear differences in chronology, ranging from mid-nineteenth-century American precursors to pioneering European countries like Sweden and Germany and to late European starters in the middle of the twentieth century.[8] In many former colonial countries organized preservation activities have barely begun.

There has been a growing awareness that the roots of preserved areas are complex indeed. Issues of nation-building and formation of "national landscapes" and various "identities" linked to collective memories have been explored in a large number of studies,[9] and the role of science and scientists has been equally thoroughly analyzed.[10] The gendered and sociologically differentiated patterns of reserves and their uses have been the object of interest both in the forming of reserves and in their use. The racial nature of parks and reserves in colonial areas has been explored, and their roots in the self-interest of colonial elites, such as hunting societies, have been underlined.[11] Issues of values and formation of values have been acknowledged at a fairly late stage in the research. Prominently, there has been dispute over the suggestion that preserved regions and areas are indeed inventions by particular social elites, an idea in the postmodern constructivist tradition that was articulated by William Cronon in his edited volume *Uncommon Ground* (1995). At about the same time Bill Adams suggested that preservation should be acknowledged as a conscious effort in forming what he called "future nature," rather than merely maintaining one that would otherwise cease.[12]

Despite increasing richness in the ways to approach, analyze, and understand preservation, there are still a number of assumed features of preservation and reserves that remain largely unquestioned even today and make up what one could call the paradigmatic understanding of preservation within the field of environmental history (EH). In this "received EH view" preservation occurred in wilderness areas and was representative, monumental, nationalist, expert based, and hierarchical. It created a dualist divide between regulated single-purpose reserves in remote areas and an outside world of vile markets and ecological decline. In that sense, despite the increasing richness, maturity, and critical perspectives of recent scholarship, there is also a striking continuity in the paradigmatic preservation narrative. What has changed is not so much its content as its valuation. The protagonists stay much the same: they are white men, and a very few women, venturing into the field; discovering the "beauty, health and permanence" of it; and starting social processes in order to preserve it.[13] The motives of these people are deconstructed in new scholarship; the gendered character of the project is penetrated, as are the complex travels of ideas from the field to the media and through to politics and decision making and the omnipresence of commercial interests. The stage, the arena of preservation, and the main tropes and actors of the narratives, however, have remained by and large the same.

There is also a phenomenal absence of theory in the mainstream preservation literature. Apart from a few major absorptions of theories of nation building, of deconstruction, and of a few standard tropes in the history of science, for example, the uses of "the field," there has been a conspicuous absence of such theory that could help explain preservation better as a social process.[14] In fact, research on preservation and on natural and national parks is among the few areas in the history of science that have been almost untouched by the rolling wave of STS work crashing on our shores since the late 1980s. The laboratory, the classroom, industrial research labs, Cold War science, the disciplines, the institutions, the hospitals, the asylums, the society of risk and danger—virtually all possible scientific environments, topics, and periods seem to have become objects of STS interest, but not nature reserves and their formation. Why is that?

## A HISTORIOGRAPHY OF ABSENCE

Notwithstanding the general lack of interaction between STS and the history of preservation and conservation, there is one feature of this "historiography of absence" that stands out as particularly striking, and that is STS and urban preservation. Is there any reason why STS and the urban, both separately and combined, should not be of interest to the historical study of preservation, a full half century after Hays's path-breaking study?

Urban preservation may not follow the same standard pattern as the formation of reserves in the so-called wilderness. The urban does not fit the stereotype partly because the protected areas are fewer, but that is deceptive: the level of protection may not be as high or as exclusive as in the remote areas, but protection nonetheless exists when social movements or other social actors manage to stave off exploitation and maintain areas as parks, commons, or green sectors. As such the history of urban preservation is rich indeed, although, interestingly, it has not been much conceived as such. Examples of struggle over urban space are countless; almost every city can cite several, and some have gone down in history as formative archetypes of the environmental movement: the formation of the "free city" Christiania in

Copenhagen in the 1970s; the "struggle of the elms" in central Stockholm in 1970; the fight to preserve People's Park in Berkeley, starting as early as the 1960s; and notably the establishment in many cities throughout the world of recreational spaces, parks, waterfront areas, outdoor landscapes, and so on. Like many environmental issues and movements, the urban green movements also saw a *Gründerzeit* in the 1960s; before then urban expansion and sprawl were by and large regarded as signs of progress and prosperity or as a necessary sacrifice at worst.[15]

Strangely, these pervasive and historically deeply rooted forces of preservation have not been much observed by environmental historians, let alone those who specialize in issues of conservation and preservation. Instead urban geographers, sociologists, architects, planners, and some social and urban historians have taken on the task of writing about them. A case in point is the compelling story of a full century of the preservation history of the San Francisco Bay Area by Berkeley geographer Richard Walker, *The Country in the City* (2007). Walker starts his story at just about the point where most previous work on California preservation left off: after the early years of the Sierra Club and the formation of significant early national parks like Yosemite, Sequoia, and Redwood. Instead he demonstrates that there was indeed an interest in the protection of urban or peri-urban nature in the Sierra Club. He then traces, decade by decade throughout the twentieth century, how new groups in the Bay Area have been drawn to the cause of protection, from local communities, especially those with educated and well-off housewives during the midwar period, to students, including Berkeley's architecture students, who envisioned green planning already in the 1930s, and postwar environmentalists, as well as, in later years, ethnic groups, communities, and business and organized interests. The main thrust of Walker's story is his focus on civic involvement and the civic spirit that he finds in almost all such protective work.[16]

His story therefore stands out from the archetypical environmental history of protection; he seeks the intersection of the social, the political, and the ecological, and he connects it to the already existing and quite large body of environmental history on cities.[17] But as Walker observes, this literature is mostly a narrative of urban decline or urban infrastructures and only marginally an attempt to explain urban green preservation. Nor could one cite, in support of his historiography, the massive literature on creative cities, edge cities, globalizing cities, high-tech cities, or resilient cities, since it almost totally neglects the preservationist aspect of events and does not develop it as a criterion of (sustainable) urban development.[18] Still, that fusion of urban innovation and urban preservation needs to take place.[19] Instead, and lacking much support from urban or environmental history, Walker turns his interest to the urban social sciences, in particular those that try to understand why citizens act, often voluntarily and in civic movements, to take responsibility for their city, for local ecosystem services and improved neighborhoods.

## RECOMBINANT CONSERVATION: STS AND THE NEW URBAN POLITICAL ECOLOGY

What factors are at play in shaping the trajectories of cities in either direction? And, assuming that preservation ambitions did exist and protective projects were attempted, how can we assert the factors that might explain why some of these projects have been successful whereas others have failed? What is it in the histories of protection that signifies success, like in the Bay Area? No typology exists, nor is there a database that is readily available that could help us sort out the success factors. There are, however, some attempts at sorting out the storyline on a metropolitan basis. There is an emerging history of the way Melbourne and Stockholm have attempted to govern ecosystem services through metropolitan-scale strategic planning and to what effect.[20] Nonetheless, we will for the time being have to rely mostly on the classical historical method of writing stories of individual cases or groups of cases and comparing those in order to sort out the forces that are at play.

However, some pattern is recognizable. Just like wilderness parks, protected urban areas are preconditioned by value-creating processes. Research on these processes highlights the importance of social articulation, that is, the work of social actors to mobilize recognition for certain places or objects.[21] Using concepts and ideas from actor-network theory, we can see this work as a *political program* that gains power as actors "pick up" artifacts or symbols from different cultural and historical contexts and translate them to fit the program, that is, linking them to places or objects in order to give weight to their program.[22] The term *artifact* is used here in its broad meaning, as "an object made by a human being" or, as elucidated by Latour, as a product or leftovers from other actors' labor. They can thus be buildings, paintings, or maps, but also scientific reports.

In the by now well-known language of Callon's classic 1986 article and its followers, we may speak of a "sociology of translation" whereby certain authoritative "actors" in the network serve as an "obligatory point of passage" and are able to align the network through four stages of action, from problematization (definition of problem), over *interessement* (connecting to other actors and gaining their commitment) and enrollment (forming and conducting the action), to the mobilization of allies (the mature stage of actor networks where communication is formalized, and networks can use "immutable mobiles" like artifacts, diagrams, maps, etc.).[23] While such networks do not possess power in any formal sense, they can achieve a lot, that is, gain power, as "communities of practice" wielding power and knowledge.[24] In this context narrative is practice and, as we shall see, a very useful practice when it comes to aligning actors in the interest of articulating values to support preservation.

However, this and similar research has focused on nonurban landscapes and has only rarely dealt with processes that have taken place in the very recent past, say since the 1980s. This means that they have not yet adequately considered the "recombinant ecology" (in Barker's terms) of what have been called "living cities" (see Hinchliffe and Whatmore), which has become the focal point of new urban and political ecologies and what we may call a "recombinant conservation" of the "more-than-human" (see Braun) urban geography, that is, one that takes the nonhuman world and entanglements seriously.[25] Likewise, and as a consequence, most historical approaches so far have largely missed out on the formation of "new social movements" or urban movements and their ability to create structural change in modern urban societies, thus also reflecting these entanglements and turning them into local politics.[26]

Observations in recent research indicate that while the general pattern of social articulation and the sociology of translation is still valid—that is, the need for construction of values and the mobilizing role of knowledge-brokering elites—the social forms and practices of nature conservation may be in a state of change or may at least demonstrate a rather broad and growing variation.[27] Classical "monumental" wilderness areas are complemented by urban "heterotopian" or "amonumental" sites that are more congenial to entanglement practices.[28] Therefore, the tools, or "artifacts," used in the "political program" are also likely to change, as will some of the content of the narrative itself. National, or nationalist, arguments that functioned to legitimate grand "green-field" sceneries are being supplemented, even replaced, with arguments advocating for "brown-field" urban nature conservation and the protection of urban green space that takes place through the agency of local movements, nongovernmental organizations (NGOs), think-tanks, and organizations of low-status groups.[29]

## TELLING AND RETELLING PROTECTIVE STORIES

A thorough shift in the view of urban nature, as "nature," was not possible until the radicalization of ecology in the 1960s and 1970s and, equally important, the emergence of conservation biology and alarming reports on biodiversity.[30] The acknowledgment of urban nature was to a high degree the work of ecologists. The concept of *urban ecology*, however, was coined by Chicago sociologists Robert Park and Ernest Burgess in 1910.[31] What these founders of the "Chicago school" had in mind was rather the city as a natural ecosystem. Their analysis of human society and behavior came out as quite reductionist and is best understood as an early contribution to the checkered history of human ecology in the twentieth century.[32] Only after World War

II was there a more active interest among ecologists in urban environments. From the 1960s onward research in urban ecology by ecologists appeared on an experimental scale, but not until the 1990s did the efforts become significant, for example, through the National Science Foundation's two urban Long Term Ecological Research (LTER) sites, in Phoenix and Baltimore.[33] In parallel, conservation biology appears as an active subdiscipline; an international organization, the Society for Conservation Biology, was founded in 1985. With its roots in animal parks and their species-oriented work for preservation, conservation biology thrived in the ever more favorable climate of concern for biodiversity.[34] Urban conservation grew into quite a movement, with particular strength in the United Kingdom, where grassroots initiatives were aligned with academic research on the plant and animal life of cities.

Conservation biologists soon found that infrastructures and urban environments stood in a complex relationship to biodiversity. On the one hand migration corridors and habitat were destroyed by urban growth, an argument advanced with other concepts by earlier generations of preservationists and advocates of the Garden City Movement since the beginning of the twentieth century. On the other hand urban environments functioned as vitalizing refuges for certain species, especially those that were victims of the mechanization and homogenization of agriculture and those that were sensitive to disturbance regimes that provoked new dynamics, species immigration, and unprecedented forms of specialization. Recent research in urban ecology has confirmed that not only urban parks, which always served a similar purpose, but also railway yards, allotment parks, golf courses, sports fields, and other typically urban zones have become the home of remarkable and unexpected varieties of biodiversity, not in spite of but rather as a consequence of human interaction and "disturbance" and with significant implications for urban planning.[35] The "urban" thus offers an entirely new ecology, where old truths and theories from "pristine" nature did not work adequately. Given that humans largely live in cities and now impact all parts of the world, it was perhaps in the city that a more general social ecology could develop.

In many synergistic ways the city, it was demonstrated, held a substantial portion of "nature"—and this nature was far more interesting than people had previously thought. These insights lay behind the emerging interest among social scientists and historians to explain what constituted protection of green areas and urban ecosystems, and their "services," a concept that gained currency from the mid-1990s.[36] The ecosystem approach was used in work on the remarkable protection of vast green areas in central Stockholm as a National Urban Park, with a legal status since 1995 on a par with classical wilderness national parks.

The connection to real-world politics is important, and that interaction can also be analyzed with STS tools. The National Urban Park in Stockholm has acquired legal status, and it has done so in an area that is fundamentally alien to the typical wilderness nature reserve. It is in a city and in an area where the pressure for economic use is extremely high and with a very high population density. One would have expected the area to rank low on a biodiversity index, but the opposite is true. The lively and varied use of the park, possibly more intense than anywhere else in Sweden, by at least sixty to seventy organizations (on some fifty square kilometers of land area), also explains the exceptional biodiversity in the park.[37] The entanglement of these actors and the green resources within these natural spaces helps explain the high level of protection. A network analysis of sixty-two organizations using the park—ranging from boating clubs to riding clubs and allotment garden associations—demonstrated frequent, intense, and sustained contacts between members of the organizations and a core "task force" of competent experts, drawn from these various organizations, which skillfully and efficiently influenced public opinion and key decision makers.[38]

The process behind the establishment of this legal innovation and its protection of urban nature on an unprecedented scale started as a movement to protect the "Eco Park," an area threatened acutely by rail and road projects that would balkanize the area into smaller parts.[39] The whole issue can be readily followed if one uses the lens of translation theory. The typical alarmists were involved, one a biologist, the other a medical researcher. Neither had any particular knowledge of conservation biology or of the new urban ecology. These were added later on as the legal issue was pending.[40] Nonetheless, these brokers and their scientific supporters, who prepared maps and scientific arguments, are congruent with Callon's translators and display a similar way of mobilizing arguments and social sympathy.

Then followed the broader social legitimization through articulation of values, with the typical mobilization of powerful institutions (the Royal Court, the Royal Swedish Academy of Sciences, famous artists, scientists, and scholars), the presence of social elites who live in the neighboring well-to-do northern and eastern areas and suburbs of Stockholm, and—crucially—the mobilization of a range of canonized "artifacts" in the park, buildings, sculptures, and famous and cherished vistas, often with royal heritage or connotation.[41] The whole sequence of events follows very closely Callon's translation process but is also in line with Latour's suggested scheme of artifacts, institutions, and designs (maps, names, images) as actants and "weights."

The above is of course both an interpretation of a historical set of events and facts and a theory-informed narrative provided to make sense of what has happened. It in turn reveals the social process, the tools, and the mechanisms that were involved (sometimes, I believe, unconsciously on the part of the "Eco Park" movement) in protecting the park. I have suggested, in collaborative work with urban political ecologist Henrik Ernstson, that we should call the strategic narrative employed by the movement a "protective story."[42] Part of that story is the social history of the protective work itself, which is typically seen among the preservationists as an integrated part of the story that they tell about the park and its venerable past.

## PRESERVATION AND THE POLITICS OF CONVIVIALITY

Apart from the interpretative framework given by ANT and STS, the theory of contemporary urban preservation proposed here also recognizes important changes in the scientific and cognitive framework. In the formation of wilderness and "nationalist" reserves in the nineteenth and twentieth centuries, legitimization often came from the geological sciences. In the country of Sweden (in this case) the geological sciences played a key role in the protective narrative explaining how the country's physical features had been formed by "ice and water" during and after the most recent glaciation and how they had been used as "weights" in the Latourian sense.[43] The nationalistic and iconic "deep time" of geophysical features was complemented by the assumed representative diversity in the form of landscape types and species diversity, offered by leading zoologists and botanists.[44] After World War II and increasingly since the definitive institutional breakthrough of scientific ecology in Sweden with the foundation of the journal *Oikos* in 1949, the integrative perspectives of ecologists have become ever more hegemonic, and this is especially true when it comes to urban nature, where geoscientists have had barely any influence at all.[45]

Ecology thus filled a void, especially as it increasingly included humans and their urban "nature practices" in their thinking. But ecology is not likely to remain the key scientific discipline to inform preservation. The encounter of society and nature in the city has already involved a wide range of scientific areas in analytical and preservation work. Through network studies in many countries anthropologists and sociologists have been able to pinpoint how different social groups, including their organizations, have been able to mobilize public protection for their own respective uses of nature, as both the Bay Area and the Stockholm cases suggest.[46] Through intricate analyses of the spatial organization of social and biological life, architects and landscape architects have corroborated Walker's dictum that there is, and should be, "nature in the city" but also that there should be a "city in nature."[47] The strong polarization between them is in fact one of the cornerstone ideas that nature preservation has contributed to cementing.

Luckily, there seem to be roads leading out of the dichotomy. "Alternative protection" with active use of maps, marked but unprepared paths, and downright "branding" of selected areas is a method that has been suggested, using Cambridge geographer William Adams's "creative conservation" as an intellectual tool.[48] Adams's concept is not unlike Alfred Schumpeter's "creative destruction." Roots should thus be sought in the modernism of breaking up toward a new future with a rich nature, perhaps even richer than today's, rather than in retrospective conservationism. Another example of a conceptual innovation is "sociotopia," that is, the topography of local places and sites that is determined by and corresponds to "experience value" and where sociological and economic circumstances influence the spatial distribution of site-specific properties.[49] Nature in a place is valuable not just because of its natural or historical properties as monuments, but because of its social embeddedness and usefulness, or rather the entanglement through which it is connected to the social. This should perhaps be obvious, but it hasn't been to most environmental historians and social scientists of any kind who are interested in preservation.

One could move one step further and suggest the possibility of setting up social programs to distribute and equalize access to urban green areas and their social use. One argument would be the positive health effects that have been demonstrated to follow protection. The most radical attempts so far might be those among geographers and sociologists who have ventured into the prospects of a "more than human nature" in the urban landscape.[50] Common uses of urban geographies have been laid as cornerstones for "cosmopolitan experiments" where new forms of "ecological politics" strive to undermine old dichotomies between man and nature, science and politics.[51] Beyond the most simple representations of nature (red listed species, additive lists of ecosystem services), some scholars try to understand, reconsider, and mutually accept and be influenced by the links and ever-deepening "entanglements" that characterize the relation between humans and the non-human world.[52] On these premises we should be able to find and articulate new political communities and rules for how to live together, humans as well as nonhumans: a "politics of conviviality."[53] Similar "more than human" visions also thrive at the intersection between art and scholarship/science. Cohabitation between animals and humans in the city has become the object of constructive experiments and keeps informing a thriving discourse on what might be called the "new boundaries of humans," stressing the mutual interdependencies and interactions of humans with domestic animals, ecological webs, and the social fabric.[54]

## PRESERVING PRESERVATION?

The preservation of nature is thus on the brink of being rescued as an intellectual, and perhaps also as a political, project by progressive currents in art, artistic research, architecture, urban ecology, and cultural and economic geography, with ecology as a still important backdrop. This preservation of preservation, as it were, when the future of preservation has long seemed very bleak,[55] is also assisted by intersectional historic rereadings of *wie es eigentlich gewesen ist*, when nature was not only separated from culture and reason in the grand modernist narrative but also separated from the city in the twentieth century, a process that is the latest in a number of such separations in late modern societies. I have previously, in collaborative work with Paul Warde, termed this process "environing."[56] Separation is important, but so is the process of healing, when nature is being reintegrated into urban settings, which luckily happens occasionally in our day and time. It might still be part of our environing, but it is a process with political dimensions quite dissimilar from the largely unsustainable practices we have pursued so far. New work on the resilience of cities suggests that access to urban nature, gardens and agriculture has provided long-term stability and an ability to survive in the face of war, siege, or natural disasters.[57] Other work suggests healing effects from green space in urban settings or integrative approaches to urban planning, where nature is made part of the built environment, and vice versa. These are forms of preservation that do not fit previous dichotomous versions of exploitation/preservation. They rather reflect the complex relationships that Darwin once saw on that "entangled bank," and it is a slight irony that his observation, more than 150 years old, returns to us by way of science and technology studies.

Some of the ideas that have been presented in this chapter are also quite novel and were almost impossible for earlier generations of nature preservationists to hold. Neither the knowledge behind them nor the conceptual innovations were available. The fact that nature protection has become an issue for cities and their populations is a result of ideas that have emerged at the intersection of research, civic organizations, and political discussion on how we can form livable and sustainable societies. "Nature" is no longer just for the small group of elites who could travel to and benefit from distant wilderness reserves. Nature should rather be seen as part of precisely that entanglement that becomes ever more characteristic of what it means to be human, which in turn means to be more and more part of the nonhuman. It is perhaps telling that the social articulation of this entanglement would start in cities, where nature and culture meet in more seamless and sprawling and discontinuous ways than in most other spaces. Cities thus form a paradoxical vanguard of nature politics and seem called to give new meanings to the old historical concept of preservation. As environmental historians we should be prepared to assume our part of this responsibility and describe, analyze, and explain this emerging and growing phenomenon, historicize it, and contextualize it.

# NOTES

## CHAPTER 1. JOINING ENVIRONMENTAL HISTORY WITH SCIENCE AND TECHNOLOGY STUDIES: PROMISES, CHALLENGES, AND CONTRIBUTIONS

I thank two anonymous reviewers, Dolly Jørgensen, Finn Arne Jørgensen, María Fernández, Durba Ghosh, Ron Kline, Laura Martin, Trevor Pinch, Rachel Prentice, Suman Seth, Marina Welker, and Wendy Wolford for their thoughtful comments on previous versions of this chapter, as well as all of the Trondheim workshop participants for their stimulating papers and insightful remarks during the conference. I also thank Robert Kulik and Connie Hsu Swenson for their editorial assistance.

1. David Bloor, *Knowledge and Social Imagery*, 2nd ed. (Chicago: University of Chicago Press, 1991), esp. chap. 1. For an overview of the principle of generalized symmetry, which extends the principle of symmetry to all dichotomies including the social and the natural, see H. M. Collins, "Science Studies and Machine Intelligence," in *Handbook of Science and Technology Studies*, 2nd ed., ed. Sheila Jasanoff, Gerald E. Markle, James C. Petersen, and Trevor Pinch (Thousand Oaks: Sage, 1995), 294–96. For an elaboration of this approach through empirical cases, see Michel Callon, "Some Elements of a Sociology of Translation: Domestication of the Scallops and the Fishermen of St. Brieuc Bay," in *Power, Action and Belief: A New Sociology of Knowledge*, ed. John Law (Boston: Routledge, 1985), 196–233; Bruno Latour, *Science in Action: How to Follow Scientists and Engineers through Society* (Cambridge: Harvard University Press, 1987). One can also see the premise at work in Latour's later book, *Aramis or the Love of Technology*, trans. Catherine Porter (Cambridge: Harvard University Press, 1996), in which different actors—"actants," in Callon and Latour's view—in the story of personalized public transportation systems in France speak, including Aramis itself.

2. For introductions and overviews to the field of STS, see Mario Biagioli, ed., *The Science Studies Reader* (New York: Routledge, 1999); Edward J. Hackett, Olga Amsterdamska, Michael Lynch, and Judy Wajcman, eds., *Handbook of Science and Technology Studies*, 3rd ed. (Chicago: University of Chicago Press, 2009); Jasanoff et al., *Handbook of Science and Technology Studies*; Sergio Sismondo, *An Introduction to Science and Technology Studies* (Malden, MA: Blackwell, 2004). For overviews of environmental history and historiographical essays at various points in the subdiscipline's history, see Kristin Asdal, "The Problematic Nature of Nature: The Post-Constructivist Challenge to Environmental History," *History and Theory* 42 (2003): 60–74; Elizabeth Ann R. Bird, "The Social Construction of Nature: Theoretical Ap-

proaches to the History of Environmental Problems," *Environmental Review* 11 (1987): 255–64; William Cronon, "Modes of Prophecy and Production: Placing Nature in History," *Journal of American History* 76 (1990): 1122–31; William Cronon, "A Place for Stories: Nature, History, and Narrative," *Journal of American History* 78 (1992): 1347–76; William Cronon, "The Uses of Environmental History," *Environmental History Review* 17 (1993): 1–22; Hugh S. Gorman and Betsy Mendelsohn, "Where Does Nature End and Culture Begin? Converging Themes in the History of Technology and Environmental History," in *The Illusory Boundary: Environment and Technology in History,* ed. Martin Reuss and Stephen H. Cutcliffe (Charlottesville: University of Virginia Press, 2010), 265–90; Arthur F. McEvoy, "Toward an Interactive Theory of Nature and Culture: Ecology, Production, and Cognition in the California Fishing Industry," in *The Ends of the Earth,* ed. Donald Worster (New York: Cambridge University Press, 1988), 211–29; Ted Steinberg, "Down to Earth: Nature, Agency, and Power in History," *American Historical Review* 107 (2002): 798–820; Jeffrey K. Stine and Joel A. Tarr, "At the Intersection of Histories: Technology and the Environment," *Technology and Culture* 39 (1998): 601–40; Douglas R. Weiner, "A Death-Defying Attempt to Articulate a Coherent Definition of Environmental History," *Environmental History* 10 (2005): 404–20; Richard White, "American Environmental History: The Development of a New Historical Field," *Pacific Historical Review* 54 (1985): 297–335; Richard White, "Environmental History, Ecology, and Meaning," *Journal of American History* 76 (1990): 1111–16; Richard White, "Discovering Nature in America," *Journal of American History* 79 (1992): 874–91; Richard White, "From Wilderness to Hybrid Landscapes: The Cultural Turn in Environmental History," *Historian* 66 (2004): 557–64; Donald Worster, "Doing Environmental History," in Worster, *Ends of the Earth,* 289–307. As White notes in several of his essays, environmental history certainly has deeper roots in related fields such as the history of the US West. The Annales School, established in France during the 1920s, is also significant.

3. Gregg Mitman, *The State of Nature: Ecology, Community, and American Social Thought, 1900–1950* (Chicago: University of Chicago Press, 1992); Robert E. Kohler, *Lords of the Fly: Drosophila Genetics and the Experimental Life* (Chicago: University of Chicago Press, 1994); Conevery Bolton Valenčius, *The Health of the Country: How American Settlers Understood Themselves and Their Land* (New York: Basic Books, 2002); Michelle Murphy, *Sick Building Syndrome and the Problem of Uncertainty: Environmental Politics, Technoscience, and Women Workers* (Durham: Duke University Press, 2006); and Linda Nash, *Inescapable Ecologies: A History of Environment, Disease, and Knowledge* (Berkeley: University of California Press, 2006). I cannot provide an exhaustive bibliography of all relevant scholarship here, but for a few works influenced by the history of science and/or science studies, see Benjamin R. Cohen, *Notes from the Ground: Science, Soil, and Society in the American Countryside* (New Haven: Yale University Press, 2009); Nancy Langston, *Forest Dreams, Forest Nightmares: The Paradox of Old Growth in the Inland West* (Seattle: University of Washington Press, 1995); Gregg Mitman, Michelle Murphy, and Christopher Sellers, eds., *Landscapes of Exposure: Knowledge and Illness in Modern Environments* (Chicago: University of Chicago Press, 2004); Jeremy Vetter, ed., *Knowing Global Environments: New Historical Perspectives in the Field Sciences* (New Brunswick: Rutgers University Press, 2010). Studies shaped by the history of technology include Finn Arne Jørgensen, *Making a Green Machine: The Infrastructure of Beverage Container Recycling* (New Brunswick: Rutgers University Press, 2011); Timothy J. LeCain, *Mass Destruction: The Men and Giant Mines That Wired America and Scarred the Planet* (New Brunswick: Rutgers University Press, 2009); basically all of the books in David E. Nye's remarkable oeuvre, including *Electrifying America: Social Meanings of a New Technology* (Cambridge: MIT Press, 1990) and *American Technological Sublime* (Cambridge: MIT Press, 1994); Sara B. Pritchard, *Confluence: The Nature of Technology and the Remaking of the Rhône* (Cambridge: Harvard University Press, 2011); Peter Thorsheim, *Inventing Pollution: Coal, Smoke, and Culture in Britain since 1800* (Ath-

ens: University of Ohio Press, 2006); Thomas Zeller, *Driving Germany: The Landscape of the German Autobahn, 1930–1970* (New York: Oxford University Press, 2007).

4. See, e.g., Mark Fiege, *Irrigated Eden: The Making of an Agricultural Landscape in the American West* (Seattle: University of Washington Press, 1999); Richard White, *The Organic Machine: The Remaking of the Columbia River* (New York: Hill and Wang, 1995); White, "From Wilderness to Hybrid Landscapes"; Brett Walker, *Toxic Archipelago: A History of Industrial Disease in Japan* (Seattle: University of Washington Press, 2010). See also Linda Nash's current research on Bureau of Reclamation engineers' work in Afghanistan after World War II. Actor-network theory may be particularly influential within environmental history because it has a materialist orientation shared by many environmental historians; it offers an(other) way to bring nature as a material actor into historical accounts. I thank one of the anonymous reviewers for highlighting this point.

5. Stéphane Castonguay, personal communication with the author, Feb. 16, 2011.

6. For instance, PhD programs in history with thematic emphases on science, technology, medicine, and the environment include Montana State University, Rutgers, and the University of Virginia. For STS programs with a significant environmental history presence, see MIT, the University of Pennsylvania, and the University of Virginia.

7. A few examples include Kohler, *Lords of the Fly;* Angela N. H. Creager, *The Life of a Virus: Tobacco Mosaic Virus as an Experimental Model, 1930–1965* (Chicago: University of Chicago Press, 2002); Scott Frickel, *Chemical Consequences: Environmental Mutagens, Scientist Activism, and the Rise of Genetic Toxicology* (New Brunswick: Rutgers University Press, 2004); Gregg Mitman, *Breathing Space: How Allergies Shape Our Lives and Landscapes* (New Haven: Yale University Press, 2007).

8. On mutual shaping, see Ronald R. Kline, *Consumers in the Countryside: Technology and Social Change in Rural America* (Baltimore: Johns Hopkins University Press, 2000); Sheila Jasanoff, ed., *States of Knowledge: The Co-Production of Science and the Social Order* (New York: Routledge, 2004).

9. For a number of historiographical reviews, see n2. For an accessible overview, see J. Donald Hughes, *What Is Environmental History?* (Cambridge: Polity, 2006).

10. On industrialization, see Theodore Steinberg, *Nature Incorporated: Industrialization and the Waters of New England* (New York: Cambridge University Press, 1991); Edmund Russell, *Evolutionary History: Uniting History and Biology to Understand Life on Earth* (New York: Cambridge University Press, 2011), chap. 9. For a sweeping reinterpretation of American history through this lens, see Mark Fiege, *Republic of Nature: An Environmental History of the United States* (Seattle: University of Washington Press, 2012); see also his earlier article, "The Atomic Scientists, the Sense of Wonder, and the Bomb," *Environmental History* 12 (2007): 578–613.

11. On the question of agency, see Linda Nash, "The Agency of Nature and the Nature of Agency," *Environmental History* 10 (Jan. 2005): 67–69. On entanglement, see Timothy Mitchell, *Rule of Experts: Egypt, Techno-Politics, Modernity* (Berkeley: University of California Press, 2002), esp. chap. 1; Walker, *Toxic Archipelago;* Paul Sutter, comment on J. R. McNeill's *Mosquito Empires: Ecology and War in the Greater Caribbean, 1620–1914* (New York: Cambridge University Press, 2010), American Society for Environmental History. On producing nature through knowledge systems, see Kim Fortun, "From Bhopal to the Informating of Environmentalism: Risk Communication in Historical Perspective," *Osiris* 19 (2004): 283–96; Weiner, "Death-Defying Attempt."

12. Sara B. Pritchard, "An Envirotechnical Disaster: Nature, Technology, and Politics at Fukushima," *Environmental History* 17 (2012): 219–43. See also the rest of this special issue on Fukushima for additional commentaries and analyses.

13. Different conceptualizations of the relationship between "natural" and "technological"

systems are encapsulated, e.g., by Thomas Parke Hughes's notion of technological systems and Charles Perrow's idea of "eco-system" (and specifically "eco-system accident"). These premises and definitions have significant implications for engineering design and practice. For an overview, see Pritchard, "Envirotechnical Disaster." As an empirical example of these issues, after Hurricane Katrina, some landscape architects, urban planners, engineers, and others have argued that New Orleans should be rebuilt according to "soft engineering" principles, which attempt to accommodate some environmental variation and processes within urban landscapes. Another timely example is illustrated by Wiebe E. Bijker's comparison of Dutch and American approaches to hydraulic engineering in "America and Dutch Coastal Engineering: Differences in Risk Conception and Differences in Technological Culture," *Social Studies of Science* 37 (Feb. 2007): 143–51.

14. On genes and evolution, see Russell, *Evolutionary History*, as well as two earlier articles by Russell: "Evolutionary History: Prospectus for a New Field," *Environmental History* 8 (2003): 204–28; and "The Garden in the Machine: Toward an Evolutionary History of Technology," in *Industrializing Organisms: Introducing Evolutionary History*, ed. Philip Scranton and Susan R. Schrepfer (New York: Routledge, 2004). For "biological" organisms, see the other essays in *Industrializing Organisms;* Ann Norton Greene, *Horses at Work: Harnessing Power in Industrial America* (Cambridge: Harvard University Press, 2008). On disease, see Alfred W. Crosby, *Ecological Imperialism: The Biological Expansion of Europe, 900–1900* (New York: Cambridge University Press, 1986); Alfred W. Crosby, "Ecological Imperialism: The Overseas Migration of Western Europeans as a Biological Phenomenon," in Worster, *Ends of the Earth*, 103–17; David Igler, "Diseased Goods: Global Commodities in the Eastern Pacific Basin, 1770–1850," *American Historical Review* 109 (2004): 693–719; J. R. McNeill, *Mosquito Empires: Ecology and War in the Greater Caribbean, 1620–1914* (New York: Cambridge University Press, 2010). On hydrology, see White, *Organic Machine;* Pritchard, *Confluence*.

15. On corn, see Deborah Fitzgerald, *The Business of Breeding: Hybrid Corn in Illinois, 1890–1940* (Ithaca: Cornell University Press, 1990); Arturo Warman, *Corn and Capitalism: How a Botanical Bastard Grew to Global Dominance* (Chapel Hill: University of North Carolina Press, 2003). For an accessible popular history of corn, see Michael Pollan, *The Omnivore's Dilemma* (New York: Penguin Press, 2006). On cotton, see Russell, *Evolutionary History*, esp. chap. 9. On flies, see Kohler, *Lords of the Fly*. On mice, see Karen Rader, *Making Mice: Standardizing Animals for American Biomedical Research, 1900–1955* (Princeton: Princeton University Press, 2004). On dogs, see Stephen Pemberton, "Canine Technologies, Model Patients: The Historical Production of Hemophiliac Dogs in American Biomedicine," in Scranton and Schrepfer, *Industrializing Organisms*. On viruses, see Creager, *Life of a Virus*.

16. On reciprocity and related ideas, see Russell, *Evolutionary History*, esp. chap. 8, "Co-evolution." His discussion of antibiotic resistance and especially the significance of how reciprocal dynamics are temporally framed is particularly compelling; see 98–101. Of course, terms such as *human world* and *nonhuman world* reproduce the very dichotomies complicated and challenged in environmental history. At the same time, actor-network theorists and other STS scholars have particularly called attention to the ways in which such terms are not only analysts' categories but also actors'. As such, they can (and should) be studied, rather than assumed as self-evident. I discuss the importance of paying attention to both dimensions later in this chapter.

17. Russell declares that "the workhorse sciences of modern environmentalism, ecology and public health, have held pride of place in environmental history as well" (*Evolutionary History*, 147).

18. For instance, the American Society for Environmental History was founded in 1977. Although declensionist narratives have repeatedly been problematized and challenged by environmental historians, they nonetheless remain influential. For several useful overviews, see

Candace Slater, "Amazonia as Edenic Narrative," Carolyn Merchant, "Reinventing Eden: Western Culture as a Recovery Narrative," and William Cronon, "The Trouble with Wilderness; or, Getting Back to the Wrong Nature," all in *Uncommon Ground: Toward Reinventing Nature,* ed. William Cronon (New York: W. W. Norton, 1995).

19. See, e.g., Jared Orsi, *Hazardous Metropolis: Flooding and Urban Ecology in Los Angeles* (Berkeley: University of California Press, 2004); Nash, *Inescapable Ecologies.* Nancy Langston advocates for an ecological understanding of health in *Toxic Bodies: Hormone Disrupters and the Legacy of DES* (New Haven: Yale University Press, 2009). It is worth noting, however, that the "ecology of health" discussed in Langston's book is also an actors' concept that develops in response to earlier bifurcations between (human) bodies and environments that emerged in association with germ theory.

20. See the subsection on constructivism below.

21. My comment here should not be conflated, however, with the attempt to write "big history." See, e.g., David Christian, *Maps of Time: An Introduction to Big History,* 2nd ed. (Berkeley: University of California Press, 2011).

22. I reference here just a few influential STS works on these themes. On imperialism, see Michael Adas, *Machines as the Measure of Men: Science, Technology, and Ideologies of Western Dominance* (Ithaca: Cornell University Press, 1989). On slavery, see Judith A. Carney, *Black Rice: The African Origins of Rice Cultivation in the Americas* (Cambridge: Harvard University Press, 2001). On capitalism, see Cori Hayden, *When Nature Goes Public: The Making and Unmaking of Bioprospecting in Mexico* (Princeton: Princeton University Press, 2003); Kaushik Sunder Rajan, *Biocapital: The Constitution of Postgenomic Life* (Durham: Duke University Press, 2006). On industrialization, see Jennifer Karns Alexander, *The Mantra of Efficiency: From Waterwheel to Social Control* (Baltimore: Johns Hopkins University Press, 2008). On the atomic age, see Itty Abraham, *Making of the Indian Atomic Bomb: Science, Secrecy, and the Postcolonial State* (New York: St. Martin's Press, 1998); Gabrielle Hecht, *Being Nuclear: Africans and the Global Uranium Trade* (Cambridge: MIT Press, 2011).

23. For key early laboratory studies, see Bruno Latour and Steve Woolgar, *Laboratory Life: The Construction of Scientific Facts,* 2nd ed. (Princeton: Princeton University Press, 1986); Karin Knorr-Cetina, "The Ethnographic Study of Scientific Work: Towards a Constructivist Sociology of Science," in *Science Observed,* ed. Karin Knorr-Cetina and Michael Mulkay (London: Sage, 1983): 115–40; Steven Shapin and Simon Schaffer, *Leviathan and the Air-Pump: Hobbes, Boyle, and the Experimental Life* (Princeton: Princeton University Press, 1985); Michael Lynch, *Art and Artifact in Laboratory Science* (London: Routledge, 1985); Sharon Traweek, *Beamtimes and Lifetimes: The World of High Energy Physics* (Cambridge: Harvard University Press, 1988). For a helpful overview of various approaches to laboratory studies in STS, see Karin Knorr-Cetina, "Laboratory Studies: The Cultural Approach to the Study of Science," in Jasanoff et al., *Handbook of Science and Technology Studies,* 140–66. For two examples of specific controversies in STS amid the genre of "controversy studies," see Brian Wynne, "May the Sheep Safely Graze? A Reflexive View of the Expert-Lay Knowledge Divide," in *Risk, Environment, and Modernity: Towards a New Ecology,* ed. Scott M. Lash, Bronislaw Szerszynski, and Brian Wynne (Thousand Oaks: Sage, 1996); Diane Vaughn, *The Challenger Launch Decision: Risky Technology, Culture, and Deviance at NASA* (Chicago: University of Chicago Press, 1997).

24. William Cronon, *Changes in the Land: Indians, Colonists, and the Ecology of New England* (New York: Hill and Wang, 1983); William Cronon, *Nature's Metropolis: Chicago and the Great West* (New York: W. W. Norton, 1991); Crosby, *Ecological Imperialism;* David Blackbourn, *The Conquest of Nature: Water, Landscape, and the Making of Modern Germany* (New York: W. W. Norton, 2006); Russell, *Evolutionary History.* By saying this, I do not mean to critique microhistories. In fact, historians have developed rich analyses of large-scale processes and issues through seemingly micro examples. Furthermore, recent scholars such as Roland Robertson

and Anna Tsing have shown, in fact, the articulation of the macro through the micro. See Roland Robertson, "Glocalization: Time-space and Homogeneity-Heterogeneity," in *Global Modernities*, ed. Mike Featherstone, Scott Lash, and Roland Robertson (London: Sage, 1995), 25–44; Anna Lowenhaupt Tsing, *Friction: An Ethnography of Global Connection* (Princeton: Princeton University Press, 2005).

25. See, e.g., Kohler, *Lords of the Fly*; Rader, *Making Mice*; Pemberton, "Canine Technologies, Model Patients"; Creager, *Life of a Virus*.

26. For an example of the materialization of gendered assumptions, which had consequences for both human bodies and nonhuman nature, see Langston, *Toxic Bodies*, 136, 146. See also Finn Arne Jørgensen's chapter in this volume on how systems and scripts materialize assumptions and attempt to enforce practices. For a theoretical discussion of historical ontology, as well as useful empirical models of this approach, see Murphy, *Sick Building Syndrome*; Mitman, *Breathing Space*. Furthermore, Gabrielle Hecht highlights the problematic borders between the cultural and material in *The Radiance of France: Nuclear Power and National Identity after World War II*, 2nd ed. (Cambridge: MIT Press, 2009), 11.

27. Conversations with Suman Seth helped me distill and organize the points in this section.

28. Most of the chapters in this collection come from larger projects—from doctoral dissertations to book manuscripts—although the role of STS theory and especially its contributions to environmental history vary in those projects. Readers interested in the specific research sites and historical analyses may want to consult the larger projects from which these chapters are drawn.

29. I am grateful to Jonathan Steinberg for the useful shorthand.

30. For instance, as feminist historians have convincingly shown, class analysis can illuminate complex (and unequal) social relations, while obscuring the gendered dimensions and dynamics of those relations. The classic essay here is Joan W. Scott, "Gender: A Useful Category of Historical Analysis," *American Historical Review* 91 (Dec. 1986): 1053–75. See also Joan Scott, *Gender and the Politics of History* (New York: Columbia University Press, 1988).

31. On narrative, see Cronon, "Place for Stories." Of course, there are many examples of historical scholarship that are both theoretically oriented and empirically rich. Studies of gender are a prime example.

32. See overviews of STS in n2 above, as well as foundational works in the sociology of scientific knowledge, including Bloor, *Knowledge and Social Imagery*. See also Trevor J. Pinch and Wiebe E. Bijker, "The Social Construction of Facts and Artifacts; or, How the Sociology of Science and the Sociology of Technology Might Benefit Each Other," in *The Social Construction of Technological Systems: New Directions in the Sociology and History of Technology*, ed. Wiebe Bijker, Thomas Hughes, and Trevor Pinch (Cambridge: MIT Press, 1987); Sal Restivo, "The Theory Landscape in Science Studies: Sociological Traditions," in Jasanoff et al., *Handbook of Science and Technology Studies*, 95–113.

33. Fiege, *Irrigated Eden*. "Nature-culture" is from Bruno Latour, *We Have Never Been Modern*, trans. Catherine Porter (Cambridge: Harvard University Press, 1993), 7. "Naturecultures" is from Donna Haraway, *The Companion Species Manifesto: Dogs, People, and Significant Otherness* (Chicago: Prickly Paradigm Press, 2003), 1; Donna Haraway, *When Species Meet* (Minneapolis: University of Minnesota Press, 2008), 16. On envirotech, a few key works include Stine and Tarr, "At the Intersection of Histories"; Scranton and Schrepfer, *Industrializing Organisms*; Reuss and Cutcliffe, *Illusory Boundary*; Pritchard, *Confluence*.

34. Of course, as all academics know, theoretical jargon can also obscure.

35. Bryan Pfaffenberger, "The Harsh Facts of Hydraulics: Technology and Society in Sri Lanka's Colonization Schemes," *Technology and Culture* 31 (1990): 361–97; Pritchard, *Confluence*, chaps. 3–5; Sarah T. Phillips, *This Land, This Nation: Conservation, Rural America, and*

*the New Deal* (New York: Cambridge University Press, 2007), epilogue. On forests as hybrid landscapes (although this term is not used), see Henry Lowood, "The Calculating Forester: Quantification, Cameral Science, and the Emergence of Scientific Forestry Management in Germany," in *The Quantifying Spirit in the Eighteenth Century,* ed. Tore Frangsmyr, J. L. Heilbron, and Robin E. Rider (Berkeley: University of California Press, 1991), 315–42; Paul W. Hirt, *A Conspiracy of Optimism: Management of the National Forests since World War Two* (Lincoln: University of Nebraska Press, 1994); Langston, *Forest Dreams, Forest Nightmares;* W. Scott Prudham, *Knock on Wood: Nature as Commodity in Douglas Fir Country* (New York: Routledge, 2005); Robert Gardner, "Constructing a Technological Forest: Nature, Culture, and Tree-Planting in the Nebraska Sand Hills," *Environmental History* 14 (2009): 275–97. On rivers, see White, *Organic Machine;* Mark Cioc, *The Rhine: An Eco-Biography, 1815–2000* (Seattle: University of Washington Press, 2002); Ari Kelman, *A River and Its City: The Nature of Landscape in New Orleans* (Berkeley: University of California Press, 2003); Pritchard, *Confluence.* Of note, there are surprisingly few comparative environmental histories. A few notable exceptions include Ian Tyrrell, *True Gardens of the Gods: Californian-Australian Environmental Reform, 1860–1930* (Berkeley: University of California Press, 1999); Kate Brown, "Gridded Lives: Why Kazakhstan and Montana Are Nearly the Same Place," *American Historical Review* 106, no. 1 (2001): 17–48; Paul Sutter, "What Can U.S. Environmental Historians Learn from Non-U.S. Environmental Historiography?" *Environmental History* 8, no. 1 (2003): 109–29; Lynne Heasley, "Reflections on Walking Contested Land: Doing Environmental History in West Africa and the United States," *Environmental History* 10, no. 3 (2005): 510–31; Mart Stewart, "If John Muir Had Been an Agrarian: American Environmental History West and South," *Environment and History* 11, no. 2 (2005): 139–62.

36. This anecdote comes from Mary Louise Roberts, now at the University of Wisconsin, when she was still teaching at Stanford University. One might combine Roberts's comment with Michelle Murphy's notion of regimes of perceptibility to argue for scholars' reflective theorization. Overall, conversations with Finn Arne Jørgensen influenced my thinking in this section.

37. See, e.g., the subsequent subsections on expertise and boundary-work for more detailed elaborations of these issues. However, the larger point I am making here is that theoretical knowledge has historically been valued more than empirical or "applied" knowledge. For an analysis of these hierarchies of knowledge in the American engineering context specifically, see Ronald Kline, "Construing 'Technology' as 'Applied Science': Public Rhetoric of Scientists and Engineers in the United States, 1880–1945," *Isis* 86 (1995): 194–221. For other important analyses of the ways in which knowledge is constructed and perceived hierarchically (including the fact that some knowledge is not perceived as knowledge at all), see Nina E. Lerman, "'Preparing for the Duties and Practical Business of Life': Technological Knowledge and Social Structure in Mid-19th-Century Philadelphia," *Technology and Culture* 38 (1997): 31–59. For an example of knowledge hierarchies in this volume, see Eunice Blavascunas's chapter.

38. For historiographical overviews of the nexus of the history of technology and environmental history, see Stine and Tarr, "At the Intersection of Histories"; Reuss and Cutcliffe, *Illusory Boundary,* especially the introduction, the afterword, and Gorman and Mendelsohn, "Where Does Nature End?" (265–90); Sara B. Pritchard, "Toward an Environmental History of Technology," in *Oxford Handbook of Environmental History,* ed. Andrew C. Isenberg (New York: Oxford University Press, forthcoming). For specific theoretical concepts, see Pritchard, *Confluence,* for envirotechnical system and envirotechnical regime. On energy landscape, see Christopher F. Jones, "A Landscape of Energy Abundance: Anthracite Coal Canals and the Roots of American Fossil Fuel Dependence, 1820–1860," *Environmental History* 15 (2010): 449–84.

39. On the interconnections between models and data (e.g., data shape the development

of models, models shape the collection of data), see Paul N. Edwards, *A Vast Machine: Computer Models, Climate Data, and the Politics of Global Warming* (Cambridge: MIT Press, 2010). Furthermore, the dynamics of empirics and theories allude to the often blurry boundaries between actors' and analysts' concepts.

40. Conversations with Djahane Salehabadi particularly inspired this important point. It is easy to see (and perhaps judge) actors' assumptions, categories, and so forth, but it is also worth considering why new theoretical approaches emerge and how they may reflect given historical, cultural, and political moments. For example, growing interest at the intersection of the history of technology and environmental history in the 1990s and 2000s and the move to question tidy divisions between nature and technology may reflect increasingly porous borders at the turn of the twenty-first century: GMOs, endocrine disruptors, global climate change, and so on. One might also contrast Donna Haraway's cyborg metaphor of the Cold War 1980s with her recent turn to companion species.

41. For an accessible overview of recent ecological research that has challenged earlier paradigms and promoted alternative possibilities such as rewilding and novel ecosystems, see Emma Marris, *Rambunctious Garden: Saving Nature in a Post-Wild World* (New York: Bloomsbury, 2011).

42. Of course, this point is not limited to these two terms. One could extend the argument to nature, ecosystem, species, and so on.

43. On "earthly politics," see Sheila Jasanoff and Marybeth Long Martello, eds., *Earthly Politics: Local and Global in Environmental Governance* (Cambridge: MIT Press, 2004). Notably, Olson's example suggests there might also be "cosmic" politics. Thus, problematizing the environment, both historically and analytically, can open up new research sites. For example, scholars working in marine environmental history have critiqued the field's tendency to focus on terrestrial environments. See W. Jeffrey Bolster, "Opportunities in Marine Environmental History," *Environmental History* 11 (July 2006): 567–97. But this example, along with Olson's above, provides another illustration of the ongoing dialogue and dynamics between actors and analysts (see also n40).

44. Although *object* tends to imply material artifact, I mean nature and knowledge as historical entities, both culturally and physically.

45. I do not have space here to discuss and differentiate these schools of thought, but instead reference several overviews here. For summaries of the strong program, social construction, and actor-network theory, see Sismondo, *Introduction to Science and Technology Studies,* esp. chaps. 5–7. For the classic essay on the social construction of technology, see Pinch and Bijker, "Social Construction of Facts and Artifacts." For a discussion of different kinds of constructivist approaches, see Sergio Sismondo, "Some Social Constructions," *Social Studies of Science* 23 (1993): 515–53. For more on social construction, see Ian Hacking, *The Social Construction of What?* (Cambridge: Harvard University Press, 1999). On mutual production or coproduction, see Jasanoff, *States of Knowledge,* esp. chaps. 1 and 2. For a concise explanation of historical ontology, see Murphy, *Sick Building Syndrome,* introduction.

46. Of course, many other scholars make parallel moves. For instance, Marxist historians use class and historians of women and gender and feminist scholars use gender as categories of analysis, even if their historical actors did not use these exact terms. Historians use such categories to help frame and organize their study, while remaining attentive to actors' own terms, definitions, assumptions, and so forth.

47. I am grateful to Benjamin R. Cohen for reminding me of this point. See his comments at "Nature and Knowledge: Conversations at the Interface of Environmental History and Science Studies," American Society for Environmental History conference, Madison, WI, Mar. 31, 2012. Furthermore, historians and philosophers of science have developed useful frameworks for historicizing how and what we know (historical epistemology and ontology), as well as the

concept of objectivity. On the latter point, see Lorraine Daston and Peter Galison, *Objectivity* (New York: Zone Books, 2007).

48. On the relationship between problem framing and implied solutions in the case of BP's Deepwater Horizon, see Christopher Jones, "Defining the Problem," posted to H-Energy, June 27, 2010, http://www.h-net.org/~energy/roundtables/Jones_Gulf.html. Environmental justice provides an instructive example. Some concerns such as lead poisoning may not be perceived as a problem if they disproportionately affect historically disadvantaged groups (the poor, people of color, etc.). Moreover, framing problems as "environmental" risks masks their consequences for humans and may particularly obscure their *differential* effects on certain social groups. In addition, on the unnatural history of "natural" disasters, see, e.g., Ted Steinberg, *Acts of God: The Unnatural History of Natural Disaster in America* (New York: Oxford University Press, 2000). Studies like Steinberg's highlight the deep sociopolitical implications of naturalizing events caused by both "environmental" and "social" processes ("nature-culture" in STS language). However, by being framed as (wholly) natural, these political, social, economic, and cultural dimensions are naturalized and thus ultimately neutralized.

49. John Law, "Technology and Heterogeneous Engineering: The Case of Portuguese Expansion," in Bijker, Hughes, and Pinch, *Social Construction of Technological Systems*, 114. See also Callon, "Some Elements of a Sociology of Translation."

50. On DDT, the classic work is Rachel Carson, *Silent Spring* (Boston: Houghton Mifflin Company, 1987). On endocrine disruptors, see Langston, *Toxic Bodies*.

51. Murphy, *Sick Building Syndrome*. See also Thorsheim, *Inventing Pollution*; Frickel, *Chemical Consequences*.

52. On the study of ignorance (and various forms thereof, from what might be called naive ignorance to willful ignorance for political and economic motives), see Robert N. Proctor and Londa Schiebinger, eds., *Agnotology: The Making and Unmaking of Ignorance* (Stanford: Stanford University Press, 2008). See also Naomi Oreskes and Erik Conway, *Merchants of Doubt: How a Handful of Scientists Obscured the Truth on Issues from Tobacco Smoke to Global Warming* (New York: Bloomsbury Press, 2010). On the political uses of uncertainty, see also Murphy, *Sick Building Syndrome*; Aaron M. McCright and Riley E. Dunlap, "Anti-reflexivity: The American Conservative Movement's Success in Undermining Climate Science and Policy," *Theory Culture Society* 27 (2010): 100–133.

53. Jones, "Defining the Problem."

54. Of course, scholars can also examine how (certain) historical actors questioned or rejected "environmental problems" as such. Gerald Markowitz and David Rosner have been two forceful (and controversial) advocates of the "deceit and denial" approach to challenging environmental problems; see *Deceit and Denial: The Deadly Politics of Industrial Pollution* (Berkeley: University of California Press, 2002). As an alternative to what might be called the "conspiracy" theory tactic, recent STS scholars have shown how the production of ignorance (or, at a minimum, uncertainty) is another strategy of undermining environmental problems. See Proctor and Schiebinger, *Agnotology*; Oreskes and Conway, *Merchants of Doubt*.

55. On "matters of fact" and "matters of concern," see Shapin and Schaffer, *Leviathan and the Air-Pump*; Bruno Latour, *Politics of Nature: How to Bring the Sciences into Democracy*, trans. Catherine Porter (Cambridge: Harvard University Press, 2004).

56. Fortun, "From Bhopal to the Informating of Environmentalism"; Weiner, "Death-Defying Attempt."

57. H. M. Collins and Robert Evans, *Rethinking Expertise* (Chicago: University of Chicago Press, 2007). STS scholars tend to focus on the sociopolitical implications of expertise: who is considered an expert, what social power such status confers, and so forth.

58. This point illustrates the notion of co- or mutual production. See Kline, *Consumers in the Countryside*; Jasanoff, *States of Knowledge*.

59. For critiques of objectivity, see Daston and Galison, *Objectivity*.

60. Latour, *Science in Action*, 70–74.

61. For one example of this, see Kevin C. Armitage's chapter in this volume for discussion of soil erosion experts during the New Deal seeking to reform farmers' practices; see also Phillips, *This Land, This Nation*; Neil M. Maher, *Nature's New Deal: The Civilian Conservation Corps and the Roots of the American Environmental Movement* (New York: Oxford University Press, 2008). For the politics of environmental expertise, see also Stephen Bocking, *Nature's Experts: Science, Politics, and the Environment* (New Brunswick: Rutgers University Press, 2004).

62. STS scholars have, however, also challenged typical representations of the relationships between "expert" and "lay" knowledge, public understandings of science, and the "communication" of science, since such terms tend to suggest a "deficiency" model of knowledge in which lay, local, indigenous, and/or other supposedly nonexpert people lack proper, correct knowledge that experts hold, rather than considering the ways in which these groups may have competing knowledge and competing knowledge systems. For useful overviews, see Bruce V. Lewenstein, "Science and the Media," and Brian Wynne, "Public Understanding of Science," both in Jasanoff et al., *Handbook of Science and Technology Studies*.

63. Murphy, *Sick Building Syndrome*; Langston, *Toxic Bodies*.

64. Thomas F. Gieryn, "Boundary-Work and the Demarcation of Science from Non-Science: Strains and Interests in Professional Ideologies of Scientists," *American Sociological Review* 48 (1983): 781–95; Thomas Gieryn, *Cultural Boundaries of Science: Credibility on the Line* (Chicago: University of Chicago Press, 1999).

65. Stine and Tarr, "At the Intersection of Histories"; Scranton and Schrepfer, *Industrializing Organisms*; Reuss and Cutcliffe, *Illusory Boundary*; LeCain, *Mass Destruction*; Pritchard, *Confluence*.

66. On nature-culture and naturecultures, see Latour, *We Have Never Been Modern*, 7; Haraway, *Companion Species Manifesto*, 1; Haraway, *When Species Meet*, 16. On sociotechnical, see Thomas Parke Hughes, *Networks of Power: Electrification in Western Society, 1880–1930* (Baltimore: Johns Hopkins University Press, 1983); Wiebe E. Bijker, "Sociohistorical Technology Studies," in Jasanoff et al., *Handbook of Science and Technology Studies*, 229–56. On technopolitics, see Hecht, *Radiance of France*.

67. On hybridity in environmental history, see Fiege, *Irrigated Eden*; White, "From Wilderness to Hybrid Landscapes." For a classic study of hybridity in science studies, see Donna Haraway, "A Cyborg Manifesto: Science, Technology, and Socialist-Feminism in the Late Twentieth Century," in *Simians, Cyborgs, and Women: The Reinvention of Nature* (New York: Routledge, 1991), 149–81. On multiplicity, see Murphy, *Sick Building Syndrome*.

68. Gieryn, "Boundary-Work and the Demarcation of Science"; Gieryn, *Cultural Boundaries of Science*.

69. Helen Rozwadowski, *Fathoming the Ocean: The Discovery and Exploration of the Deep Sea* (Cambridge: Harvard University Press, 2005). See also Robert E. Kohler, *Landscapes and Labscapes: Exploring the Lab-Field Border in Biology* (Chicago: University of Chicago Press, 2002).

70. Although many of the subsequent chapters in this collection emphasize the historicity and politics of categories such as *science, technology*, and so on, at times conference participants slipped into unreflective uses of the terms. Clapperton Mavhunga repeatedly flagged this slippage and its vast implications, reminding many of us to use the more inclusive term *knowledge*. Otherwise, we, as analysts, reproduce the historic devaluation of certain areas of knowledge and ways of knowing. That is, we risk replicating our actors' categories and hierarchies of knowledge and thus the very concepts and hierarchies that most of us have been trying to write against. It is also worth noting, however, that while *knowledge* has the potential to include many more types of knowledge and ways of knowing, the binary of *knowledge* and *not knowledge* is still freighted with the politics of inclusion (and hence exclusion).

71. Pritchard, *Confluence*, chap. 2.

72. Cronon, "Trouble with Wilderness"; Louis S. Warren, *The Hunter's Game: Poachers and Conservationists in Twentieth-Century America* (New Haven: Yale University Press, 1997); Roderick P. Neumann, *Imposing Wilderness: Struggles over Livelihood and Nature Preservation in Africa* (Berkeley: University of California Press, 1998); Mark David Spence, *Dispossessing Wilderness: Indian Removal and the Making of the National Parks* (New York: Oxford University Press, 1999); Karl Jacoby, *Crimes against Nature: Squatters, Poachers, Thieves, and the Hidden History of American Conservation* (Berkeley: University of California Press, 2001); Mark Dowie, *Conservation Refugees: The Hundred-Year Conflict between Global Conservation and Native Peoples* (Cambridge: MIT Press, 2009).

73. For a few examples, see n72.

74. For example, see the recent special issue "Theorized History," ed. Paul Friedland and Mary Louise Roberts, *French Historical Studies* 35 (Spring 2012).

75. Alan Brinkley, "Half a Mind Is a Terrible Thing to Waste," *Newsweek Magazine*, Nov. 13, 2009, http://www.thedailybeast.com/newsweek/2009/11/13/half-a-mind-is-a-terrible-thing-to -waste.html.

76. Latour, *Politics of Nature*; Shapin and Schaffer, *Leviathan and the Air-Pump*.

## CHAPTER 2. THE NATURAL HISTORY OF EARLY NORTHEASTERN AMERICA: AN INEXACT SCIENCE

1. As the title indicates, Dwight also wrote about eastern New York. He explains in the preface that although he originally intended to describe only the New England states, he added eastern New York because it was "intimately connected with New England by business, intercourse, and attachments." See Timothy Dwight, *Travels in New England and New York*, 4 vols. (New Haven, 1821), 1: 9–12, 104–5, 2: 213 (dismal), 340 (rough, lean, solitary). For a close reading of *Travels in New England and New York* that emphasizes how Dwight's descriptions of the landscape reflected a struggle to assimilate his sometimes contradictory theological and political positions, see Jane Kamensky, "'In These Contrasted Climes, How Chang'd the Scene': Progress, Declension, and Balance in the Landscapes of Timothy Dwight," *New England Quarterly* 63, no. 1 (Mar. 1990): 80–108.

2. Early Americans were typically inattentive to the ongoing dynamics of Native land use that predated colonization; see William Cronon, *Changes in the Land: Indians, Colonists, and the Ecology of New England* (1983; New York: Hill and Wang, 2003). On scientific perceptions of and interventions in rapidly changing colonial ecologies in tropical islands, see Richard Grove, *Green Imperialism: Colonial Expansion, Tropical Island Edens, and the Origins of Environmentalism, 1600–1860* (New York: Cambridge University Press, 1996).

3. On imperfection and inconclusiveness as a central characteristic of a range of Enlightenment literary genres, including natural history narratives, see Joanna Stalnaker, *The Unfinished Enlightenment: Description in the Age of the Encyclopedia* (Ithaca: Cornell University Press, 2010). On the tension between the growing emphasis on precision and new geographical knowledge resulting from imperial exploration, see Michael Bravo, "Precision and Curiosity in Scientific Travel: James Rennell and the Orientalist Geography of the New Imperial Age, 1760–1830," in *Voyages and Visions: Towards a Cultural History of Travel*, ed. Jas Elsner and J. P. Rubies (London: Reaktion Books, 1999), 162–83. On the need to more thoroughly integrate early modern history with science studies, see the provocative essay by Lorraine Daston, "Science Studies and the History of Science," *Critical Inquiry* 35, no. 4 (Summer 2009): 798–815.

4. For a range of approaches to the culture of precision and accuracy and debates about the subtle distinction between the two terms, see M. Norton Wise, ed., *The Values of Precision* (Princeton: Princeton University Press, 1995); Tore Frängsmyr, J. L. Heilbron, and Robin E. Rider, eds., *The Quantifying Spirit in the Eighteenth Century* (Berkeley: University of California Press,

1990). See also Theodore M. Porter, *Trust in Numbers: The Pursuit of Objectivity in Science and Public Life* (Princeton: Princeton University Press, 1995); and Ian Hacking, *The Emergence of Probability: A Philosophical Study of Early Ideas about Probability, Induction and Statistical Inference* (1975; New York: Cambridge University Press, 2006). Natural history was of course not an inherently imprecise practice. For example, air temperatures and other meteorological data were recorded using exact measuring techniques and contributed to improvements in quantification. See Theodore S. Feldman, "Late Enlightenment Meteorology," in Frängsmyr, Heilbron, and Rider, *Quantifying Spirit*, 143–78.

5. George B. Goode, "The Beginnings of Natural History in America," in *The Origins of Natural Science in America: The Essays of George Brown Goode*, ed. Sally Gregory Kohlstedt (Washington, DC: Smithsonian Institution Press, 1991), 23–89. For the social world of eighteenth-century North American naturalists, see Joyce E. Chaplin, *The First Scientific American: Benjamin Franklin and the Pursuit of Genius* (New York: Basic Books, 2006); and Susan Scott Parrish, *American Curiosity: Cultures of Natural History in the Colonial British Atlantic World* (Chapel Hill: University of North Carolina Press, 2006).

6. For the conventional periodization of the history of science in the colonial, revolutionary, and early national eras, see Parrish, *American Curiosity;* Andrew J. Lewis, "A Democracy of Facts, An Empire of Reason: Swallow Submersion and Natural History in the Early American Republic," *William and Mary Quarterly* 62, no. 4 (Oct. 2005): 663–96; Joyce E. Chaplin, "Nature and Nation: Natural History in Context," in *Stuffing Birds, Pressing Plants, Shaping Knowledge: Natural History in North America, 1730–1860*, ed. Sue Ann Prince (Philadelphia: American Philosophical Society, 2003); John C. Greene, *American Science in the Age of Jefferson* (Ames: Iowa State University Press, 1984); Raymond P. Stearns, *Science in the British Colonies of America* (Urbana: University of Illinois Press, 1970); Brooke Hindle, *The Pursuit of Science in Revolutionary America* (Chapel Hill: University of North Carolina Press, 1956).

7. Nicholas Dew and James Delbourgo, eds., *Science and Empire in the Atlantic World* (New York: Routledge, 2008); *Itineraries of Atlantic Science—New Questions, New Approaches, New Directions*, special issue, *Atlantic Studies: Literary, Cultural and Historical Perspectives* 7, no. 4 (2010).

8. Joseph Banks (JB) to Benjamin Waterhouse (BW), n.d., Benjamin Waterhouse Papers (HMS c16.4), Harvard Medical Library in the Francis A. Countway Library of Medicine (hereafter BW-HMS c16.4); BW to JB, Dec. 20, 1793, BL Add. Mss. 8098.305–306.

9. On the empiricist orientation and "philosophical modesty" of colonial American science, see James Delbourgo, *A Most Amazing Scene of Wonders: Electricity and Enlightenment in Early America* (Cambridge: Harvard University Press, 2006), 282–83.

10. Patricia Cline Cohen, *A Calculating People: The Spread of Numeracy in Early America* (1982; New York: Routledge, 1999).

11. "Astronomical Lectures Read in the Chapel of Harvard College at Cambridge in America, 1780–1781," box 1, folder 2 (underline [sic] per original), "Change of Climate in North America and Europe," n.d., box 1, folder 10, both Samuel Williams Papers, Harvard University Archives. For more on Williams's ideas about climate change and their reception, see Jan Golinski, "American Climate and the Civilization of Nature," in Dew and Delbourgo, *Science and Empire in the Atlantic World*, 153–74; and James R. Fleming, *Historical Perspectives on Climate Change* (New York: Oxford University Press, 1998), 11–32.

12. Jeremy Belknap, *The History of New-Hampshire*, 3 vols. (1784; Boston, 1792), 3: 40–41, 49.

13. David Bosse, "'To Promote Useful Knowledge': An Accurate Map of the Four New England States by John Norman and John Coles," *Imago Mundi* 52 (2000): 143–57.

14. Belknap, *History of New-Hampshire*, vol. 3, *Containing a Geographical Description of the State; with Sketches of its Natural History, Productions, Improvements, and Present State of Society and Manners, Laws and Government*, 96, 120. I do not mean to imply that qualitative data were

or came to be considered as inherently imprecise, but rather that the achievement of a consensus about precise quantification was a more prominent dispute.

15. J. C. Greene, *American Science in the Age of Jefferson;* Stearns, *Science in the British Colonies of America;* Hindle, *Pursuit of Science in Revolutionary America.*

16. "Lecture: Introductory on Natural History, October 12, 1810," BW-HMS c16.4; Spary, *Utopia's Garden,* 13; Fredrik Albritton Jonsson, "Rival Ecologies of Global Commerce: Adam Smith and the Natural Historians," *American Historical Review* 115, no. 5 (Dec. 2010): 1342–63; Michael Dettelbach, "'A Kind of Linnaean Being': Forster and Eighteenth-Century Natural History," in *Observations Made during a Voyage Round the World,* by Johann Reinhold Forster, ed. Nicholas Thomas et al. (Manoa: University of Hawai'i Press, 1996), lv–lxxiv. American historians looking for the intellectual counterpart to the Industrial Revolution in the North or for the origins of the international prominence of American Cold War science have been dismayed by the prevalence of amateurs and the utilitarian focus of eighteenth-century science. See Hindle, *Pursuit of Science in Revolutionary America;* Stearns, *Science in the British Colonies of America;* and J. C. Greene, *American Science in the Age of Jefferson.*

17. Cronon, *Changes in the Land,* 19–22.

18. Matthew H. Edney, "The Irony of Imperial Mapping," in *The Imperial Map: Cartography and the Mastery of Empire,* ed. James R. Akerman (Chicago: University of Chicago Press, 2009), 28–29. On the cartographical history of Massachusetts Bay Colony's expansionist aims, see Matthew H. Edney, "Printed but Not Published: Limited-Circulation Maps of Territorial Disputes in Eighteenth-Century New England," in *Mappae antiqua: Liber amicorum Günter Schilder: Vriendenboek ter gelegenheid van zijn 65ste verjaardag¾Essays on the Occasion of His 65th Birthday,* ed. Paula van Gestel–van het Schip and Peter van der Krogt ('t Goy-Houten: Hes and De Graaf, 2007), 147–58.

19. William Bradford, *Of Plymouth Plantation, 1620–1647,* ed. Samuel Eliot Morison (New York: Knopf, 1952), quote on 76; William Douglass, *A Summary Historical and Political, Of the First Planting, Progressive Improvements, and Present State of the British Settlements in North America,* 2 vols. (Boston: Rogers and Fowle, 1747–52), 1: 7; Arthur Young, *Annals of Agriculture and Other Useful Arts* 1, no. 1 (1784); 13; H. Klockhoff, "A Chorographical Map of the Northern Department of North America," (Amsterdam, 1780).

20. Mary Pedley, "Map Wars: The Role of Maps in the Nova Scotia/Acadia Boundary Disputes of 1750," *Imago Mundi* 50 (1998): 96–104. On the role of maps in border and property disputes in New York, see Sara Stidstone Gronim, "Geography and Persuasion: Maps in British Colonial New York," *William and Mary Quarterly* 58, no. 2 (Apr. 2001): 373–402. On early-seventeenth-century English attempts to assimilate New Netherlands into maps of New England, see Benjamin Schmidt, "Mapping an Empire: Cartographic and Colonial Rivalry in Seventeenth-Century Dutch and English North America," *William and Mary Quarterly* 54, no. 3 (July 1997): 549–78.

21. Paul Mascarene, Description of Nova Scotia, 1720–21, MG1, v. 1520, folder P, Mascarene (Engineer), Nova Scotia Archives and Records Management (hereafter NSA). On the wavering commitment of the British to Nova Scotia before the 1740s, see Geoffrey G. Plank, *An Unsettled Conquest: The British Campaign against the Peoples of Acadia* (Philadelphia: University of Pennsylvania Press, 2001); and John G. Reid, *Acadia, Maine, and New Scotland: Marginal Colonies in the Seventeenth Century* (Buffalo: University of Toronto Press, 1981).

22. Jedidiah Morse, *The American Geography: or, A View of the Present Situation of the United States of America,* 2nd ed. (1789; London: 1792), iv, vi, 140, 193. Morse may have been unaware of the recent boundary settlement between the District of Maine and New Brunswick in 1784, but maps through the 1790s continued to show New England encroaching well into New Brunswick; see Barbara B. McCorkle, *New England in Early Printed Maps, 1513–1800: An Illustrated Carto-Bibliography* (Providence: John Carter Brown Library, 2001), 114. In the 1780s

there had also been plans to establish a New Ireland colony in the area between the Penobscot and St. Croix rivers, a project that was nullified during the Treaty of Paris negotiations. See David Demeritt, "Representing the 'True' St. Croix: Knowledge and Power in the Partition of the Northeast," *William and Mary Quarterly* 54, no. 3 (July 1997): 516.

23. Also see James Sullivan, *History of the District of Maine* (Boston, 1795). The major American model for these works was Thomas Jefferson's *Notes on the State of Virginia* (1781–84).

24. On the practice of chorography and its ideological aspects, see Charles W. J. Withers, "Reporting, Mapping, Trusting: Making Geographical Knowledge in the Late Seventeenth-Century," *Isis* 90, no. 3 (Sept. 1999): 497–521; and Richard Helgerson, "The Land Speaks: Cartography, Chorography, and Subversion in Renaissance England," *Representations* 16 (Autumn 1986): 50–85.

25. David Ramsay to Jedidiah Morse, Nov. 30, 1787, in Robert L. Brunhouse, "David Ramsay, 1749–1815: Selections from His Writings," *Transactions of the American Philosophical Society* 55, no. 4 (1965): 1–250, letter on 116–17.

26. Morse, *American Geography*, 5–8. Morse did not provide a table of the southern equatorial zone.

27. Dwight, *Travels in New England*, 4: 233; C. F. Volney, A *View of the Soil and Climate of the United States of America* (London, 1804), 1–2, 6–7.

28. Thomas Pennant, *Arctic Zoology*, 1st ed. (London: Henry Hughes, 1785), quotes on 3.

29. MC to Professor Samuel Williams, June 20, 1780, in *Life, Journals, and Correspondence of Reverend Manasseh Cutler, LLD, By His Grandchildren*, ed. William Cutler and Julia Cutler, 2 vols. (Cincinnati: Robert Clarke and Co., 1888), 1: 80–82.

30. Manasseh Cutler, *An Account of Some of the Vegetable Productions, Naturally Growing in this Part of America, Botanically Arranged* (Boston, 1785), 401; Jacob Bigelow, *Florula bostoniensis: A Collection of Plants of Boston and its Environs, with their generic and specific characters, synonyms, descriptions, places of growth, and time of flowering, and occasional remarks* (Boston: Cummings and Hilliard, 1814), vi–vii.

31. Thomas Caulfield to Board of Trade, Nov. 21, 1715, MG1 v. 1520, folder K, NSA.

32. Belknap, *History of New-Hampshire*, 3: 120.

33. D. Graham Burnett, "Hydrographic Discipline among the Navigators: Charting an Empire of Commerce and Science in the Nineteenth-Century Pacific," in Akerman, *Imperial Map*, 247. On the fragmented geography of imperial legal sovereignty on the ground rather than the territorial sweep of imperial maps based on abstract claims to land, see Lauren Benton, *A Search for Sovereignty: Law and Geography in European Empires, 1400–1900* (New York: Cambridge University Press, 2009).

34. Bernard Bailyn, *Voyagers to the West: A Passage in the Peopling of America on the Eve of the Revolution* (New York: Vintage, 1986), 10–11.

35. Belknap, *History of New-Hampshire*, 3: 47, 97; Russell M. Lawson, *Passaconaway's Realm: Captain John Evans and the Exploration of Mount Washington* (Hanover: University Press of New England, 2002). Well-documented surveys of the range were also completed in 1774, 1804, and 1816. On economic tourism as a practice of natural history, see Fredrik A. Jonsson, "The Enlightenment in the Highlands: Natural History and Internal Colonization in the Scottish Enlightenment, 1760–1830" (PhD diss., University of Chicago, 2005), 81–100.

36. Jeremy Belknap, "Description of the White Mountains in New-Hampshire, (Read 15 October 1784)," *Transactions of the American Philosophical Society* 2 (1786): 42–49, quotes on 43, 46; *Life, Journals and Correspondence of Reverend Manasseh Cutler*, 2: 98–99, 103; Belknap, *History of New-Hampshire*, 3: 48–51; Jeremy Belknap, *Journal of a Tour to the White Mountains in July, 1784* (Boston: Massachusetts Historical Society, 1876), 16. Mount Washington is the highest peak north of the Carolinas and east of the Mississippi River, and it has been described

as an "arctic island in the temperate zone," comparable to northern Labrador and western Greenland. See Charles P. Alexander, "The Presidential Range of New Hampshire as a Biological Environment," *American Midland Naturalist* 24, no. 1 (1940): 104.

37. Belknap, *History of New-Hampshire*, 3: 48–51; Belknap, *Journal of a Tour to the White Mountains*, 16; *Life, Journals, and Correspondence of Reverend Manasseh Cutler*, 2: 98–99, 103.

38. *Life, Journals, and Correspondence of Reverend Manasseh Cutler*, 2: 221.

39. Belknap, "Description of the White Mountains," 43, 49.

40. Cutler to Belknap, Feb. 28, 1785, in *Life, Journals, and Correspondence of Reverend Manasseh Cutler*, 2: 227.

41. Morse, *American Geography*, 164.

42. P. Medows, J. Bruce, and J. Merrill to Board of Trade, June 22, 1717, MG1, v. 1520, folder M, NSA.

43. Dwight, *Travels in New England*, 2: 142.

44. Samuel Williams, *The Natural and Civil History of Vermont* (Walpole, NH, 1794), xi.

45. Joyce E. Chaplin, *An Anxious Pursuit: Agricultural Innovation and Modernity in the Lower South, 1730–1815* (Chapel Hill: University of North Carolina Press, 1993), 77; P. J. Marshall and Glyndwr Williams, *The Great Map of Mankind: British Perceptions of the World in the Age of Enlightenment* (Cambridge: Harvard University Press, 1982); Dwight, *Travels in New England*, 2: 142.

46. For example, see the various editions of Morse's *American Geography*, first published in 1789 and annually revised throughout the 1790s.

## CHAPTER 3. FARMING AND NOT KNOWING: AGNOTOLOGY MEETS ENVIRONMENTAL HISTORY

1. Frank Uekotter, *The Age of Smoke: Environmental Policy in Germany and the United States, 1880–1970* (Pittsburgh: University of Pittsburgh Press, 2009), 201.

2. Donald Worster, *A River Running West: The Life of John Wesley Powell* (Oxford: Oxford University Press, 2001).

3. Naomi Oreskes and Erik M. Conway, *Merchants of Doubt: How a Handful of Scientists Obscured the Truth on Issues from Tobacco Smoke to Global Warming* (New York: Bloomsbury Press, 2010).

4. Robert N. Proctor, "Agnotology: A Missing Term to Describe the Cultural Production of Ignorance (and Its Study)," in *Agnotology: The Making and Unmaking of Ignorance*, ed. Robert N. Proctor and Londa Schiebinger (Stanford: Stanford University Press, 2008), 3 (emphasis in original).

5. Proctor, "Agnotology," 1.

6. Wilbert E. Moore and Melvin M. Tumin, "Some Social Functions of Ignorance," *American Sociological Review* 14 (1949): 787–95.

7. Moore and Tumin, "Some Social Functions," 787. See also Vannevar Bush, *Science, the Endless Frontier: A Report to the President* (Washington, DC: US Government Printing Office, 1945).

8. See Peter Wehling, "Jenseits des Wissens? Wissenschaftliches Nichtwissen aus soziologischer Perspektive," *Zeitschrift für Soziologie* 30, no. 6 (2001): 465–84; and S. Holly Stocking, "On Drawing Attention to Ignorance," *Science Communication* 20, no. 1 (1998): 165–78.

9. Robert N. Proctor, *Cancer Wars: How Politics Shapes What We Know and Don't Know about Cancer* (New York: Basic Books, 1995); Robert N. Proctor, *The Nazi War on Cancer* (Princeton: Princeton University Press, 1999).

10. Proctor and Schiebinger, *Agnotology*, 16.

11. Robert N. Proctor, "Unwissen ist Macht," *Süddeutsche Zeitung*, Jan. 2, 2008.

12. Proctor, *Agnotology*, 17.

13. Naomi Oreskes and Erik M. Conway, "Challenging Knowledge: How Climate Science Became a Victim of the Cold War," in Proctor and Schiebinger, *Agnotology*, 55–89. See also Oreskes and Conway, *Merchants of Doubt*.

14. P. H. Gleick et al., "Climate Change and the Integrity of Science," *Science* 328 (2010): 689–90.

15. Deborah Fitzgerald, *The Business of Breeding: Hybrid Corn in Illinois, 1890–1940* (Ithaca: Cornell University Press, 1990), 1.

16. Thomas Miedaner, *Von der Hacke bis zur Gen-Technik. Kulturgeschichte der Pflanzenproduktion in Mitteleuropa* (Frankfurt: DLG-Verlag, 2005), 115.

17. Landwirtschaftskammer Nordrhein-Westfalen, *Zahlen zur Landwirtschaft in Nordrhein-Westfalen 2004* (Münster and Bonn: Landwirtschaftskammer NRW, 2004), n30.

18. Rembert Unterstell, "Auf der Suche nach 'Miss World Energiemais,'" *Forschung Spezial Energie. Magazin der Deutschen Forschungsgemeinschaft*, June 2010, 28–33, 30.

19. On corn in Africa, see James C. McCann, *Maize and Grace: Africa's Encounter with a New World Crop, 1500–2000* (Cambridge and London: Harvard University Press, 2005).

20. Thomas Nipperdey, *Deutsche Geschichte 1866–1918. Bd. 1: Arbeitswelt und Bürgergeist* (Munich: C. H. Beck, 1990), 193.

21. Ludwig Niggl, *Die Geschichte der Deutschen Grünlandbewegung 1914–1945* (n.p., n.d. [ca. 1953]), n79. See also Wilhelm Opitz von Boberfeld, "Zur Historik grünlandwissenschaftlicher Strukturen des 20. Jahrhunderts und ableitbare Perspektiven," *Berichte über Landwirtschaft* 77 (1999): 469–78.

22. Heinz Haushofer, *Die Furche der DLG 1885 bis 1960* (Frankfurt: DLG-Verlag, 1960), 72.

23. Hans Buß, *Der Mais, eine wichtige landwirtschaftliche Kulturpflanze. Beobachtungen auf dem Gesamtgebiete des Maisbaues, der Maissilage und der Maiszüchtung in Ungarn und Rumänien und ihre Nutzanwendung auf deutsche Verhältnisse* (Berlin: Arbeiten der Deutschen Landwirtschafts-Gesellschaft Heft 372, 1929).

24. Buß, *Der Mais*, 46.

25. See *Mitteilungen der Deutschen Landwirtschafts-Gesellschaft* 42 (1927): 79–82; Ruths, "Zweckmäßiger Silobau für Mais und Erfordernisse zur Gewinnung einer guten Maissilage," *Mitteilungen der Deutschen Landwirtschafts-Gesellschaft* 42 (1927): 82–86; Ludwig Niggl, "Der Maisbau vom betriebswirtschaftlichen Standpunkte aus," *Mitteilungen der Deutschen Landwirtschafts-Gesellschaft* 42 (1927): 86–89; Richard Lieber, "Praktische Durchführung und Entwicklungsmöglichkeiten des Maisanbaues in Deutschland," *Mitteilungen der Deutschen Landwirtschafts-Gesellschaft* 44 (1929): 130–35.

26. Johannes Zscheischler et al., *Handbuch Mais. Umweltgerechter Anbau, wirtschaftliche Verwertung*, 4th ed. (Frankfurt: DLG-Verlag, 1990), 20; Staatsarchiv Münster Landwirtschaftliche Kreisstellen no. 627, Maisanbau-Gesellschaft to Landwirtschaftsschule und Wirtschaftsberatungsstelle Soest in Westfalen, Nov. 8, 1937.

27. Hans Buß, "Körnermaisbau in Deutschland. Klarheit in allen wichtigen technischen Einzelfragen," *Mitteilungen für die Landwirtschaft* 51 (1936): 67–69, 98–100.

28. Carl Schrimpf, *Mais. Anbau und Düngung* (Bochum: Ruhr-Stickstoff AG, 1960).

29. For this story, see Fitzgerald, *Business of Breeding*.

30. See Thomas Wieland, "'Wir beherrschen den pflanzlichen Organismus besser . . . ,'" in *Wissenschaftliche Pflanzenzüchtung in Deutschland, 1889–1945* (Munich: Deutsches Museum, 2004). See also the classic Jack Ralph Kloppenburg Jr., *First the Seed: The Political Economy of Plant Biotechnology*, 2nd ed. (Madison: University of Wisconsin Press, 2004).

31. Landwirtschaftskammer Nordrhein-Westfalen, *Zahlen zur Landwirtschaft 2004*, S. 34.

32. M. Baur, "Die Sorte dem Standort anpassen. Zum Anbau von Silomais," *Landwirtschaftliches Wochenblatt für Westfalen und Lippe* 123, no. 7 (Feb. 17, 1966): ed. A: 20.

33. M. Baur, "Hat der Mais, was er zum Wachsen braucht? Sechs Fragen für den Mais-bauer," *Landwirtschaftliches Wochenblatt Westfalen-Lippe* 127, no. 11 (Mar. 12, 1970): ed. A: 42.

34. Gustav Aufhammer, *Neuzeitlicher Getreidebau*, 2nd ed. (Frankfurt: DLG-Verlag, 1963), 137.

35. Zscheischler et al., *Handbuch Mais*, 259.

36. Johannes Zscheischler and Friedrich Groß, *Mais. Anbau und Verwertung* (Frankfurt: DLG-Verlag, 1966), 55.

37. Carina Weber, "Pestizide," in *Landwirtschaft 1993. Der kritische Agrarbericht*, ed. Agrar-bündnis (Rheda-Wiedenbrück: Arbeitsgemeinschaft bäuerliche Landwirtschaft Bauernblatt, n.d.), 147.

38. Günter Spielhaus, "Gülle paßt gut zu Mais," *Landwirtschaftliches Wochenblatt West-falen-Lippe* 137, no. 13 (Mar. 27, 1980), ed. B: 26.

39. Zscheischler and Groß, *Mais*, 36.

40. Jürgen Rimpau, "Düngung und ökologische Auswirkungen. Berichterstattung," in *Mit welcher Düngungsintensität in die 90er Jahre? Vorträge und Ergebnisse des DLG-Kolloquiums am 13. und 14. Dezember 1988 in Bad Nauheim* (Frankfurt: DLG-Verlag, 1989), 58.

41. Zscheischler et al., *Handbuch Mais*, 107.

42. George Orwell, *Nineteen Eighty-Four. With a Critical Introduction and Annotations by Bernard Crick* (New York: Oxford University Press, 1984).

43. Moore and Tumin, "Some Social Functions," 795.

## CHAPTER 4. ENVIRONMENTALISTS ON BOTH SIDES: ENACTMENTS IN THE CALIFORNIA RIGS-TO-REEFS DEBATE

The epigraph is from Brock Bernstein's oral testimony with slides during the panel discussion "Oil Platform Decommissioning Study," California Ocean Protection Council, June 25, 2010, http://www.cal-span.org/cgi-bin/archive.php?owner=COPC&date=2010–06–25 (video clip).

1. Bernstein, oral testimony, June 25, 2010.

2. California Senate Bill 1 (2000–2001), Veto Statement, Oct. 13, 2001.

3. Donna Schroeder and Milton Love, "Ecological and Political Issues Surrounding De-commissioning of Offshore Oil Facilities in the Southern California Bight," *Ocean and Coastal Management* 47, no. 1 (2004): 21–48; Dan Rothbach, "Rigs-to-Reefs: Refocusing the Debate in California," *Duke Environmental Law and Policy Forum* 17 (2006): 283–95.

4. Annemarie Mol, *The Body Multiple: Ontology in Medical Practice* (Durham: Duke University Press, 2002); John Law, *Aircraft Stories: Decentering the Object in Technoscience* (Durham: Duke University Press, 2002); John Law, "Enacting Naturecultures: A Note from STS" (Lancaster: Centre for Science Studies, Lancaster University, 2004), http://www.lancs .ac.uk/fass/sociology/papers/law-enacting-naturecultures.pdf; Christopher Gad and Casper Bruun Jensen, "On the Consequences of Post-ANT," *Science, Technology and Human Values* 35 (2010): 55–80.

5. Mol, *Body Multiple*, 43.

6. Mol, *Body Multiple*, 176.

7. Wiebe Bijker, *Of Bicycles, Bakelites, and Bulbs: Toward a Theory of Sociotechnical Change* (Cambridge: MIT Press, 1997).

8. See the essays in H. Tristram Engelhardt Jr. and Arthur L. Caplan, eds., *Scientific Con-troversies: Case Studies in the Resolution and Closure of Disputes in Science and Technology* (Cam-bridge: Cambridge University Press, 1987).

9. Leland Glenna, "Value-Laden Technocratic Management and Environmental Conflicts: The Case of the New York City Watershed Controversy," *Science, Technology and Human Values* 35 (2010): 81–112.

10. Mol, *Body Multiple*, 104.

11. National Oceanic and Atmospheric Association (NOAA), *National Artificial Reef Plan*, comp. Richard B. Stone, NOAA Technical Memorandum NMFS OF-6 (US Department of Commerce, 1985).

12. Louisiana Department of Wildlife and Fisheries, *Louisiana Artificial Reef Plan*, prep. Charles A. Wilson, Virginia R. Van Sickle, and David L. Pope, Technical Bulletin No. 41 (1987); Texas Parks and Wildlife Department, *Texas Artificial Reef Fishery Management Plan*, prep. C. Dianne Stephan et al., Fishery Management Plan Series, No. 3 (1990).

13. Committee on Disposition of Offshore Platforms, *Disposal of Offshore Platforms*, prepared for the National Research Council (Washington, DC: National Academy Press, 1985), 9. About half of the Gulf facilities are located in extremely shallow water (less than fifty feet), making them quite different from the much larger, deeper installations along the Pacific coast. In 1983, there were 4,056 offshore structures in the Gulf of Mexico, versus only 24 offshore structures in California and 14 in Alaska.

14. "Louisiana Adds New Reef Sites for Storm-Damaged Structures," *Oil and Gas Journal*, June 11, 2007.

15. Dolly Jørgensen, "An Oasis in a Watery Desert? Discourses on an Industrial Ecosystem in the Gulf of Mexico Rigs-to-Reefs Program," *History and Technology* 25 (2009): 343–64.

16. Bob Williams, "An Angler's Tale of Two States," *Oil and Gas Journal*, Oct. 17, 1994. The position of the DFG against rigs-to-reefs appears confirmed in a newspaper article from 1986 that quotes state marine biologist John Grant as saying that the DFG was opposed to the program and had strict requirements for artificial reef material other than rock and concrete: Shari Roan, "Refuse or Refuge? Environmentalists Watch for Pollution," *Orange County Register*, Nov. 5, 1986.

17. Kermit Pattison, "Conservationists Envision Offshore Oil Rigs as Artificial Reefs," *Los Angeles Daily News*, Oct. 25, 1994.

18. Interagency Decommissioning Working Group, "Action Plan for Addressing Pacific Region Offshore Oil and Gas Facility Decommissioning Issues," July 12, 1999, 3.

19. SB 2173 (1997–98), introduced Feb, 20, 1998, amended May 4, 1998, and May 22, 1998. Chevron had presented rigs-to-reefs as its preferred decommissioning option for five platforms (Gail, Grace, Harvest, Hermosa, and Hidalgo) in late 1997 to the County of Santa Barbara Planning and Development Department, so the timing of the legislation coincides with these efforts.

20. In a confusing (but not untypical) move, the number SB 2173 was reused by the author for an employment bill later that same session. That bill was eventually passed by the Senate. So when looking at the history of this bill, only the history through May 27 applies to the artificial reef issue.

21. Nora Lynn (former aid to Senator Alpert and lead on the legislation), interview with the author, Sept. 11, 2008, Sacramento, CA.

22. SB 241 (1999–2000), as introduced, Jan. 26, 1999.

23. Kelly Gilleland, "Rigs-to-Reef Feud Grips California," *Upstream*, Nov. 5, 1999.

24. Sally J. Holbrok et al., *Ecological Issues Related to Decommissioning of California's Offshore Production Platforms*, report to the California Marine Council (2000).

25. Terry Rodgers, "Alpert's 'Rigs to Reefs' Measure Is Making Waves," *San Diego Union-Tribune*, Sept. 18, 2000. The conflict with the governor's office appears to have focused on the makeup of the board of trustees who would oversee the funding, with the governor wanting direct control of appointments.

26. SB 1 (2000–2001), as introduced, Dec. 4, 2000.

27. Barry Broad, Pete Price, and Shane Gusman to All Members of the Senate Natural Resources Committee, Dec. 13, 1999, Senate Committee on Natural Resources and Wildlife

file for SB 241, Legislative Papers, LP383:334, California State Archives, Office of the Secretary of State, Sacramento. All legislative papers cited in this article come from the same archive.

28. Associated Press, "Debates Surround Plan to Convert Oil Platforms into Artificial Reefs," Dec. 4, 1999.

29. See, e.g., Associated Press, "A Plan to Turn Oil Rigs into Fish Reefs," *San Francisco Chronicle*, Oct. 17, 1994; Dave Strege, "United Anglers Has a Plan for Discarded Oil Rigs," *Orange County Register*, Oct. 23, 1994; Pattison, "Conservationists Envision."

30. Dick Long, President, San Diego Oceans Foundation, to Tom Hayden, Chair of Senate NR&W committee, Jan. 7, 2000, Senate Committee on Natural Resources and Wildlife file for SB 241, Legislative Papers, LP383:334. See D. Jørgensen, "Oasis in a Watery Desert," for a discussion of the oasis metaphor.

31. Jim Hill, *CNN World View*, July 17, 2000.

32. Senate NR&W Committee meeting, Apr. 24, 2001, video, tape #122, California State Archives, Office of the Secretary of State, Sacramento.

33. Jerry Gandy, "Plan Calls for Donation of Oil Rigs," *Contra Costa Times*, Jan. 20, 2000.

34. Pattison, "Conservationists Envision."

35. Broad, Price, and Gusman to All Members of the Senate Natural Resources Committee, Dec. 13, 1999.

36. Milton Love quoted in Stephanie Greenman, "From Rockfish to Rigfish," *California Wild* (Fall 2001), http://researcharchive.calacademy.org/calwild/2001fall/stories/habitats.html.

37. K. C. Bishop, Chevron, to Members of Senate, Jan. 25, 2000, Senate Committee on Natural Resources and Wildlife file for SB 241, Legislative Papers, LP383:334.

38. Robert Lindsey, "Offshore Oil Rigs Prove Fertile Farm for Mussels on the Coast," *New York Times*, Nov. 5, 1985.

39. Jim Hill, *CNN World View*, July 17, 2000. Other examples of using the 1969 spill as a framework for media coverage of rigs-to-reefs includes Kermit Pattison, "California's Offshore Rigs: Safe Enough to Be Reefs?" *Christian Science Monitor*, Aug. 22, 1996; Ryck Lydecker, "Fish Dig Big Rigs," *BOAT/U.S. Magazine*, Sept. 2007, 24; Scott LaFee, "From Rigs to Riches? After the Oil's Gone, Structures May Help Fish Species Survive, Thrive," *San Diego Union-Tribune*, Oct. 18, 2007.

40. Senate Floor Debate SB 241 (Alpert), Jan. 31, 2000, transcript of the proceedings, Senate Committee on Natural Resources and Wildlife file for SB 241, Legislative Papers, LP383:333.

41. W. F. "Zeke" Grader Jr., PCFFA, to Sheila Kuehl, Chair of Senate NR&W committee, Mar. 29, 2001, Senate Committee on Natural Resources and Wildlife file for SB 1, Legislative Papers, LP383:385.

42. EDC to Tom Hayden, Chair of Senate NR&W committee, Dec. 22, 1999, Senate Committee on Natural Resources and Wildlife file for SB 241, Legislative Papers, LP383:334. The wording of the EDC letter is duplicated verbatim in several other opposition letters, including California Coastkeeper to Tom Hayden, Jan. 7, 2000, Citizens for Goleta Valley to Tom Hayden, Jan. 10, 2000, and Eve Kliszewski, Environmental Director, Surfrider Foundation, to Tom Hayden, Jan. 6, 2000, all in the same file.

43. EDC to Sheila Kuehl, Chair of Senate NR&W committee, Mar. 30, 2001, Senate Committee on Natural Resources and Wildlife file for SB 1, Legislative Papers, LP383:385.

44. Robert Sollen to Tom Hayden, Chair of Senate NR&W committee, Jan. 2, 2000, Senate Committee on Natural Resources and Wildlife file for SB 241, Legislative Papers, LP383:334; Robert Sollen, Offshore Oil Policy coordinator, Sierra Club, to Senator Alpert, Dec. 18, 1988, Senate Committee on Natural Resources and Wildlife file for SB 241, Legislative Papers, LP383:333; Linda Krop, EDC, to Tom Hayden, Chair of Senate NR&W committee, Apr. 10, 1998, Senate Committee on Natural Resources and Wildlife file for SB 2173, Legislative Pa-

pers, LP383:300; Pamela Marshall Heatherington, Environmental Center of San Luis Obispo County, to Sheila Kuehl, Chair of Senate NR&W committee, Mar. 16, 2001, Senate Committee on Natural Resources and Wildlife file for SB 1, Legislative Papers, LP383:385.

45. Pattison, "California's Offshore Rigs."

46. Southern California Trawlers Association to Senator Alpert, Dec. 28, 1998, Senate Committee on Natural Resources and Wildlife file for SB 241, Legislative Papers, LP383:333.

47. Pattison, "California's Offshore Rigs."

48. Hannah-Beth Jackson to Governor Gray Davis, Sept. 26, 2001, SB 1 Veto file, California State Archives, Office of the Secretary of State, Sacramento.

49. Bill Allayaud, Legislative Director, Sierra Club California, to Governor Gray Davis, Sept. 19, 2001, SB 1 Veto file.

50. See, e.g., Pattison, "California's Offshore Rigs"; Keith Lair, "Channel Islands Rockfish Are Finding a Home on Oil Platforms," *Los Angeles Daily News*, July 6, 2000; Greenman, "From Rockfish to Rigfish."

51. Dick Long, President, San Diego Oceans Foundation, to Tom Hayden, Chair of Senate NR&W committee, Jan. 7, 2000, Senate Committee on Natural Resources and Wildlife file for SB 241, Legislative Papers, LP383:334.

52. Senator Dede Alpert to Governor Gray Davis, Sept. 24, 2001, Deidre Alpert Papers, SB 1 Bill File, LP384:261.

53. Pattison, "California's Offshore Rigs."

54. Keith Lair, "Bill Would Make Reefs of Oil Rigs," *Los Angeles Daily News*, Mar. 18, 1999.

55. Strege, "United Anglers Has a Plan."

56. D. Jørgensen, "Oasis in a Watery Desert."

57. EDC clearly provided the template for many of the opposition letters submitted for both SB 241 and SB 1. After the Decommissioning Report was issued, the EDC correspondence typically includes a statement that the report found that there is "no sound scientific evidence that platforms contribute to regional fish stock." See, e.g., Linda Krop, Environmental Defense Center, to Sheila Kuehl, Chair of Senate NR&W committee, Mar. 30, 2001, Committee on Natural Resources and Wildlife file for SB 1, Legislative Papers, LP383:385.

58. Jim Hill, *CNN World View*, July 17, 2000.

59. Holbrok et al., *Ecological Issues*, 20, 32.

60. Leon Drouin Keith, "Bill Would Require Mercury Testing around Oil Rigs," Associated Press, Apr. 2002.

61. Mike McCorkle, President, Southern California Trawlers Association, to Tom Hayden, Chair of Senate NR&W committee, Jan. 6, 2000, Senate Committee on Natural Resources and Wildlife file for SB 241, Legislative Papers, LP383:334.

62. Susan Rose, Board of Supervisors, County of Santa Barbara, to Tom Hayden, Chair of Senate NR&W committee, Jan. 4, 2000, Senate Committee on Natural Resources and Wildlife file for SB 241, Legislative Papers, LP383:334.

63. D. Jørgensen, "Oasis in a Watery Desert."

64. Pattison, "California's Offshore Rigs."

65. Kristin Valette, Project Aware, to Senate NR&W Committee, Apr. 13, 2001, Deidre Alpert Papers, Bill files—SB 1 (2001–2002), LP384:263.

66. Greenman, "From Rockfish to Rigfish."

67. Mike McCorkel, Southern California Trawlers' Association, to Rick Van Nieuwburg, Feb. 17, 1998, Senate Committee on Natural Resources and Wildlife file for SB 2173, Legislative Papers, LP383:300.

68. Mike McCorkle, Southern California Trawlers Association, quoted in "Enviros, Oil Firms Debate 'Rigs To Reefs,'" *Daily Energy Briefing*, June 21, 1999.

69. Some examples: J. E. Caselle et al., "Trash or Habitat? Fish Assemblages on Offshore Oilfield Seafloor Debris in the Santa Barbara Channel, California," *ICES Journal of Marine Science* 59 (2002): S258–65; M. Helvey, "Are Southern California Oil and Gas Platforms Essential Fish Habitat?" *ICES Journal of Marine Science* 59 (2002): S266–71; M. S. Love, D. M. Schroeder, and W. H. Lenarz, "Distribution of Bocaccio (*Sebastes paucispinis*) and Cowcod (*Sebastes levis*) around Oil Platforms and Natural Outcrops Off California with Implications for Larval Production," *Bulletin of Marine Science* 77 (2005): 397–408; M. S. Love and A. York, "The Relationships between Fish Assemblages and the Amount of Bottom Horizontal Beam Exposed at California Oil Platforms: Fish Habitat Preferences at Man-Made Platforms and (by Inference) at Natural Reefs," *Fishery Bulletin* 104 (2006): 542–49; M. S. Love et al., "Potential Use of Offshore Marine Structures in Rebuilding an Overfished Rockfish Species, *Bocaccio* (Sebastes paucispinis)," *Fishery Bulletin* 104 (2006): 384–90.

70. Although I have not reviewed the full letters of support and opposition, the summaries extracted from them for the Senate bill analysis indicates that much of the language used in the earlier debates has been repeated. Senate Committee, Bill Analysis of AB 2503 (2009–2010), June 25, 2010, http://www.leginfo.ca.gov/pub/09-10/bill/asm/ab_2501-2550/ab_2503_cfa_20100625_180535_sen_comm.html.

71. This was clearly the attitude of Linda Krop when the author interviewed her in Sept. 2008.

## CHAPTER 5. THE BACKBONE OF EVERYDAY ENVIRONMENTALISM: CULTURAL SCRIPTING AND TECHNOLOGICAL SYSTEMS

1. See, e.g., Frank Ackerman, *Why Do We Recycle? Markets, Values, and Public Policy* (Washington, DC: Island Press, 1996). See also Susan Strasser, *Waste and Want: A Social History of Trash* (New York: Owl Books, 2000); Carl Zimring, *Cash for Your Trash: Scrap Recycling in America* (New Brunswick: Rutgers University Press, 2005).

2. Zsuzsa Gille, *From the Cult of Waste to the Trash Heap of History: The Politics of Waste in Socialist and Postsocialist Hungary* (Bloomington: Indiana University Press, 2007); Martin Melosi, *Garbage in the Cities: Refuse Reform and the Environment* (Pittsburgh: Pittsburgh University Press, 2004); Benjamin Miller, *Fat of the Land: Garbage in New York: The Last Two Hundred Years* (New York: Four Walls Eight Windows, 2000); Heather Rogers, *Gone Tomorrow: The Hidden Life of Garbage* (New York: New Press, 2005); Elizabeth Royte, *Garbage Land: On the Secret Trail of Trash* (New York: Little, Brown and Company, 2005); Adam S. Weinberg, David N. Pellow, and Allan Schnaiberg, *Urban Recycling and the Search for Sustainable Community Development* (Princeton: Princeton University Press, 2000).

3. Industrial recycling, on the other hand, still remains all about materials reclamation and resource conservation. At the same time, many businesses use the symbolic environmental value of recycling for all it is worth.

4. While the recycling rates are not as high worldwide as they are in Norway and a few other well-organized Western European countries, packaging recycling is still perhaps the most common "environmentalist" action in the world.

5. Resirk, "Fakta og tall," 2009, http://resirk.no/Fakta-og-tall-52.aspx. In addition to these single-use containers, reusable glass and PET bottles still have a significant share of the Norwegian beverage container market, although the total number is unavailable.

6. Norwegian Pollution Control Authority, "Vedtak om returandel for retursystem for emballasje til drikkevarer for 2009-Bryggeri-og Drikkevareforeningens pante- og retursystem," letter to Bryggeri-og drikkevareforeningen, June 12, 2009.

7. Note that these are the average numbers for the United States and include both deposit and nondeposit states. Container Recycling Institute, "Aluminum, Plastic, and Glass Recy-

cling Rates, 1986–2006," http://www.container-recycling.org/facts/all/data/recrates-3mats
.htm (accessed Feb. 14, 2012).

8. See, e.g., Robert Gottlieb, *Forcing the Spring: The Transformation of the American Environmental Movement* (Washington, DC: Island Press, 2005); Char Miller, *Gifford Pinchot and the Making of Modern Environmentalism* (Washington, DC: Island Press, 2004); Paul Sutter, *Driven Wild: How the Fight against Automobiles Launched the Modern Wilderness Movement* (Seattle: University of Washington Press, 2004).

9. John Tierney, "Recycling Is Garbage," *New York Times*, June 30, 1996; Heather Rogers, *Green Gone Wrong: How Our Economy Is Undermining the Environmental Revolution* (New York: Scribner, 2010). See also Donna Green and Liz Minchin, *Screw Light Bulbs: Smarter Ways to Save Australians Time and Money* (Perth: University of Western Australia Press, 2010), for a good discussion of the significance of small environmentalist actions.

10. Siegfried Giedion laments the same situation for historians in general in his classic *Mechanization Takes Command: A Contribution to Anonymous History* (New York: Oxford University Press, 1948).

11. Wiebe E. Bijker, Thomas P. Hughes, and Trevor J. Pinch, eds., *The Social Construction of Technological Systems: New Directions in the Sociology and History of Technology* (Cambridge: MIT Press, 1987); Wiebe E. Bijker and John Law, eds., *Shaping Technology/Building Society: Studies in Sociotechnical Change* (Cambridge: MIT Press, 1992).

12. "Where Are the Missing Masses? The Sociology of a Few Mundane Objects," in Bijker and Law, *Shaping Technology/Building Society*, 225–58.

13. Madeleine Akrich, "The De-Scription of Technical Objects," in Bijker and Law, *Shaping Technology/Building Society*, 217.

14. Marit Hubak suggests this division of scripts into physical and sociotechnical scripts in "The Car as a Cultural Statement," in *Making Technology Our Own? Domesticating Technology into Everyday Life*, ed. Merete Lie and Knut H. Sørensen (Oslo: Scandinavian University Press, 1996), 175.

15. Akrich, "De-Scription of Technical Objects," 217.

16. Kjetil Fallan, "De-scribing Design: Appropriating Script Analysis to Design History," *Design Issues* 4 (2008): 61–75.

17. Akrich, "De-Scription of Technical Objects," 216.

18. Bruno Latour, *Science in Action: How to Follow Scientists and Engineers through Society* (Milton Keynes: Open University Press, 1987), 259.

19. As argued by a variety of scholars, e.g., Ronald Kline and Trevor Pinch, "Users as Agents of Technological Change: The Social Construction of the Automobile in the Rural United States," *Technology and Culture* 4 (1996): 763–95. See also Nelly Oudshoorn and Trevor Pinch, *How Users Matter: The Co-Construction of Users and Technologies* (Cambridge: MIT Press, 2003), for a collection of articles exploring the relationship among users, designers, and technologies.

20. For a more in-depth history of Scandinavian and American beverage container recycling, see my book *Making a Green Machine: The Infrastructure of Beverage Container Recycling* (New Brunswick: Rutgers University Press, 2011). The book gives a considerably more detailed empirical analysis of the development of beverage container recycling but does not duplicate the explicit discussion of STS methodology in this article.

21. This setup is made possible by standardized, interchangeable, and reusable bottles shared by all the brewers on the Norwegian market. The first such bottle was developed by the Swedish bottler Anders Bjurholm and the cork-cap factory owner Gustaf Emil Boëthius in the 1880s. See Samuel E. Bring, *Anders och Pehr Bjurholms bryggerier* (Stockholm: Stockholms bryggerier, 1949), 116–17.

22. Chr. P. Killengreen, *Den Norske Bryggeriforening: Jubileumsskrift i anledning foreningens 25 aars bestaaende* (Oslo: Centraltrykkeriet, 1926), 25.

23. See, e.g., Peter Bohm, *Deposit-Refund Systems: Theory and Applications to Environmental, Conservation, and Consumer Policy* (Baltimore: Johns Hopkins University Press, 1981).

24. E. C. Dahls bryggeri, *Aksjeselskapet E.C. Dahls bryggeri 1856–1956* (Oslo, 1956), 104–5.

25. Numerous letters to Norwegians newspapers in the late 1960s and 1970s testify to the growing frustration with littering. See, e.g., "Knuste tomflasker," *VG*, Oct. 3, 1972, 30; and "Nei takk til plastflasker," *VG*, Oct. 10, 1972, 30.

26. Øystein Øystå, *Brygg, brus og bruduljer. Bryggeri- og mineralvannbransjen i Norge 100 år* (Oslo: Bryggeri-og mineralvannforeningen, 2001), 152.

27. "Ad: Engangsemballasje. Medlemsforslag i Nordisk Råd om ensartede prinsipper for lovgivning.—Deres j.nr. 20466/70 H.5," Mineralvannindustriens landslag to Det kongelige sosialdepartement, Jan. 7, 1971, subfolder "Korrespondanse fra Industriforbundet til Sosial-departementet."

28. Øystå, *Brygg, brus og bruduljer*, 150.

29. Again I refer to my book *Making a Green Machine* for those interested in the rest of the story.

30. The following description of the early RVMs integration in Norwegian grocery stores is primarily based on the author's interviews with the grocer Aage Fremstad, owner of the first store to install a Tomra I RVM (pictured in fig. 5.1), and the Planke brothers.

31. Tore Planke, "Apparat for automatisk mønstergjenkjenning og registering av tom-flasker," Norwegian Patent No. 126900, filed Dec. 14, 1971, issued Apr. 9, 1973.

32. Tore Planke, interview with the author, Jan. 17, 2006, Oslo, digital recording.

33. "Om midlertidig lov om panteordninger for emballasje til øl, mineralvann og andre leskedrikker," Norwegian Parliament, Ot.prp. nr. 61, 1973–74, passed May 27, 1974.

34. A few other RVM manufacturers existed, but their market share was very small.

35. Part of this discussion is based on Erik Røsrud's unpublished history of Resirk, *RESIRK historien, 1989–1999* (Oslo: Resirk, 1999), as well as interviews with Røsrud, Tore Planke, and Resirk's first CEO, Jarle Grytli. I wish to thank Røsrud for sharing his manuscript with me.

36. Resirk, *Pante-og retursystemer på drikkevaresektoren i Norge* (Oslo: Resirk, 1990).

37. Various draft proposals for an aluminum-can recycling system had circulated since 1984 but did not gain approval. The discussion of disposable containers peaked in a massive controversy in the mid-1990s, involving bottlers, labor unions, environmentalists, the European Union, and others. Tomra attempted to keep a low profile in the proposal, but Tore Planke still came to play a key role in the public debate over the new system.

38. Norsk Resirk, "Framtidens påskeføre," 2007, http://www.pant.no/reklamearkiv/files/pdf/Paaske2007_Orken_146x365.pdf.

39. See Ginger Strand, "The Crying Indian: How an Environmental Icon Helped Sell Cans—and Sell Out Environmentalism," *Orion Magazine*, Nov.–Dec. 2008; Finis Dunaway, "Gas Masks, Pogo, and the Ecological Indian," *American Quarterly*, no. 1 (2008): 67–97. The full video of the ad can be seen on the Ad Council's official Youtube page: http://www.youtube.com/watch?v=862cXNfxwmE (accessed Feb. 14, 2012).

40. In the decades following the Crying Indian ad, however, KAB began organizing local clean-up events that definitely had a visible impact on American landscapes.

41. Ted Williams, "The Metamorphosis of Keep America Beautiful," *Audubon: The Magazine of the National Audubon Society* 2 (1990): 124–34.

42. Dunaway, "Gas Masks," 88.

43. Finn Arne Jørgensen, "Keep America Beautiful," in *Encyclopedia of American Environmental History*, ed. Kathleen Brosnan (New York: Facts on File, 2010).

CHAPTER 6. THE SOIL DOCTOR: HUGH HAMMOND BENNETT, SOIL CONSERVATION, AND THE SEARCH FOR A DEMOCRATIC SCIENCE

Epigraph: Franklin D. Roosevelt, "A Presidential Statement on Signing the Soil Conservation and Domestic Allotment Act" (Mar. 1, 1936), in *The Public Papers and Addresses of Franklin D. Roosevelt,* ed. Samuel Rosenman, 13 vols. (New York: Random House, 1938), 5: 97.

1. Hugh Hammond Bennett, "Science in Soil Conservation," address delivered to the Northeast Section of the American Society of Agronomy, American Association for the Advancement of Science, Durham, New Hampshire, June 26, 1941, p. 24.

2. Bennett, "Science in Soil Conservation," 25.

3. Hugh Hammond Bennett, "The Increased Cost of Erosion," *Annals of the American Academy of Political and Social Science* 142 (Mar. 1929): 170.

4. Hugh Hammond Bennett and William Clayton Pryor, *This Land We Defend* (New York: Longmans, Green and Co., 1942), 79.

5. Hugh Hammond Bennett, "The Development of Natural Resources: The Coming Technological Revolution on the Land," address delivered to USDA, Soil Conservation Service, before Engineering and Human Affairs conference, Princeton University Bicentennial Conference, Princeton, NJ, Oct. 2, 1946, USDA, Natural Resource Conservation Service, http://www.nrcs.usda.gov/about/history/speeches/19461002.html.

6. Bennett, "Science in Soil Conservation," 10.

7. Bennett, "Science in Soil Conservation," 13.

8. Hugh Hammond Bennett, "The Wasting Heritage of a Nation," *Scientific Monthly* 27, no. 2 (Aug. 1928): 97.

9. Wellington Brink, *Big Hugh: The Father of Soil Conservation* (New York: MacMillan Company, 1951), 28.

10. Erving Goffman, *Frame Analysis: An Essay on the Organization of Experience* (New York: Harper Colophon, 1974), 21.

11. Robert D. Benford and David A. Snow, "Framing Processes and Social Movements: An Overview and Assessment," *Annual Review of Sociology* 26 (2000): 614.

12. Wiebe Bijker, "The Social Construction of Bakelite: Toward a Theory of Invention," in *The Social Construction of Technological Systems: New Directions in the Sociology and History of Technology,* ed. Wiebe Bijker, Thomas Hughes, and Trevor Pinch (Cambridge: MIT Press, 1987), 169.

13. Milton Whitney, *Soils of the United States* (Washington, DC: USDA, Bureau of Soils, Bulletin no. 55, 1909), 66.

14. Quoted by Stanford Martin, *And History Is Already Shining on Home: Some Impressions of Hugh H. Bennett, Father of Soil Conservation* (Washington, DC: American Potash Institute, n.d.).

15. Brink, *Big Hugh,* 75–77.

16. Bennett and Pryor, *This Land We Defend,* 6, 9.

17. Bennett, "Wasting Heritage of a Nation," 102.

18. Hugh Hammond Bennett, "Science and Soil," radio address, *Frontiers of Democracy,* Columbia Broadcasting System, New York City, Oct. 17, 1938.

19. Bennett, "Science in Soil Conservation," 10.

20. Bennett, "Science and Soil."

21. Bennett, "Science in Soil Conservation," 23.

22. Bennett, "Science in Soil Conservation," 14–15.

23. Bennett, "Development of Natural Resources."

24. Bennett, "Wasting Heritage of a Nation," 118.

25. John Dewey, *The Later Works*, vol. 2, ed. Jo Ann Boydston (Carbondale and Edwardsville: Southern Illinois University Press, 1988), 319, 320.

26. Bennett, "Increased Cost of Erosion," 176.

27. H. H. Bennett, "Facing the Erosion Problem," *Science* 81, no. 2101 (Apr. 5, 1935): 321–26, esp. 323.

28. Douglas Helms, "Hugh Hammond Bennett and the Creation of the Soil Erosion Service," *Journal of Soil and Water Conservation* 64, no. 2 (Mar.–Apr. 2009): 72A.

29. Brink, *Big Hugh*, 83.

30. Memorandum in regard to erosion control project of the US Department of Agriculture under authority of the National Industrial Recovery Act, July 31, 1933, Soil Erosion Central Classified Files, Record Group 48, file 1–275, College Park, MD, National Archives and Records Administration, quoted in Helms, "Hugh Hammond Bennett," 73A.

31. Neil Maher, *Nature's New Deal: The Civilian Conservation Corps and the Roots of the American Environmental Movement* (New York: Oxford University Press, 2009), 63.

32. "Dust Storm Adds New Crop Menace," *New York Times*, May 11, 1934, 38.

33. Bennett and Pryor, *This Land We Defend*, 80, 94 (emphasis in original).

34. Hugh Hammond Bennett, "Erosion and Rural Relief," statement of H. H. Bennett, Chief, Soil Conservation Service, before Special Senate Committee to Investigate Unemployment and Relief, Mar. 9, 1938, USDA, Natural Resource Conservation Service. http://www.nrcs.usda.gov/about/history/speeches/19380309.html.

35. Bennett, "Development of Natural Resources."

36. Bennett and Pryor, *This Land We Defend*, 83.

37. Maher, *Nature's New Deal*, 64.

38. Bennett, "Wasting Heritage of a Nation," 122.

39. Special Committee on Survey of Land and Water Policies of the U.S., "Hearings before a Special Committee on Survey of Land and Water Policies of the U.S," 74th Congress, 1st session, Aug. 21, 1935, 11.

40. Bennett, "Science in Soil Conservation," 7–8.

41. Bennett and Pryor, *This Land We Defend*, 67.

42. Maher, *Nature's New Deal*, 67.

43. Bennett and Pryor, *This Land We Defend*, 62, 69.

44. Bennett and Pryor, *This Land We Defend*, 35, 36, 38.

45. Bennett and Pryor, *This Land We Defend*, 28–29.

46. Donald Worster, *Dust Bowl* (New York: Oxford University Press, 2004), 7, 8.

47. Hugh Hammond Bennett, address delivered to the American Association for the Advancement of Science, Richmond, VA, Dec. 29, 1938, USDA, Natural Resource Conservation Service. http://www.nrcs.usda.gov/about/history/speeches/19461002.html (my emphasis).

## CHAPTER 7. COMMUNICATING KNOWLEDGE: THE SWEDISH MERCURY GROUP AND VERNACULAR SCIENCE, 1965–1972

I am grateful to Johan Gran for translating many of the Swedish newspaper articles used in this piece.

1. Hans Ackefors, Nils-Erik Landell, Göran Löfroth, and Carl-Gustaf Rosén, oral history with the author, Stockholm, Sweden, Aug. 7, 2009. Mercury is a naturally occurring element, but it rarely occurs independently in nature without human intervention. Rather, it is trapped in coal and other mineral deposits and freed into air, soil, and water through such human activities as waste incineration, coal combustion, and other industrial practices. The environ-

mental hazard posed by mercury stems from organic forms of the element such as those used as fungicides in treating seedgrain. These methylmercuries are very dangerous neurotoxins, which accumulate in the soil, water, and living organisms.

2. The core Mercury Group was originally founded by three young scientists, Carl-Gustaf Rosén, Hans Ackefors, and Robert Nilsson. In 1965, Göran Löfroth joined the group after returning to Sweden from a postdoctoral fellowship at Michigan State University. Nils-Erik Landell was a relative outsider but also participated in making the more technical literature accessible to a public audience. Though a little older, Claes Ramel was also a peripheral member, who provided valuable statistical data. The Mercury Group originally met in a biocide seminar at the University of Stockholm. Rosén and Löfroth had both studied under Jos Arenberg; Arenberg had encouraged taking a broader view to social and scientific issues and had dramatized the scientific effects of detergents a generation earlier.

3. Because environmental narratives "belong as much to rhetoric and human discourse as to ecology and nature . . . [environmental historians] cannot escape confronting the challenge of multiple competing narratives in our efforts to understand both nature and the human past." See William Cronon, "A Place for Stories: Nature, History, and Narrative," *Journal of American History* 78 (Mar. 1992): 1367.

4. As early as 1912, German chemists had experimented with organic mercury compounds but found them to be too expensive. For a brief history of seed treatment, see D. E. Mathre, R. H. Johnston, and W. E. Grey, "Small Grain Cereal Seed Treatment," *Plant Health Instructor* (2001), http://www.apsnet.org/edcenter/advanced/topics/Pages/CerealSeedTreatment.aspx.

5. Göran Löfroth and Margaret E. Duffy, "Birds Give Warning," *Environment*, May 1969, 10–17, quotation on 10. In more than four hundred tests performed in Sweden between 1933 and 1963, mercury-treated seedgrain demonstrated average crop yields of 6–16 percent for oats, winter rye, barley, and winter wheat. See I. Granhall, in *Kvicksilverfrågan i Sverige* (Stockholm: Holmqvist, 1965), 92–108; and Stig Tejning, "Mercury in Pheasants (*Phasianus colchius* L.) Deriving from Seed Grain Tested with Methyl and Ethyl Mercury Compounds," *Oikos* 18 (1967): 334–44.

6. Claes Bernes and Lars J. Lundgren, *Use and Misuse of Nature's Resources: An Environmental History of Sweden* (Stockholm: Swedish Environmental Protection Agency, 2009), 80.

7. K. Borg et al., "Mercury Poisoning in Swedish Wildlife," *Journal of Applied Ecology* 3 (June 1966): 171–72. See also K. Borg et al., "Alkylmercury Poisoning in Terrestrial Swedish Wildlife," *Viltrevy* 6 (1969): 301. Borg's earlier work, in Swedish, had been published in 1963 and 1965.

8. Arne Jernelöv, oral history with the author, Stockholm, Sweden, Aug. 6, 2009.

9. A. G. Johnels and T. Westermark analyzed goshawk (*Accipter gentilis*) feathers, drawing samples from over one hundred years. Prior to 1940, their mean value was 2,200 ng/g of mercury; after 1940, their mean value had climbed to 29,000 ng/g. See A. G. Johnels and T. Westermark, "Mercury Contamination of the Environment in Sweden," in *Chemical Fallout: Current Research on Persistent Pesticides*, ed. Morton W. Miller and George G. Berg (Springfield, IL: Charles C. Thomas, 1969), 221–41. Their report was derived from studies from a 1965 conference on mercury in the environment, convened by the Royal Commission on Natural Resources: A. G. Johnels, "Preliminär rapport rörande Hg-analyser på havsörn och berguv," in *Kvicksilverfrågan i Sverige*, 165–68; and T. Westermark, "Kvicksilver hos vattenlevande organismer," in *Kvicksilverfrågan i Sverige*, 25–76. See also W. Berg et al., "Mercury Content in Feathers of Swedish Birds from the Past 100 Years," *Oikos* 17 (1966): 71–83.

10. For an overview, see Lundqvist, *The Case of Mercury Pollution in Sweden: Scientific Information and Public Responses* (Stockholm: Swedish Natural Science Council, 1974), 12–25. See also Katherine Montague and Peter Montague, *Mercury* (San Francisco: Sierra Club Books,

1971), 34–37; and Patricia A. D'Itri and Frank M. D'Itri, *Mercury Contamination: A Human Tragedy* (New York: John Wiley and Sons, 1977), 29–34.

11. A good summary is J. E. Larsson, *Environmental Mercury Research in Sweden* (Stockholm: Swedish Environmental Protection Board, 1970).

12. D'Itri and D'Itri, *Mercury Contamination*, 30.

13. Lundqvist, *Case of Mercury Pollution in Sweden*, 29.

14. "Industrirök ger kvicksilver i fisk," *Dagens Nyheter*, Feb. 5, 1965, 28. Johnels and Westermark analyzed for total mercury—not methylmercury—using a nuclear activation technique developed in Westermark's laboratory.

15. "Toppmöte om gifter: Statsråd informeras," *Dagens Nyheter*, Sept. 8, 1965, 26.

16. Carl-Gustaf Rosén, "Biociddebatt i bakvatten," *Svenska Dagbladet*, Sept. 16, 1965, 4.

17. Lundqvist, *Case of Mercury Pollution in Sweden*, 18.

18. "Inget förbud mot kvicksilver: Jordbruksministern: Giftnämndsbeslutet gäller även för oss," *Dagens Nyheter*, Sept. 10, 1965, 3. See also "Anpassad betning: Oenig giftnämnd föregrep expertis," *Dagens Nyheter*, Sept. 9, 1965, 1. The Poisons Board, however, was not unanimous in its decision. Karl Borg represented a dissenting view, writing a separate letter to the government and urging a complete ban on all mercury-based fungicides by Jan. 1, 1966.

19. Rosén, "Biociddebatt i bakvatten," 4.

20. Ackefors et al., oral history, Aug. 7, 2009.

21. Ackefors et al., oral history, Aug. 7, 2009.

22. *Dagens Nyheter*, Feb. 24, 1964.

23. Barbara Soller, "Forskarkrav: Ny granskning av kvicksilverrön," *Dagens Nyheter*, Sept. 25, 1965, 20.

24. Soller, "Forskarkrav," 1; and Gunnel Westöö, "Kvicksilver i ägg," *Vår Föda* 17 no. 5 (1965): 1–8. Eggs from six continental countries averaged only seven parts per billion. See Gunnel Westöö, "Methylmercury Compounds in Animal Foods," in Miller and Berg, *Chemical Fallout*, 75–93. Westöö notes that eggs in Norway also had heightened levels of mercury, which she attributes to Sweden's and Norway's extensive use of methylmercury dicyandiamide as a fungicide. See also Löfroth and Duffy, "Birds Give Warning," 13; and D'Itri and D'Itri, *Mercury Contamination*, 30.

25. Ackefors et al., oral history, Aug. 7, 2009.

26. C-G. Rosen et al., *Svensk Kemisk Tidskrift* 78 (1966), 8, cited in Göran Löfroth, "Methylmercury: A Review of Health Hazards and Side Effects Associated with the Emissions of Mercury Compounds into Natural Systems" (Swedish Natural Science Research Council, 1970), 8.

27. Rosén, "Biociddebatt i bakvatten," 4.

28. M. Lagervall and Gunnel Westöö, *Vår Föda* 21 (1969): 9. In light of the food scare, on Feb. 1, 1966, the Poisons and Pesticides Board ultimately revoked the license for all agricultural alkylmercury dressings. Infected seeds could still be treated, but only with the less toxic methoxyethylmercury, Panogen Metox. Prior to the growing concern over mercury in agriculture, 80 percent of seeds sown in the spring were treated with methylmercury; in 1967, only 12 percent were treated, and those were treated with Panogen Metox. Studies conducted after 1965 suggested that bird populations were recovering, mercury in eggs was substantially reduced, and the mercury content in meat had declined. By 1968, the mercury residues in meat and eggs were similar to those in the rest of Europe. See Löfroth and Duffy, "Birds Give Warning," 13; and Gunnel Westöö, "Totalkvicksilver: och metylkvicksilverhalter i ägg köpta i Sverige, Juni 1966–September 1967," *Vår Föda* 19 (1967): 121–24.

29. Rosén, "Biociddebatt i bakvatten," 4.

30. Soller, "Forskarkrav," 1, 20.

252 NOTES TO PAGES 111–115

31. Ackefors et al., oral history, Aug. 7, 2009.

32. Lundqvist, *Case of Mercury Pollution in Sweden*, 20, 35. Lundqvist tallied articles from *Arbetet, Dagens Nyheter, Svenska Dagbladet,* and *Sydsvenska Dagbladet/Snällposten*. Between 1965 and 1968, as the number of mercury articles increased, so too did the number of authors. Well over fifty different people contributed news or editorial pieces on mercury to Swedish newspapers.

33. Ackefors et al., oral history, Aug. 7, 2009.

34. Ackefors et al., oral history, Aug. 7, 2009.

35. Gunnel Westöö, "Methylmercury Compounds in Fish: Identification and Determination," *Acta Chemica Scandinavica* 20 (1966): 2131–37; Gunnel Westöö, "Kvicksilver i fisk," *Vår Föda* 19 (1967): 1–11; and K. Norén and Gunnel Westöö, "Metylkvicksilver i fisk," *Vår Föda* 19 (1967): 13–17.

36. *Dagens Nyheter,* Jan. 21, 1967. See also Nils-Erik Landell, "Miljödebatten har borrats i sänk," *Svenska Dagbladet,* July 31, 2001; H. O. Bouveng, "Control of Mercury Effluents from Chlorine Plants," *Pure and Applied Chemistry* 29, no. 1 (1972): 75–91; P. Wihlborg and Å. Danielsson, "Half a Century of Mercury Contamination in Lake Vänern (Sweden)," *Water, Air, and Soil Pollution* (2006): 285–300; and Lennart Lindeström, "Mercury in Sediment and Fish Communities of Lake Vänern, Sweden: Recovery from Contamination," *Ambio* (Dec. 2001): 538–44.

37. F. Berglund and A. Wretlind, "Toxikologisk värdering av kvicksilverhalter i svensk Fisk," *Vår Föda* 19 (1967): 9–11, trans. in Lundqvist, *Case of Mercury Pollution,* 31.

38. The standard account of this discovery also credits Torbjörn Westermark as having made the same discovery independently. According to Göran Löfroth and other members of the Mercury Group, the Biochemistry Department was not quiet about its findings. Through a younger relative in the lab, this information was passed along to Westermark, who was the first to share it with the Public Health Institute. Ackefors et al., oral history, Aug. 7, 2009.

39. *Sydsvenska Dagbladet Snällposten,* Feb. 22, 1967, May 20, 1967; and *Dagens Nyheter,* May 20, 1967, Oct. 25, 1967.

40. *Dagens Nyheter,* Jan. 31, 1967; *Svenska Dagbladet,* Feb. 1, 1967; *Arbetet,* Feb. 1, 1967. See also Löfroth and Duffy, "Birds Give Warning," 15–16. The new methodology was based on one man's experience of eating roughly 1 mg/kg of fish daily for over a year and showing no symptoms of mercury poisoning.

41. Ackefors et al., oral history, Aug. 7, 2009.

42. Lundqvist, *Case of Mercury Pollution in Sweden,* 31. See also Nils-Erik Landell, *Fågeldöd—Fiskhot—Kvicksilver* (Stockholm: Aldus/Bonnier, 1970), 88; and *Sydsvenska Dagbladet Snällposten,* Sept. 13, 1968.

43. In 1968, the National Board on Social Welfare changed its name to the National Board of Medicine.

44. Lundqvist, *Case of Mercury Pollution in Sweden,* 41.

45. Å. W. Edfeldt, *Kvicksilvergäddan* (Stockholm: Tiden/Folksam, 1969), 34. See also *Svenska Dagbladet,* Apr. 26, 1967.

46. Landell, *Fågeldöd—Fiskhot—Kvicksilver,* 88–89. See also *Dagens Nyheter,* Feb. 14, 1967.

47. Edfeldt, *Kvicksilvergäddan,* 31.

48. *Svenska Dagbladet,* Nov. 15, 1967; *Dagens Nyheter,* Nov. 15, 1967; *Dagens Nyheter,* Nov. 29, 1967.

49. *Dagens Nyheter,* Oct. 26, 1967.

50. *Svenska Dagbladet,* Feb. 3, 1968, quoted in Lundqvist, *Case of Mercury Pollution,* 43.

51. Lundqvist, *Case of Mercury Pollution in Sweden,* 43.

52. Chandra Mukerji, *A Fragile Power: Scientists and the State* (Princeton: Princeton University Press, 1989).

53. Julie Thompson Klein, *Crossing Boundaries: Knowledge, Disciplinarities, and Interdisciplinarities* (Charlottesville: University of Virginia Press, 1996), 4.

54. Michael Soulé, "What Is Conservation Biology?" *Bioscience* 35 (1985): 727–34.

55. Brian Wynne and Sue Mayer, "How Science Fails the Environment," *New Scientist* 5 (1993): 33–35.

56. S. O. Funtowicz and J. R. Ravetz, "Three Types of Risk Assessment and the Emergence of Post-Normal Science," in *Social Theories of Risk*, ed. S. Krimsky and D. Golding (Westport, CT: Praeger, 1992), 251–73, quotation on 254.

57. H. M. Collins and Robert Evans, "The Third Wave of Science Studies: Studies of Expertise and Experience," *Social Studies of Science* 32 (Apr. 2002): 235–96, quotation on 235.

58. Bruno Latour, *The Pasteurization of France* (Cambridge: Harvard University Press, 1993). See also Sheila Jasanoff, "Beyond Epistemology: Relativism and Engagement in the Politics of Science," *Social Studies of Science* 26 (1996): 393–418; David H. Guston, "Boundary Organizations in Environmental Policy and Science: An Introduction," *Science, Technology, and Human Values* 26 (Autumn 2001): 399–408; and Christian Pohl, "From Science to Policy through Transdisciplinary Research," *Environmental Science and Policy* 11 (2008): 46–53.

59. Collins and Evans, "Third Wave of Science Studies," 281.

60. Marybeth Long Martello and Sheila Jasanoff, "Globalization and Environmental Governance," in *Earthly Politics: Local and Global in Environmental Governance*, ed. Sheila Jasanoff and Marybeth Long Martello (Cambridge: MIT Press, 2004), 1–29. See also Bruno Latour, *We Have Never Been Modern* (Cambridge: Harvard University Press, 1993), 1–5; and Peter Groenewegen, "Accommodating Science to External Demands: The Emergence of Dutch Toxicology," *Science, Technology, and Human Values* 27 (Autumn 2002): 479–98.

## CHAPTER 8. SIGNALS IN THE FOREST: CULTURAL BOUNDARIES OF SCIENCE IN BIAŁOWIEŻA, POLAND

1. The demographics in Białowieża have changed considerably since I conducted research in 2005–7. A new group of residents live in Białowieża. These people run the burgeoning tourist industry and work at hotels. Also, a different group of biologists now work in Białowieża and do not necessarily share the same position as the biologists I write about, who were in their twenties and thirties in the early 1990s.

2. Since 1994 Belarus has been run by President Alexander Lukashenko, now in his fourth five-year term. His regime is widely condemned in the Western press as "the last dictatorship in Europe."

3. On the cultural boundaries of science see Susan Leigh Star and J. R. Griesemer, "Institutional Ecology, 'Translations' and Boundary Objects: Amateurs and Professionals in Berkeley's Museum of Vertebrate Zoology, 1907–39," *Social Studies of Science* 19, no. 4 (1989): 387–420. See also Donna Haraway, *Primate Visions: Gender, Race, and Nature in the World of Modern Science* (London: Routledge, 1989); Karen Barad, "Agential Realism: Feminist Interventions in Understanding Scientific Practices," in *The Science Studies Reader*, ed. M. Biagioli (New York: Routledge, 1998), 1–11; Cori Hayden, *When Nature Goes Public: The Making and Unmaking of Bioprospecting in Mexico* (Princeton: Princeton University Press, 2003).

4. Gieryn, *Cultural Boundaries of Science*, 7.

5. See Chris M. Hann, *Not the Horse We Wanted! Postsocialism, Neoliberalism, and Eurasia* (Picataway, NJ: Transaction Publishers, 2006).

6. As of Dec. 2012 local governing councils had rejected every proposal since 2001 to expand the national park. Greenpeace led a campaign in Poland in 2010–11 to overturn the law on nature protection, collecting more than 250,000 signatures.

7. See James Scott, *Seeing like a State: How Certain Schemes to Improve the Human Condition Have Failed* (New Haven: Yale University Press, 1998).

8. Anna Lawrence has argued that Central and East European traditions of forestry all emerged in the age of empires and feudalism so that laws encoding forest use and rights are deeply engrained in local rural cultures: Anna Lawrence, "Forests in Transition: Negotiated Expertise in Post-Socialist Europe" (paper presented at Scientific Framework of Environmental and Forest Governance—The Role of Discourses and Expertise conference, Göttingen, Germany, 2007).

9. For a discussion of Polish foresters' involvement in the 1863 uprising, see Simon Schama, *Landscape and Memory* (New York: A. A. Knopf, 1995). Ravi Rajan's monograph provides important insights into the spread and application of German scientific forestry throughout the nineteenth and twentieth centuries. See S. Ravi Rajan, *Modernizing Nature: Forestry and Imperial Ecodevelopment 1800–1950* (New York: Oxford University Press, 2006).

10. See Simona Kossak, *The Saga of the Bialowieza Forest* (Bialystok: Bialydruk, 2001), 391–407.

11. Kossak, *Saga*, 311.

12. Borys Nikitiuk, *Z dziejów Hajnówki i jej okolic (1915–1939)* (Hajnówka: Starostwo Powiatowe w Hajnówce, 2004).

13. Bison reconstitution is considered partially successful. The herd consists of approximately five hundred individuals for both the Polish and Belarusian part of the woodland, but the population is seriously compromised genetically. For a scientific explanation, see E. Wołk and M. Krasińska, "Has the Condition of European Bison Deteriorated over Last Twenty Years?" *Acta Theriologica* 49 (2004): 405–18.

14. Polish biologists claim that Polish scientists had much more autonomy during the socialist period than those in the Soviet Union. Paul Josephson's work details Soviet applications of science under Khrushchev, roughly the same time period as Białowieża's scientific institutes began. Soviet science at the outposts of Lake Baikal was supposed to accelerate communism. See Paul Josephson, *New Atlantis Revisited: Akademgorodok, the Siberian City of Science* (Princeton: Princeton University Press, 2007). Douglas Weiner has argued, however, that Russian biologists possessed a great deal of freedom under the Soviet system. See Douglas Weiner, *A Little Corner of Freedom: Russian Nature Protection from Stalin to Gorbachev* (Berkeley: University of California Press, 2002). See Jane Dawson's work on science and environmentalism in post-Socialist Europe for a discussion of science in creating new social relations: Jane Dawson, *Eco-Nationalism: Anti-Nuclear Activism and National Identity of Russia, Lithuania and Ukraine* (Durham: Duke University Press, 1996).

15. For an analysis of the protest events, see Stuart Franklin, "Bialowieza Forest, Poland: Representation, Myth, and the Politics of Dispossession," *Environment and Planning A* 34, no. 8 (2002): 459–85.

16. Biologists and other forest activists repeatedly drew my attention to a "corrupting" gift to then–prime minister Wlodzimierz Cimoszewicz. Foresters gifted Cimoszewicz with a historic cottage deep in the forest at the same time Cimoszewicz expressed disapproval for the national park expansion. Cimoszewicz was also an avid hunter and was seen to have received special privileges in this regard. Upon a personal visit to Cimoszewicz's cottage, I observed two hunting towers and cleared field within two hundred meters of his home.

17. Much ecotourism to the area came in the form of large hotels, including a Best Western that could accommodate hundreds of people. The village of Białowieża transformed from a small village with very specific and limited tourism in the 1990s to a site of large, mass tourism, with forays to the forest organized by tourist offices throughout Europe. Also, conference tourism marked a whole new kind of tourism in the village and was very much part of ecotourism.

18. Etienne Benson explains how radio telemetry has many meanings embedded in it that are often ignored by scholars. He argues that the technology is not just a form of surveillance

and control, but a technology where scientists can become intimately connected with particular animals and landscapes. See Etienne Benson, *Wired Wilderness: Technologies of Tracking and the Making of Modern Wildlife* (Baltimore: John Hopkins University, 2010).

19. Data from historical periods comes from two sources: royal censuses of animals and foresters' animal counts. Since the nineteenth century censuses have been conducted in roughly the same manner. Animals were driven by groups of beaters encroaching from two sides and then were respectively counted.

20. Although the term *Pan* (the formal masculine *you*) is placed before most names, when it appears behind an occupation, such as *Pan Leśniczy* or *Pan Lekarz* (Sir Doctor), an added form of respect is imparted, marking deference for a person of a higher social class. It was common in the nineteenth century for peasants to address any member of the intelligentsia as *Pan* and *Pani*, but for the forester to emphasize the continuation of this relationship in the present signals foresters' attachments to their transhistorical relationship with local people.

21. This biologist is often viewed as responsible for starting the animal torture/radio telemetry controversy. Because the details of this are complex and there is not full room for discussion here, I suggest reading my third dissertation chapter: see Eunice Blavascunas, "The Peasant and Communist Past in the Making of an Ecological Region: Podlasie, Poland" (PhD diss., University of California Santa Cruz, 2008).

22. See Joseph Dumit for a discussion of objectifying technologies: Joseph Dumit, "A Digital Image of the Category of the Person," in *Cyborgs and Citadels: Anthropological Interventions in Emerging Sciences and Technologies*, ed. Joseph Dumit (Seattle: University of Washington Press, 1997).

23. Scott, *Seeing like a State*.

24. See Gregg Mitman, *Reel Nature: America's Romance with Wildlife on Films* (Cambridge: Harvard University Press, 1999). See also Haraway, *Primate Visions*.

## CHAPTER 9. THE PRODUCTION AND CIRCULATION OF STANDARDIZED KARAKUL SHEEP AND FRONTIER SETTLEMENT IN THE EMPIRES OF HITLER, MUSSOLINI, AND SALAZAR

1. Himmler to Körner, May 1944, Bundesarchiv Berlin (BA), Persönlicher Stab Reichsführer-SS, NS 19/2596, "Unterbringung des Gestüts und der Karakul-Schafherde von Dr. Schäfer in Ungarn."

2. The literature on the Nazi occupation of Eastern Europe is of course enormous. The following references are especially useful to understand the imperial regime at work: Mark Mazower, *Hitler's Empire: How the Nazis Ruled Europe* (New York: Penguin Press, 2008); Götz Aly and Susanne Heim, *Vordenker der Vernichtung. Auschwitz und die deutschen Pläne für eine neue europäische Ordnung* (Hamburg: Hoffman und Campe, 1991); Michael Thad Allen, *The Business of Genocide: The SS, Slave Labor and the Concentration Camps* (Chapel Hill: University of North Carolina Press, 2002); Wendy Lower, *Nazi Empire-Building and the Holocaust in Ukraine* (Chapel Hill: University of North Carolina Press, 2005); Czeslaw Madajczyk, ed., *Vom Generalplan Ost zum Generalsiedlungsplan* (Munich: Saur, 1994).

3. This straightforward methodology of following sheep around is much inspired by Sarah Franklin, *Dolly Mixtures: The Remaking of Genealogy* (Durham: Duke University Press, 2007). If this is an obvious reference for sheep in STS, environmental history also has very good examples of narratives built around sheep, namely Elinor G. K. Melville, *A Plague of Sheep: Environmental Consequences of the Conquest of Mexico* (Cambridge: Cambridge University Press, 1994).

4. Environmental historians have offered us famous accounts of the relevance of such shifts, as in the exemplary narrative of the conversion of the buffalo range into ranchland. See William Cronon, *Nature's Metropolis: Chicago and the Great West* (New York: W. W. Norton,

1991). We also have other sets of narratives that have paid attention to hunting practices on the frontier and their connections to conservationist measures, such as the 1900 Convention for the Preservation of Wild Animals, Birds and Fish in Africa. See William Beinart and Peter Coates, *Environment and History: The Taming of Nature in the USA and South Africa* (New York: Routledge, 1995), 17–33; William Kelleher Storey, *Guns, Race, and Power in Colonial South Africa* (Cambridge: Cambridge University Press, 2008).

5. L. Adametz, *El Carnero Karakul* (Buenos Aires: Talleres de Publicación de la Dirección Meteorológica, 1914), 3–5.

6. For the importance of expansionist ambitions for fascism, see Aristotle A. Kallis, *Fascist Ideology: Territory and Expansionism in Italy and Germany, 1922–1945* (London: Routledge, 2000).

7. Robert Lewis Koehl, *RKFDV: German Resettlement and Population Policy, 1939–1945: A History of the Reich Commission for the Strengthening of Germandom* (Cambridge: Harvard University Press, 1957); Bruno Wasser, *Himmlers Raumplanung im Osten: Der Generalplan Ost in Polen 1940–1944* (Berlin: Birkhäuser Verlag, 1993); Metchild Rössler and Sabine Schleiermacher, eds., *Der Generalplan Ost: Hauptlinien der nationalsozialistischen Planungs- und Vernichtungspolitik* (Berlin: Akademie Verlag, 1993); Aly and Heim, *Vordenker*, n2.

8. Joachim Wolschke-Bulmahn, "Violence as the Basis of National Socialist Landscape Planning in the 'Annexed Eastern Areas,'" in *How Green Were the Nazis? Nature, Environment, and Nation in the Third Reich*, ed. Franz-Josef Brüggemeier, Mark Cioc, and Thomas Zeller (Athens: Ohio University Press, 2005), 243–56.

9. Heather Pringle, *The Master Plan: Himmler's Scholars and the Holocaust* (London: Harper Perennial, 2006).

10. For an exploration of the comparison between the American frontier and the role of the "'Wild East'" in Nazi ideology, see David Blackbourn, *The Conquest of Nature: Water, Landscape and the Making of Modern Germany* (London: Random House, 2006), 280–96.

11. Henrik Eberle, *Die Martin-Luther-Universität in der Zeit des Nationalsozialismus 1933–45* (Halle [Saale]: mdv Mitteldeutscher, 2002), 234–35.

12. For a description of German colonial rule in the Ukraine, see Lower, *Nazi Empire-Building*.

13. Debórah Dwork and Robert Jan van Pelt, *Auschwitz, 1270 to the Present* (New York: W. W. Norton, 1996), 140.

14. This technoscientific infrastructure in the occupied regions was not created ex nihilo. Much of it was based on taking over previous Soviet agriculture research institutes, as made clear by Susanne Heim, *Kalorien, Kautschuk, Karrieren. Pflanzenzüchtung und landwirtschaftliche Forschung in Kaiser-Wilhelm-Instituten 1933–1945* (Göttingen: Wallstein, 2003), 40–49. Considering the highly developed exploration of karakul sheep by Soviet animal breeders and the previous role of the Ukraine as one of the Soviet areas where karakul were intensively explored, it seems reasonable to suppose that the German Kriwoj Rog experiment station was based on preexistent Soviet research efforts. See Gustav Frölich and Hans Hornitschek, *Das Karakulschaf und seine Zucht* (Munich: F. C. Mayer Verlag, 1942).

15. Heim, *Kalorien*, 60n14; Eberle, *Die Martin-Luther-Universität*, 234–35n11.

16. Frölich and Hornitschek, *Das Karakulschaf,* n14. The first edition of the work was published in 1928 and was authored only by Frölich. On Frölich, see Heim, *Kalorien*, 39n14. Also important is his Festschrift for his sixtieth anniversary in the special issue of *Kühn Archiv* 52 (1939).

17. Hans Hornitschek, "Bau und Entwicklung der Locke des Karakulschafes," *Kühn-Archiv* 47 (1938): 81–175.

18. Jonathan Harwood, *Styles of Scientific Thought: The German Genetics Community 1900–1933* (Chicago: University of Chicago Press, 1993), 54.

19. Hornitschek, "Bau und Entwicklung," 81n16.

20. Gustav Frölich, "Die Zucht des Karakulschafes am Tierzuchtinstitut der Universität Halle als Versuchs- und Forschungsobjekt," *Kühn-Archiv* 18 (1928): i–xiii.

On model organisms in the biological sciences, see Hans-Jörg Rheinberger, *An Epistemology of the Concrete: Twentieth Century Histories of Life* (Durham: Duke University Press, 2010); Robert E. Kohler, *Lords of the Fly: Drosophila Genetics and the Experimental* Life (Chicago: University of Chicago Press, 1994); Karen A. Rader, *Making Mice: Standardizing Animals for American Biomedical Research, 1900–1955* (Princeton: Princeton University Press, 2004); Angela N. H. Creager, *The Life of a Virus: Tobacco Mosaic Virus as an Experimental Model, 1930–1965* (Chicago: University of Chicago Press, 2002); Angela N. H. Creager, Elizabeth Lunbeck, and M. Norton Wise, introduction, in *Science without Laws: Model Systems, Cases, Exemplary Narratives* (Durham: Duke University Press, 2007), 1–20.

21. Gustav Frölich and Hans Hornitshek, *El Karakul. Su cría, explotación y selección* (Madrid: Sindicato Nacional de Ganadería, 1946), 249.

22. Leopold Adametz, *Studien über die Mendelsche Vererbung der wichtigsten Rassenmerkmale der Karakulschafe* (Leipzig: Borntraeger, 1917).

23. Frölich and Hornitschek, *El Karakul,* 227–38n21.

24. Leopold Adametz, "Über die Eignung verschiedener Landschafrassen zur Kreuzung mit Karakuls," *Zeitschrift für Tierzucht und Züchtungsbiologie* 45 (1940): 71–97.

25. Frölich, "Die Zucht des Karakulschafes," n20.

26. Brenda Bravenboer, *Karakul: Gift from the Arid Land: Namibia 1907–2007* (Windhoek: Karakul Board of Namibia and Karakul Breeder's Society of Namibia, 2007); Jacobus Andreas Nel, "A Critical Study of the Development, Breeding and Care of the Neudam Karakul Stud" (MSc thesis, University of Stellenbosch, 1950).

27. Bravenboer, *Karakul,* 46–47.

28. Jürgen Zimmerer, "Krieg, KZ und Völkermord in Südwestafrika. Der erste deutsche Genozid," in *Völkermord in Deutsch-Südwestafrika. Der Kolonialkrieg (1904–1908) in Namibia und seine Folgen,* ed. Jürgen Zimmerer and Joachim Zeller (Berlin: Christoph Links, 2003), 45–63.

29. Robert Gerwarth and Stephan Malinowski, "Hannah Arendt's Ghosts: Reflections on the Disputable Paths from Windhoek to Auschwitz," *Central European History* 42 (2009): 279–300.

30. Helmuth Stoecker, ed., *German Imperialism in Africa: From the Beginnings until the Second World War* (Oxford: Oxford University Press, 1986), 62.

31. Guido G. Weigend, "German Settlement Patterns in Namibia," *Geographical Review* 75 (1985): 156–69.

32. D. C. Krogh, "Economic Aspects of the Karakul Industry in South West Africa," *South African Journal of Economics* 23 (1955): 99–113.

33. Heim, *Kalorien,* 35–46n14.

34. Heim, *Kalorien,* 39.

35. Max Planck Society (MPS) Archive, Abt. 1, rep. 1A, no. 2887/1–6 Vertrag für Übernahme der Karakulherde.

36. Marion Kazemi, "Das Kaiser-Wilhelm-Institut für Tierzuchtforschung in Rostock und Dummerstorf 1939–1945," *Dahlemer Archivgespräche* 8 (2002): 137–63; *Jahrbuch der Kaiser-Wilhelm-Gesellschaft* (1941): 47–58.

37. Francesco Maiocco, "Coniglicoltura e allevamento animali da pelliccia al Consiglio Zootecnico," *Rivista di Coniglicoltura e Allevamento Animali da Pelliccia* 3, no. 4 (1931): 2–19.

38. On the connections between genetics and autarky policies during the Italian fascist regime, see Tiago Saraiva, "Fascist Labscapes: Geneticists, Wheat, and the Landscapes of Fascism in Italy and Portugal," *Historical Studies in the Natural Sciences* 40 (2010): 457–98.

39. For general discussions of the political economy of the Italian fascist regime, the following sources are particularly valuable: Domenico Petri, *Economia e istituzioni nello stato fascista* (Rome: Riunti, 1980); Charles S. Maier, *In Search of Stability: Explorations in Historical Political Economy* (Cambridge: Cambridge University Press, 1987), 70–120; Adrian Lyttelton, *The Seizure of Power: Fascism in Italy, 1919–1929* (London: Weidenfeld and Nicolson, 1973), 333–63; A. James Gregor, *Italian Fascism and Developmental Dictatorship* (Princeton: Princeton University Press, 1979), 127–71.

40. Francesco Maiocco," "I circoli di allevamento," *Rivista di Coniglicoltura e Allevamento Animali da Pelliccia* 3, no. 1 (1931): 7–16.

41. "Sul piano straordinario di azione a favore della coniglicoltura," *Rivista di Coniglicoltura e Allevamento Animali da Pelliccia* 13, no. 1 (1941): 1–12.

42. *Rivista di Coniglicoltura e Allevamento Animali da Pelliccia* 11, no. 11 (1939).

43. *Rivista di Coniglicoltura e Allevamento Animali da Pelliccia* 11, no. 5 (1939).

44. "Per l'allevamento Karakul in italia," *Rivista di Coniglicoltura e Allevamento Animali da Pelliccia* 3, no. 8 (1931): 5.

45. "Per l'allevamento Karakul in italia."

46. Consiglio Nazionale delle Ricerche, *Istituti e Laboratori Scientifici Italiani* (Rome: Presso Il Consiglio Nazionale delle Ricerche, 1940), 2: 619–20.

47. For a general overview of Italian imperialism in the fascist years, the following sources are particularly helpful: Ruth Ben-Ghiat and Mia Fuller, eds., *Italian Colonialism* (New York: Palgrave, 2005); Alberto Sbacchi, *Il Colonialismo Italiano in Etiopia, 1935–1940* (Milan: Mursia, 1980); Claudio G. Segrè, *L'Italia in Libia: Dall'età giolittiana a Gheddafi* (Milan: Feltrinelli, 1978); Ricardo Bottoni, ed., *L'impero fascista. Italia ed Etiopia (1935–1941)* (Bologna: Il Mulino, 2008).

48. For the arguments of Armando Maugini, probably the most quoted expert on Italian settlement policies, see Armando Maugini, "Colonizzazione Borghese e colonizzazione contadina nella Libia," *Agricoltura Coloniale* 29 (1935): 113–23; Armando Maugini, *Agricoltura indigena e colonizzazione agricola nell' AOI* (Rome: Sindacato Nazionale Fascista Tecnici Agricoli, 1936); Franco Cardini e Isabella Gagliardi, "Verso un nuovo impero," in *L'Istituto Agronomico per L'Oltremare. La Sua Storia* (Florence: Masso delle Fate Edizioni, 2007), 201–12.

49. See n47.

50. Nicola Labanca, "Italian Colonial Internment," in Ben-Ghiat and Fuller, *Italian Colonialism*, 27–36.

51. Claudio G. Segrè, *Italo Balbo: A Fascist Life* (Berkeley: University of California Press, 1987).

52. E. J. Russell, "Agricultural Colonization in the Pontine Marshes and Libya," *Geographical Journal* 94 (1939): 273–89.

53. "Disciplinare per l'allevamento degli ovini karakul," 1939, Istituto Agronomico per l'Oltremare Archive (IAOA), Fasc. 1972.

54. "Relazione sull'allevamento degli ovini karakul nella Libia Occidentale," 1937, IAOA, Fasc. 529.

55. "Relazione sull'allevamento degli ovini karakul nella Libia Occidentale," 1937.

56. Francesco Maiocco, "L'Allevamento della Pecora Caracul per la Produzione della Pelliccia 'Agnellino di Persia' e sue Possibilitá di Sviluppo nell' Africa Italiana. Necessitá del suo Allevamento al Fini Autarchici dell' Industria Italiana dell' Abbigliamento," paper presented at the VIII Congresso Internazionale di Agricoltura Tropicale e Subtropicale, Tripoli, Mar. 1939.

57. Maiocco, "L'Allevamento della Pecora Caracul," 6–9.

58. *XXV° Anno dalla fondazione dello Istituto Sperimentale Italiano 'L. Spallanzani' per la Fecondazione Artificiale* (Aggregato all'Universitá degli Sudi di Milano, 1962); Telesforo Bonadonna, *La Fecondazione artificiale degli animali domestici. Problema zootecnico e problema*

*igienico-sanitario nazionale e coloniale. Risultati Sperimentali* (Pavia: Federazione Fasci Combattimento, 1937).

59. Telesforo Bonadonna, "Fecondazione strumentale e zootecnia coloniale," *L'Agricoltura Coloniale* 32 (1938): 193–98.

60. G. Schultze, "Allevamento delle pecore Karakul e del bestiame in generale. Rapporto sul mio viaggio di studio attraverso il Territorio dell'Impero," 1938–39, IAOA, Fasc. 553.

61. Jill Dias, "Angola," in *O Império Africano: 1825–1890*, ed. Valentim Alexandre and Jill Dias (Lisbon: Editorial Estampa), 319–556.

62. For a general overview of the relations between Salazar's dictatorship and the colonial undertaking, see Valentim Alexandre, "Ideologia, economia e política: A questão colonial na implementação do Estado Novo," *Análise Social* 28 (1993): 117–36; Luís Reis Torgal, *Estados Novos, Estado Novo. Ensaios de História Política e Cultural* (Coimbra: Imprensa da Universidade de Coimbra, 2009): 1: 467–98; Cláudia Castelo, *Passagens para África. O Povoamento de Angola e Moçambique com Naturais da Metrópole (1920–1974)* (Porto: Afrontamento, 2007).

63. René Pélissier, *História das Campanhas de Angola. Resistência e Revoltas 1845–1941* (Lisbon: Estampa, 1997), 267–75.

64. M. Santos Pereira, *Plano de fomento do Karakul de Angola* (Lisbon: self-published, 1953); M. Santos Pereira, "Situação do caraculo Angolano," *Boletim Geral do Ultramar* 406 (1959): 27–56; Relazione sul PEK (Posto Experimental do Karakul), "Osservazioni del Dott. Mario Garutti,: 1951, IAOA, Fasc. 1027.

65. Realzione sul PEK, n63.

66. Ruy Duarte de Carvalho, *Vou lá visitar pastores: exploração epistolar de um percurso angolano em território Kuvale (1992–1997)* (Lisbon: Cotovia, 2000).

67. Ruy Duarte de Carvalho, *Como se o Mundo não tivesse Leste* (Lisbon: Cotovia, 2008), 25–27.

68. Salvador de Figueiredo, *Angola, o último café* (Torres Vedras, 2006), 150–57.

69. For a general discussion of the role of nonhuman animals in environmental history, see, e.g., Harriet Ritvo, "Beasts in the Jungle (or Wherever)," *Daedalus* 137, no. 2 (2008): 22–30.

70. Kohler, *Lords of the Fly*; Rader, *Making Mice*; Creager, *Life of a Virus*. An important exception is Angela H. Creager, "The Industrialization of Radioisotopes by the U.S. Atomic Energy Commission," in *The Science-Industry Nexus: History, Policy, Implications*, ed. Karl Grandin, Nina Wormbs, and Sven Widmalm (Sagamore Beach, MA: Science History Publications/USA and Nobel Foundation, 2004), 141–68. But although the author delves into the new forms of cooperation among research laboratories, hospitals, state agencies, and industry, the emphasis of the text is above all on the industrial nature of biological research.

Historians of science dealing with standardization procedures in other scientific realms have been more successful in both horizontal and vertical circulation. See, e.g., Simon Schaffer, "Accurate Measurement Is an English Science," in *The Values of Precision*, ed. M. Norton Wise (Princeton: Princeton University Press, 1995), 135–72; M. Norton Wise, "Precision: Agent of Unity and Product of Agreement. Part II—The Age of Steam and Telegraphy" in Wise, *Values of Precision*, 222–36.

A striking contrasting example is Robert E. Kohler's reflections on laboratory history for a recent focus section of the journal *Isis*. In all the suggestions he offers on the ways of following the relevance of laboratories for modern life, he never mentions the dispersion of material laboratory artifacts in society. His stronger case, on the importance of laboratories for modern states, is made by invoking the role of laboratories in mass public education. See Robert E. Kohler, "Lab History: Reflections," *Isis* 99 (2008): 761–68.

71. Christophe Bonneuil develops a similar argument when at the end of an excellent article on vertical circulation of standardized life-forms he asks for more detailed attention to "the transformations of the material practices of observation, recording, book-keeping, processing

and manipulating." See Christophe Bonneuil, "Producing Identity, Industrializing Purity," in *A Cultural History of Heredity IV: Heredity in the Century of the Gene,* ed. Staffan Müller-Wille, Hans-Jörg Rheinberger, and John Dupré (Berlin: Max Planck Institute for the History of Science, 2008), 81–110, esp. 105–6.

72. See, e.g., Susan R. Schrepfer and Philip Scranton, eds., *Industrializing Organisms: Introducing Evolutionary History* (New York: Routledge, 2004). Many of the articles in this groundbreaking volume are very good at exploring the circulation of industrialized organisms while tending to ignore the procedures by which they have become industrialized. The most important exception to such criticism is Deborah Fitzgerald's work on hybrid corn in Illinois. The present text has drawn obvious inspiration from her discussion of "how hybrid corn came to exist, how the agricultural college got involved in commodification, and how the research-and-development interests of agribusiness have changed the way the agricultural college conceives of its mission." But it is nevertheless discouraging that her 1990 lament about the remarkably scant attention given to the historical relationship between biology and agriculture is still apt. See Deborah Fitzgerald, *The Business of Breeding Hybrid Corn in Illinois, 1890–1940* (Ithaca: Cornell University Press, 1990).

## CHAPTER 10. TRADING SPACES: TRANSFERRING ENERGY AND ORGANIZING POWER IN THE NINETEENTH-CENTURY ATLANTIC GRAIN TRADE

1. The growing subfield of "envirotech" exemplifies the possibilities of such scholarship. See in particular Philip Scranton and Susan R. Schrepfer, *Industrializing Organisms: Introducing Evolutionary History* (New York: Routledge, 2004); Timothy LeCain, *Mass Destruction: The Men and Giant Mines That Wired America and Scarred the Planet* (New Brunswick: Rutgers University Press, 2009); Martin Reuss and Stephen H. Cutcliffe, eds., *The Illusory Boundary: Environment and Technology in History* (Charlottesville: University of Virginia Press, 2010); Sara B. Pritchard, *Confluence: The Nature of Technology and the Remaking of the Rhône* (Cambridge: Harvard University Press, 2011). The relationship between energy and power has already been isolated as a place of potential overlap between the history of technology and environmental history. See Edmund Russell et al., "The Nature of Power: Synthesizing the History of Technology and Environmental History," *Technology and Culture* 52, no. 2 (2011): 246–59.

2. This chapter reflects a broader discussion of energy and power within the University of Virginia's Committee for the History of the Environment, Science and Technology (CHEST). In particular, the insights of Jaime Allison, Jack Brown, Bernie Carlson, Bart Elmore, Laura Kolar, and Ed Russell have shaped the argument. Many of its arguments can be found in further detail in the author's forthcoming doctoral dissertation, "Harvesting Power: American Agriculture and British Industry, 1776–1918" (University of Virginia).

3. Michel Callon, "Some Elements of a Sociology of Translation: Domestication of the Scallops and the Fishermen of St. Brieuc Bay," *Sociological Review Monograph* 32 (1986): 196–233; John Law, "Technology and Heterogeneous Engineering: The Case of Portuguese Expansion," in *The Social Construction of Technological Systems: New Directions in the Sociology and History of Technology* (Cambridge: MIT Press, 1987); Bruno Latour, *Science in Action: How to Follow Scientists and Engineers through Society* (Cambridge: Harvard University Press, 1987).

4. Thomas P. Hughes, *Networks of Power: Electrification in Western Society, 1880–1930* (Baltimore: Johns Hopkins University Press, 1983); Thomas Hughes, *Rescuing Prometheus,* 1st ed. (New York: Pantheon Books, 1998); Thomas Hughes, *Human-Built World: How to Think about Technology and Culture* (Chicago: University of Chicago Press, 2004).

5. Hughes, *Networks of Power,* 15–16.

6. The subfield of anthropology termed *energetics* has long recognized that there is a correlation between social structure and energy use. See L. A. White, *The Evolution of Culture* (New York: McGraw-Hill, 1959); L. A. White, "Energy and the Evolution of Culture," in *An-*

*thropology in Theory: Issues in Epistemology* (Malden, MA: Blackwell Publishing, 2006); R. N. Adams, *Energy and Structure: A Theory of Social Power* (Austin: University of Texas Press, 1975); R. N. Adams, *Paradoxical Harvest: Energy and Explanation in British History, 1870–1914* (Cambridge: Cambridge University Press, 1982).

7. This is a generally agreed-upon definition of power in sociology. Most sociologists note that power is socially constructed. While there can be many manifestations, power necessitates some control (potential or actual) over energy and/or resources. See Marvin Olsen, *Power in Societies* (New York: Macmillan, 1970).

8. The Second Law of Thermodynamics describes the process by which closed systems break down over time due to entropy, the loss of heat during energy conversion. *Entropy* is the term that describes the tendency of closed systems to move from higher to lower states of organization. See H. T. Odum, *Environment, Power, and Society for the Twenty-First Century: The Hierarchy of Energy* (New York: Columbia University Press, 2007), 31–35.

9. Brian Mitchell and Phyllis Deane, *Abstract of British Historical Statistics* (Cambridge: Cambridge University Press, 1962), 298–300.

10. While this system was not controlled by one organizational entity, as in the case of Hughesian systems, the practices of futures trading and grading, the power of the railroads, and the widespread acceptance of free-trade doctrines nonetheless produce an internal coherence that was not present a generation earlier within the grain trade.

11. Michel Callon describes this process as translation, which is "never a completed project, and it may fail." See Callon, "Some Elements of a Sociology of Translation," 1.

12. For generalized discussions of the Corn Laws and their importance, see P. A. Pickering and A. Tyrrell, *The People's Bread: A History of the Anti-Corn Law League* (London: Leicester University Press, 2000); Susan Fairlie, "The Corn Laws and British Wheat Production, 1829–76," *Economic History Review* 22, no. 1, new series (Apr. 1, 1969): 88–116, http://www.jstor.org/stable/2591948; N. McCord, *The Anti-Corn Law League, 1838–1846*, vol. 2 (London: Unwin University Books, 1968); Anna Gambles, "Rethinking the Politics of Protection: Conservatism and the Corn Laws, 1830–52," *English Historical Review* 113, no. 453 (Sept. 1998): 928–52, http://www.jstor.org/stable/578662; J. R. Wordie, "Perceptions and Reality: The Effects of the Corn Laws and Their Repeal in England, 1815–1906," in *Agriculture and Politics in England, 1815–1839* (London: Macmillan Press, 2000); T. R. Malthus, *Observations of the Effects of the Corn Laws, and of a Rise or Fall in the Price of Corn on the Agriculture and General Wealth of the Country* (St. Paul's Church-Yard: J. Johnson and Co., 1814); C. Schonhardt-Bailey, *From the Corn Laws to Free Trade: Interests, Ideas, and Institutions in Historical Perspective* (Cambridge: MIT Press, 2006); C. Schonhardt-Bailey, *Free Trade: The Repeal of the Corn Laws* (Bristol: Thoemmes Press, 1996).

13. Rudolph Peterson, *Wheat: Botany, Cultivation and Utilization* (London: L. Hill Books, 1965), 288–94.

14. Robert Albion, *The Rise of New York Port, 1815–1860* (New York: Scribner, 1970), 38–95; J. G. Clark, *The Grain Trade in the Old Northwest* (Westport, CT: Greenwood Press, 1966).

15. William Rathbone VI to Rathbone Brothers and Company, Dec. 9, 1848, Rathbone Business Papers, XXIV 2.4, University of Liverpool Special Collections.

16. Busk and Jevons to Rathbone Brothers and Company, Sept. 5, 1868, Rathbone Business Papers, XXIV 2.24.

17. William Rathbone VI to William Rathbone V, Oct. 28–31, 1848, Rathbone Business Papers, XXIV 2.4.

18. R. W. Hidy, *The House of Baring in American Trade and Finance: English Merchant Bankers at Work, 1763–1861* (Cambridge: Harvard University Press, 1949), 82–84.

19. Eustace Greg to Rathbone Brothers and Company, Feb. 15, 1861, Rathbone Business Papers, XXIV 2.17.

20. Morton Rothstein, "Antebellum Wheat and Cotton Exports: A Contrast in Marketing Organization and Economic Development," *Agricultural History* 40, no. 2 (1966): 91.

21. B. Mitchell and Deane, *British Historical Statistics*, 97–98.

22. Percy Bidwell and John I. Falconer, *History of Agriculture in the Northern United States, 1620–1860* (Washington, DC: Carnegie Institution of Washington, 1925), 196–203, 259–65, 321–38.

23. See Brian Donahue, *The Great Meadow: Farmers and the Land in Colonial Concord* (New Haven: Yale University Press, 2004).

24. Henry Tudor, *Narrative of a Tour in North America . . .* , 2 vols. (London: James Duncan, 1834), 231; William Cronon, *Changes in the Land: Indians, Colonists, and the Ecology of New England* (New York: Hill and Wang, 1983), 116–17.

25. U. P. Hedrick, *A History of Agriculture in the State of New York* (New York: Hill and Wang, 1933), 338.

26. William Rathbone VI to William Rathbone V and James Powell, June 2, 1842, Rathbone Business Papers, XXIV 2.3.

27. Henry Gair to Rathbone Brothers and Company, Apr. 2, 1855, Rathbone Business Papers, XXIV 2.12.

28. Eleanor Rathbone, *William Rathbone: A Memoir*, 1st ed. (London and New York: Macmillan, 1905), 145.

29. See Hughes, *Networks of Power*, 14–15.

30. J. R. Busk to Rathbone Brothers and Company, Jan. 16, 1868, Rathbone Business Papers, XXIV 2.27.

31. Busk to Rathbone Brothers and Company, Jan. 16, 1868.

32. Henry Gair to Rathbone Brothers and Company, Dec. 4, 1855, and Henry Gair to Rathbone Brothers and Company, Aug. 21, 1855, both Rathbone Business Papers, XXIV 2.12.

33. Sheila Marriner, "Rathbones' Trading Activities in the Middle of the Nineteenth Century," *Transactions of the Historic Society of Lancashire and Cheshire*, no. 108 (1957): 119. By the late 1860s, Rathbone Brothers employed agents in New York City to keep them abreast of potential investments and to advise them on the reliability of certain merchants.

34. Marriner, "Rathbones' Trading Activities," 119.

35. Busk and Jevons to Rathbone Brothers and Company, Oct. 4, 1873, Rathbone Business Papers, XXIV 2.32.

36. See Walter Prescott Webb, *The Great Plains* ([Boston]: Ginn and Co., 1931), 366–74; Morton Rothstein, "America in the International Rivalry for the British Wheat Market, 1860–1914," *Mississippi Valley Historical Review* 47, no. 3 (1960): 401; Allan Bogue, *From Prairie to Corn Belt: Farming on the Illinois and Iowa Prairies in the Nineteenth Century* (Chicago: University of Chicago Press, 1963); Hiram Drache, *The Challenge of the Prairie: Life and Times of Red River Pioneers* (Fargo: North Dakota Institute for Regional Studies, 1970); William Cronon, *Nature's Metropolis: Chicago and the Great West*, 1st ed. (New York: W. W. Norton, 1991), 55–147; E. West, *The Contested Plains: Indians, Goldseekers, and the Rush to Colorado* (Lawrence: University of Kansas Press, 1998), 237–70; Hiram M. Drache, *The Day of the Bonanza: A History of Bonanza Farming in the Red River Valley of the North* (Fargo: North Dakota Institute for Regional Studies, 1964); Theodore Steinberg, *Down to Earth: Nature's Role in American History* (New York: Oxford University Press, 2002), 116–37; Sterling Evans, *Bound in Twine: The History and Ecology of the Henequen-Wheat Complex for Mexico and the American and Canadian Plains, 1880–1950*, 1st ed. (College Station: Texas A&M University Press, 2007).

37. J. W. Lee, *History of Hamilton County, Iowa* (Chicago: S. J. Clarke Publishing Co., 1912), 135–43, 246.

38. Benjamin Horace Hibbard, *The History of Agriculture in Dane County, Wisconsin* (Madison: University of Wisconsin, 1904), 126–27, http://search.lib.virginia.edu/catalog/u901951.

39. Drache, *Day of the Bonanza.*

40. Samuel G. Rathbone to Rathbone Brothers and Company, June 23, 1884, Rathbone Business Papers, XXIV 1.28.

41. This is the same shift that American agrarian movements reacted to following the American Civil War. It revolved around price structures that favored large, long-distance shipments over smaller, shorter-distance freight. See H. M. Larson, *The Wheat Market and the Farmer in Minnesota* (New York: Longmans, Green and Co., 1926); Henrietta M. Larson, *Jay Cooke, Private Banker,* Harvard Studies in Business History II (Cambridge: Harvard University Press, 1936); Joseph Nimmo, *Railroad Federations and the Relation of the Railroads to Commerce* (Washington, DC: Government Printing Office, 1885); D. North, "Ocean Freights and Economic Development 1750–1913," *Journal of Economic History* 18, no. 4 (Dec. 1958): 537–55.

42. Cronon, *Nature's Metropolis,* 55–147; Alfred D. Chandler, *The Visible Hand: The Managerial Revolution in American Business* (Cambridge: Belknap Press, 1977). Richard White has recently complicated the notion that railroads were the harbingers of ordered modernity. See Richard White, *Railroaded: The Transcontinentals and the Making of Modern America,* 1st ed. (New York: W. W. Norton and Co., 2011).

43. An Imperial Quarter is roughly equivalent to eight bushels. See B. Mitchell and Deane, *British Historical Statistics,* 5, 95.

44. The standard measurement of weight for grain became the cental in 1858; see B. Mitchell and Deane, *British Historical Statistics,* 9, 99.

45. A cental is equivalent to four Imperial Quarters.

46. "Report of the Committee on Canals of New York State, 1900," New York State Assembly Document no. 79 (Albany: James B. Lyon, 1900), 220.

47. B. Mitchell and Deane, *British Historical Statistics,* 98–102.

48. These changes were also part and parcel of a dramatic change in the British diet. From 1800 to 1900, the average British diet became more reliant on wheat and less reliant on other grains such as rye and barley. See D. J. Oddy, "Food in Nineteenth Century England: Nutrition in the First Urban Society," *Proceedings of the Nutrition Society* 29, no. 1 (1970): 150–57; D. J. Oddy, "Working-Class Diets in Late Nineteenth-Century Britain," *Economic History Review* 23, no. 2 (Aug. 1970): 314–23; C. Petersen, *Bread and the British Economy, 1770–1870* (Hants, UK: Scolar Press, 1995).

49. Two environmental histories that come close to achieving an integrated assessment of energy and power are Richard White, *The Organic Machine* (New York: Hill and Wang, 1995); and West, *Contested Plains.*

## CHAPTER 11. SITUATED YET MOBILE: EXAMINING THE ENVIRONMENTAL HISTORY OF ARCTIC ECOLOGICAL SCIENCE

1. Trevor Levere, *Science and the Canadian Arctic: A Century of Exploration, 1818–1918* (Cambridge: Cambridge University Press, 1993).

2. Sherrill Grace, *Canada and the Idea of North* (Montreal and Kingston: McGill-Queen's University Press, 2002).

3. Diarmid Finnegan, "The Spatial Turn: Geographical Approaches in the History of Science," *Journal of the History of Biology* 41 (2008): 369–88; Christopher Henke and Thomas Gieryn, "Sites of Scientific Practice: The Enduring Importance of Place," in *The Handbook of Science and Technology Studies,* ed. Edward J. Hackett, Olga Amsterdamska, Michael Lynch, and Judy Wajcman, 3rd ed. (Chicago: MIT Press, 2008), 353–76.

4. Steven Shapin, "Placing the View from Nowhere: Historical and Sociological Problems in the Location of Science," *Transactions of the Institute of British Geographers* 23 (1998): 5–12.

5. Robert Kohler, *Landscapes and Labscapes: Exploring the Lab-Field Border in Biology* (Chicago: University of Chicago Press, 2002).

6. Peder Anker, *Imperial Ecology: Environmental Order in the British Empire, 1895–1945* (Cambridge: Harvard University Press, 2001); Stephen Bocking, *Ecologists and Environmental Politics: A History of Contemporary Ecology* (New Haven: Yale University Press, 1997).

7. Denis Cosgrove, *Apollo's Eye: A Cartographic Genealogy of the Earth in the Western Imagination* (Baltimore: Johns Hopkins University Press, 2001); Gregg Mitman, "When Nature Is the Zoo: Vision and Power in the Art and Science of Natural History," *Osiris* 11 (1996): 117–43.

8. Christopher Henke, "Making a Place for Science: The Field Trial," *Social Studies of Science* 30, no. 4 (2000): 483–511.

9. Sheila Jasanoff and Marybeth Martello, eds., *Earthly Politics: Local and Global in Environmental Governance* (Cambridge: MIT Press, 2004); James Scott, *Seeing like a State: How Certain Schemes to Improve the Human Condition Have Failed* (New Haven: Yale University Press, 1998).

10. Jasanoff and Martello, *Earthly Politics.*

11. Trevor Barnes, "Placing Ideas: Genius Loci, Heterotopia and Geography's Quantitative Revolution," *Progress in Human Geography* 28, no. 5 (2004): 565–95; David Livingstone, *Putting Science in Its Place: Geographies of Scientific Knowledge* (Chicago: University of Chicago Press, 2003); Simon Naylor, "Introduction: Historical Geographies of Science—Places, Contexts, Cartographies," *British Journal for the History of Science* 38, no. 1 (2005): 1–12.

12. Matthew Evenden, *Fish versus Power: An Environmental History of the Fraser River* (Cambridge: Cambridge University Press, 2004); Christine Keiner, *The Oyster Question: Scientists, Watermen, and the Maryland Chesapeake Bay since 1880* (Athens: University of Georgia Press, 2009); Harold Platt, *Shock Cities: The Environmental Transformation and Reform of Manchester and Chicago* (Chicago: University of Chicago Press, 2005).

13. Gregg Mitman, *Breathing Space: How Allergies Shape Our Lives and Landscapes* (New Haven: Yale University Press, 2007); Linda Nash, *Inescapable Ecologies: A History of Environment, Disease, and Knowledge* (Berkeley: University of California Press, 2007).

14. Warwick Anderson and Vincanne Adams, "Pramoedya's Chickens: Postcolonial Studies of Technoscience," in Hackett et al., *Handbook of Science and Technology Studies,* 181–204; Kapil Raj, "Introduction: Circulation and Locality in Early Modern Science," *British Journal for the History of Science* 43, no. 4 (2010): 513–17.

15. Anker, *Imperial Ecology;* William Beinart and Lotte Hughes, *Environment and Empire* (Oxford: Oxford University Press, 2007); Richard Grove, *Green Imperialism: Colonial Expansion, Tropical Island Edens, and the Origins of Environmentalism, 1600–1860* (Cambridge: Cambridge University Press, 1996); S. Ravi Rajan, *Modernizing Nature: Forestry and Imperial Eco-Development 1800–1950* (Oxford: Clarendon Press, 2006).

16. Anderson and Adams, "Pramoedya's Chickens."

17. Christopher Sellers, *Hazards of the Job: From Industrial Disease to Environmental Health Science* (Chapel Hill: University of North Carolina Press, 1999).

18. Matthew Farish, "Frontier Engineering: From the Globe to the Body in the Cold War Arctic," *Canadian Geographer* 50, no. 2 (2006): 177–96; Edward Jones-Imhotep, "Communicating the North: Scientific Practice and Canadian Postwar Identity," *Osiris* 24, 2nd series (2009): 144–64.

19. Magda Havas and Tom Hutchinson, "The Smoking Hills: Natural Acidification of an Aquatic Ecosystem," *Nature* 301 (1983): 23–27.

20. Harold Welch, "Metabolic Rates of Arctic Lakes," *Limnology and Oceanography* 19 (1974): 65–73.

21. Kohler, *Landscapes and Labscapes.*

22. James Woodford, *The Violated Vision: The Rape of Canada's North* (Toronto: McClelland and Stewart, 1972), 44.

23. Max Dunbar, *Ecological Development in Polar Regions: A Study in Evolution* (Englewood Cliffs: Prentice-Hall, 1968), 1, 5.

24. Suzanne Zeller, "Classical Codes: Biogeographical Assessments of Environment in Victorian Canada," *Journal of Historical Geography* 24, no. 1 (1998): 20–35; Suzanne Zeller, "Humboldt and the Habitability of Canada's Great Northwest," *Geographical Review* 96, no. 3 (2006): 382–98.

25. Jessica Shadian and Monica Tennberg, eds., *Legacies and Change in Polar Sciences: Historical, Legal and Political Reflections on the International Polar Year* (Farnham: Ashgate, 2009).

26. David Downie and Terry Fenge, *Northern Lights against POPs: Combatting Toxic Threats in the Arctic* (Montreal and Kingston: McGill-Queen's University Press, 2003); Farish, "Frontier Engineering."

27. Stephen Bocking, "Science and Spaces in the Northern Environment," *Environmental History* 12 (2007): 868–95.

28. Stephen Bocking, "A Disciplined Geography: Aviation, Science, and the Cold War in Northern Canada, 1945–1960," *Technology and Culture* 50 (2009): 320–45.

29. Nicholas Polunin, "The Real Arctic: Suggestions for Its Delimitation, Subdivision and Characterization," *Journal of Ecology* 39, no. 2 (1951): 308–15; J. A. Downes, "What Is an Arctic Insect?" *Canadian Entomologist* 94 (1962): 143–62; Louis-Edmond Hamelin, *Canadian Nordicity: It's Your North, Too* (Montreal: Harvest House, 1979).

30. William Fuller and Peter Kevan, eds., "Productivity and Conservation in Northern Circumpolar Lands," IUCN Publications, new series, no. 16 (1970).

31. Bocking, "Science and Spaces"; Dennis Chitty, *Do Lemmings Commit Suicide? Beautiful Hypotheses and Ugly Facts* (New York: Oxford University Press, 1996); Charles Elton, *Voles, Mice and Lemmings: Problems in Population Dynamics* (Oxford: Clarendon Press, 1942).

32. Charles Krebs, "The Lemming Cycle at Baker Lake, Northwest Territories, during 1959–63," Arctic Institute of North America Technical Paper No. 13 (1964); Frank Pitelka, "Ecological Studies on the Alaskan Arctic Slope," *Arctic* 22, no. 3 (1969): 333–40.

33. Krebs, "Lemming Cycle at Baker Lake," 51.

34. Stephen Pyne, *Awful Splendour: A Fire History of Canada* (Vancouver: University of British Columbia Press, 2008).

35. George Scotter, "Effects of Forest Fires on the Winter Range of Barren-Ground Caribou in Northern Saskatchewan," Canadian Wildlife Service, *Wildlife Management Bulletin*, no. 18, 1st series 1 (1964): 1–111.

36. Edward Johnson and J. Stan Rowe, "Fire and Vegetation Change in the Western Subarctic," ALUR Report 1975–76–61 (Ottawa, 1977), 18.

37. Ross Wein, "Recovery of Vegetation in Arctic Regions after Burning," Environmental-Social Program, Northern Pipelines, Rept. No. 74–6 (Ottawa, 1974); K. A. Kershaw and W. R. Rouse, "The Impact of Fire on Forest and Tundra Ecosystems, Final Report 1975," ALUR Report 1975–76–63 (Ottawa, 1976).

38. Edward Johnson and J. Stan Rowe, "Fire in the Subarctic Wintering Ground of the Beverley Caribou Herd," *American Midland Naturalist* 94 (1975): 1–14.

39. Henry Lewis, "Maskuta: The Ecology of Indian Fires in Northern Alberta," *Western Canadian Journal of Anthropology* 7, no. 1 (1977): 15–52.

40. William Pruitt, *Boreal Ecology* (Edward Arnold, 1978); William Pruitt, "Life in the Snow," *Nature Canada* 4, no. 4 (1975): 40–44; William Pruitt, "Snow and Living Things," in *Northern Ecology and Resource Management: Memorial Essays Honouring Don Gill*, ed. R. Olson, R. Hastings and F. Geddes (Edmonton: University of Alberta Press, 1984), 51–77.

41. A. N. Formozov, "Snow Cover as an Integral Factor of the Environment and Its Importance in the Ecology of Mammals and Birds," Boreal Institute for Northern Studies, Occas.

Publ. No. 1 (1964) (originally published in Russian: Moscow, 1946). Ties between Soviet and Canadian boreal ecology culminated in a bilingual (English and Russian) book coauthored by Pruitt: William Pruitt and L. M. Baskin, *Boreal Forest of Canada and Russia* (Sofia: Pensoft Publishers, 2004).

42. Pruitt, *Boreal Ecology*.

## CHAPTER 12. WHITE MOUNTAIN APACHE BOUNDARY-WORK AS AN INSTRUMENT OF ECOPOLITICAL LIBERATION AND LANDSCAPE CHANGE

1. D'Arcy McNickle, "Address," in *The Indian and State Government, 6th Annual Conference on Indian Affairs* (Vermillion: Institute of Indian Studies, University of South Dakota, 1960), 91–100, quote on 98.

2. Stephen Cornell, Miriam Jorgensen, Joseph Kalt, and Katherine Spilde, *Seizing the Future: Why Some Native Nations Do and Others Don't* (Tucson: Native Nations Institute, 2005).

3. David Rich Lewis, "American Indian Environmental Relations," in *A Companion to American Environmental History*, ed. Douglass Sackman (Hoboken: Wiley-Blackwell, 2010).

4. Frederick Hoxie, *The Campaign to Assimilate the Indians* (Lincoln: University of Nebraska Press, 1984); Vine Deloria Jr., *Custer Died for Your Sins* (Norman: University of Oklahoma Press, 1970).

5. In 1870 Fort Apache was much bigger and was called the White Mountain Reservation. In 1897 this larger reservation was divided into the San Carlos and Fort Apache reservations.

6. Thomas Gieryn, *Cultural Boundaries of Science: Credibility on the Line* (Chicago: University of Chicago Press, 1999).

7. Adele Clarke, *Disciplining Reproduction* (Berkeley: University of California Press, 1998).

8. Steven Epstein, *Impure Science* (Berkeley: University of California Press, 1996); Barbara Allen, *Uneasy Alchemy* (Cambridge: MIT Press, 2003); Reid Helford, "Rediscovering the Presettlement Landscape,'" *Science, Technology, and Human Values* 24 (1999): 55–79.

9. Kathryn Mutz, Gary Bryner, and Douglas Kenney, eds., *Justice and Natural Resources* (Washington, DC: Island Press, 2002); Richmond Clow and Imre Sutton, eds., *Trusteeship in Change* (Boulder: University of Colorado Press, 2001).

10. On cultural appropriation, see Ron Eglash, "Appropriating Technology: An Introduction," in *Appropriating Technology: Vernacular Science and Social Power*, ed. Ron Eglash et al. (Minneapolis: University of Minnesota Press, 2004), vii–xxi.

11. On creating new knowledge systems, see David Turnbull, *Masons, Tricksters and Cartographers* (New York: Routledge, 2000).

12. Kenneth Philp, *John Collier's Crusade for Indian Reform, 1920–1954* (Tucson: University of Arizona Press, 1977).

13. William Donner, "Fort Apache Enrollee Training Program, 1940–1941," United States National Archives, Record Group 75, Fort Apache Indian Reservation (hereafter RG75 FAIR).

14. For more on the Indian New Deal, see David Tomblin, *Managing Boundaries, Healing the Homeland: Ecological Restoration and the Revitalization of the White Mountain Apache* (Boulder: University of Colorado Press, forthcoming), chaps. 2–4.

15. Resolution of the White Mountain Apache Tribe, No. 53–50, RG75 FAIR.

16. "Plan of Operations for the White Mountain Recreational Enterprise," RG75 FAIR.

17. F. M. Haverland, "Memorandum for the Records," Dec. 13, 1955, RG75 FAIR.

18. Clinton Kessay to Glenn Emmons, Jan. 8, 1957, RG75 FAIR.

19. WMAT Council Resolution No. 57–31, Apr. 15, 1957, RG75 FAIR.

20. F. M. Haverland to Commissioner, May 15, 1958, RG75 FAIR.

21. Richard Dunlop, "Apache Camp-Out," *Saturday Evening Post*, n.d., RG75 FAIR.

22. L. D. Arnold to Ralph Gelvin, Mar. 12, 1953, RG75 FAIR.

23. R. D. Holtz to John Crow, Mar. 21, 1952, RG75 FAIR.

24. Resolution of the WMAT, No. 52–9, RG75 FAIR.

25. Arizona Commission of Indian Affairs, *Control of Natural Resources: Report of the 7th Annual Indian Town Hall* (Phoenix: ACIA, 1980), 33–36.

26. Arizona Commission of Indian Affairs, *Control of Natural Resources*, 33–36.

27. The reason I characterize Apache knowledge during this time as "publicly vague" is because due to the nature of my research methods, which were largely archival, the extent to which Apache knowledge is articulated in the nonpublic sphere (e.g., Apache management agencies) is not clear. More detailed ethnographic work should be employed to construct a better picture of the evolution of Apache local knowledge during this period.

28. "Elk Survey Provides Game Management Information," *Fort Apache Scout*, June 10, 1983.

29. "Elk Survey Provides Game Management Information."

30. "Cow Elk Moved as Part of Game Management," *Fort Apache Scout*, June 16, 1989.

31. "Controlled Burning Reduces Fire Hazard, Increases Game Forage," *Fort Apache Scout*, Nov. 4, 1988.

32. Peter Friederici, ed., *Ecological Restoration of Southwestern Ponderosa Pine Forest* (Washington, DC: Island Press, 2003).

33. Game and Fish Department, "Wildlife and Fire," *Fort Apache Scout*, June 1, 1990.

34. "Council Approves Tribal Member Hunting Fees, Cow Elk Hunt," *Fort Apache Scout*, June 30, 1989.

35. Punt Cooley, "The Many Faces of Maverick," *Fort Apache Scout*, Mar. 22, 2002.

36. Bennett Cosay, "While the Rest of Us Hustle and Bustle through Our Daily Lives, Somewhere Up in the Mountains a Small Group of People are Having the Time of Their Lives," *Fort Apache Scout*, Oct. 14, 1994.

37. "Trophy Elk Hunt," *Fort Apache Scout*, Oct. 20, 1989.

38. John Reiger, *American Sportsmen and the Origins of Conservation* (Norman: University of Oklahoma Press, 1985); Louis S. Warren, *The Hunter's Game* (New Haven: Yale University Press, 1997).

39. *White Mountain Apache Tribe v. the United States*, 11 Cl. Ct. 614 (1987); "White Mountain Apache Land Restoration Code," http://thorpe.ou.edu/codes/wmtnapache/Land RestorationCode.html (accessed Jan. 3, 2013).

40. On program damages, see *White Mountain Apache Tribe v. the United States*; Jonathan Long, "Cibecue Watershed Projects," in *Land Stewardship in the 21st Century*, coord. P. Ffolliot et al. (Fort Collins: USDA, Forest Service, Rocky Mountain Research Station, 2000), 227–33.

41. Keith Basso, *Wisdom Sits in Places* (Tucson: University of Arizona Press, 1996).

42. Greenville Goodwin, *Social Organization of the Western Apache* (Tucson: University of Arizona Press), 152, 574.

43. Jonathan Long and Candy Lupe, "A Process for Planning and Evaluating Success of Riparian-Wetland Projects on the Fort Apache Indian Reservation," *Hydrology and Water Resources in Arizona and the Southwest* 28 (1998): 68–74; Jonathan Long, Aregai Tecle, and Benrita Burnette, "Marsh Development at Restoration Sites on the White Mountain Apache Reservation, Arizona," *Journal of the American Water Resources Association* (2004): 1345–59.

44. Long, "Cibecue Watershed Projects."

45. Ronnie Lupe, "Comments of the Chairman of the White Mountain Apache Tribal Council," in *Indian Water: 1997 Trends and Directions in Federal Water Policy*, ed. Ted Olinger (Boulder: Report to the Western Water Policy Review Advisory Commission, 1997), 38–44.

46. Jonathan Long, Delbin Endfield, Candy Lupe, and Mae Burnette, "Battle at the Bridge: Using Participatory Approaches to Develop Community Researchers in Ecological Management," in *Proceedings of the Fifth Biennial Conference on University Education in Natural Resources*, ed. Thomas Kolb (Logan: Quinney Library, Utah State University, 2004), 29–44, quote on 33–34.

47. B. Mae Burnette, "Studying the Rivers: A Great Adventure," *Fort Apache Scout*, Nov. 22, 1996.

48. For information on cultural heritage programs, see John Welch, "Reconstructing an Ndee Sense of Place," in *The Archaeology of Meaningful Places*, ed. Brenda Bowser and Maria Nieves Zedeno (Salt Lake City: University of Utah Press, 2009), 149–62.

49. Arizona Commission of Indian Affairs, *Report on Proceedings of the 5th Annual Indian Town Hall: Tribal Water Rights, Today's Concern* (Phoenix: ACIA, 1978), 1–4; "Recreation Enterprises Legal Battle Continues," *Fort Apache Scout*, Mar. 7, 1980; Ronnie Lupe, "Chairman's Corner: Managed Tourism Essential to Tribe's Economic Development," *Fort Apache Scout*, Aug. 15, 1997.

50. US Commission on Civil Rights, *A Quiet Crisis: Federal Funding and Unmet Needs in Indian Country* (Washington, DC: Government Printing Office, 2003).

51. John Welch, "The Rodeo-Chediski Fire and Cultural Resources," *Arizona Archaeological Council Newsletter* 26 (2002): 1–3; Mary Stuever, "BAER Fairs Showcase Burn Restoration," *Fort Apache Scout*, Feb. 18, 2005; Gail Pechuli, "Wildfire Disaster Shakes Apache's Economy," *Indian Country Today*, Jan. 29, 2003.

52. "Council Approves Tribal Member Hunting Fees, Cow Elk Hunt," *Fort Apache Scout*, June 30, 1989.

53. Sandra Harding, *Is Science Multicultural?* (Bloomington: Indiana University Press, 1998).

54. On creating knowledge assemblages, see Turnbull, *Masons, Tricksters and Cartographers*.

55. On the interpretive flexibility of technology, see Andrew Feenberg, *Questioning Technology* (New York: Routledge, 1999).

## CHAPTER 13. NEOECOLOGY: THE SOLAR SYSTEM'S EMERGING ENVIRONMENTAL HISTORY AND POLITICS

1. Peter Redfield, *Space in the Tropics: From Convicts to Rockets in French Guiana* (Berkeley: University of California Press, 2000).

2. See Sheila Jasanoff, "Heaven and Earth: The Politics of Environmental Images," in *Earthly Politics: Local and Global in Environmental Governance*, ed. Sheila Jasanoff and Marybeth Long Martello (Cambridge: MIT Press, 2004), 31–54. See also Myanna Lahsen, "Transnational Locals: Brazilian Experiences of the Climate Regime," in Jasanoff and Martello, *Earthly Politics*, 151–72.

3. Tim Ingold, "Globes and Spheres: The Topology of Environmentalism," in *Environmentalism: The View from Anthropology*, ed. Kay Milton (London: Routledge, 1993), 31–42.

4. Stephen Pyne, "Forum: Extreme Environments," *Environmental History* 15 (2010): 510.

5. Ingold, "Globes and Spheres," 1.

6. Susan L. Star and James R. Griesemer, "Institutional Ecology, 'Translations,' and Boundary Objects: Amateurs and Professionals in Berkeley's Museum of Vertebrate Zoology," *Social Studies of Science* 19, no. 4 (1989): 387–420.

7. The founding work on this topic is by Thomas F. Gieryn, "Boundary-Work and the Demarcation of Science from Non-Science: Strains and Interests in Professional Ideologies of Scientists," *American Sociological Review* 48 (1983): 781–95.

8. In a review article and introduction to the first part of a special issue of *Revue d'Anthropologie des Connaissances* revisiting the "boundary object" concept as a way to understand shared objects within environmental management issues, Trompette calls attention to the concept's genesis as a response to the delineation of "ecologies" more broadly thought of as "worlds" than institutions. However, Trompette and the contributors involved in this special issue still focus on the ecological properties of objects at work, and not on the constitution of

ecologies qua ecologies. See Pascale Trompette, "Revisiting the Notion of Boundary Object," trans. Dominick Vinck, *Revue d'anthropologie des connaissances* 3 (2009): 3–25.

9. Bruno Latour, "Why Has Critique Run Out of Steam? From Matters of Fact to Matters of Concern," *Critical Inquiry B* 30 (2004): 225–48.

10. Jasanoff and Martello, *Earthly Politics*.

11. Hans-Jörg Rheinberger, *Toward a History of Epistemic Things: Synthesizing Proteins in the Test Tube* (Stanford: Stanford University Press, 1997).

12. Patricia L. Barnes-Svarney, *Asteroid: Earth Destroyer or New Frontier?* (New York: Plenum Press, 1996).

13. Steven J. Edberg and David H. Levy, *Observing Comets, Asteroids, Meteors, and the Zodiacal Light*, Practical Astronomy Handbook Series 5 (Cambridge and New York: Cambridge University Press, 1994).

14. Lisa R. Messeri, "The Problem with Pluto: Conflicting Cosmologies and the Classification of Planets," *Social Studies of Science* 40 (2010): 187–214.

15. Peter Galison, *Image and Logic: A Material Culture of Microphysics* (Chicago: University of Chicago Press, 1997).

16. Luis W. Alvarez, F. Asaro, and H. V. Michel, "Extraterrestrial Cause for the Cretaceous-Tertiary Extinction," *Science* 208 (1981): 1095–108.

17. Dandridge Cole, "$50,000,000,000,000 from the Asteroids," *Space World* 4 (1963): 1–8.

18. John S. Lewis, *Mining the Sky: Untold Riches from the Asteroids, Comets, and Planets* (Reading: Addison-Wesley, 1996).

19. See Joseph Masco, *The Nuclear Borderlands: The Manhattan Project in Post–Cold War New Mexico* (Princeton: Princeton University Press, 2006), for a discussion of nuclear bombs as objects that remade modern cosmology and rescaled imaginaries of environmental security regimes.

20. See inter alia Alan J. Stern, "Myriad Planets in Our Solar System and Copernicus Smiled," *Space Daily*, Sept 11, 2002, http://www.spacedaily.com/reports/Myriad_Planets_In _Our_Solar_System_And_Copernicus_Smiled_999.html.

21. Donald K. Yeomans, "Killer Rocks and the Celestial Police: The Search for Near-Earth Asteroids," *Planetary Report* 11 (1991): 4–7.

22. Dave Morrison, developer of the Ames Research Center's "Asteroid and Comet Impact Hazards" project (http://impact.arc.nasa.gov/index.cfm), reports this story in a "News Archive" section entitled "Edward Teller (1908–2003) and Defense against Asteroids" and indicates that these notes on "Teller at Los Alamos, January 1992," are "adapted from an unpublished book manuscript."

23. See Felicity Mellor, "Colliding Worlds: Asteroid Research and the Legitimization of War in Space," *Social Studies of Science* 37 (2010): 499–531. Mellor discusses the ways that asteroid research and military cooperation legitimate the idea of war in space and the development of space-based weapons and platforms for testing Strategic Defense Initiative systems otherwise opposed politically. Certainly, these proposals bring space scientist and military groups together and perpetuate "the politics of fear" at the heart of US defense policy. However, as I argue further on, there is also scientific and political pushback against the utility of such weapons. Proposals to develop nonweaponized NEO mitigation strategies, and to secure international collaboration to do so, provide counternarratives that also call for greater understanding of technology impacts on the environment.

24. Peter Garretson and Douglias Kaupa, "Planetary Defense: Potential Department of Defense Mitigation Roles," *Air and Space Power Journal* 22, no. 3 (2008); http://www.airpower .au.af.mil/airchronicles/apj/apj08/fal08/garretson.html (accessed Dec. 2009).

25. David Morrison, "The Spaceguard Survey: Report of the NASA International Near-

Earth-Object Detection Workshop," NASA International Near-Earth-Object Detection Workshop (Colorado, 1992), http://impact.arc.nasa.gov/downloads/spacesurvey.pdf.

26. This phrase is used in the 2010 US space policy to include both natural and "manmade" threats: http://www.whitehouse.gov/sites/default/files/national_space_policy_6-28-10.pdf.

27. Arthur C. Clarke, *Rendezvous with Rama* (London: Pan Books, 1974).

28. David Morrison, "The Impact Hazard: Advanced NEO Surveys and Societal Responses," in *Comet/Asteroid Impacts and Human Society: An Interdisciplinary Approach,* ed. Peter T. Bobrowsky and Hans Rickman (New York, Berlin: Springer, 2007), 163–73.

29. Morrison, "Impact Hazard."

30. Bobrowsky and Rickman, *Comet/Asteroid Impacts and Human Society,* vi.

31. Bobrowky and Rickman, *Comet/Asteroid Impacts and Human Society,* vi.

32. Ulrich Beck, *Risk Society: Towards a New Modernity* (London and Newbury Park: Sage Publications, 1992).

33. See B612 Foundation Website, http://www.b612foundation.org/about/welcome.html (accessed July 2010).

34. Stephen J. Collier and Andrew Lakoff, "The Vulnerability of Vital Systems: How 'Critical Infrastructure' Became a Security Problem," in *Securing "The Homeland": Critical Infrastructure, Risk, and (In)security,* ed. M. Dunn Cavelty and K. S. Kristensen (New York: Routledge, 2008). This quote is available on p. 30 of their prepublished online document, http://anthropos-lab.net/wp/publications/2008/01/collier-and-lakoff.pdf.

35. Russell L. Schweickart, Thomas D. Jones, Franz von der Dunk, and Sergio Camacho-Lara, "Asteroid Threats: A Call for Global Response," in "Association of Space Explorers Committee on Near-Earth Objects," 2008, 46, http://www.space-explorers.org/committees/NEO/ASE_NEO_Final_Report_excerpt.pdf.

36. Schweickart et al., "Asteroid Threats."

37. Schweickart et al., "Asteroid Threats."

38. Russell L. Schweickart, "The Real Deflection Dilemma," in *Planetary Defense Conference: Protecting Earth from Asteroids* (Orange County, CA: 2004).

39. Schweickart et al., "Asteroid Threats," 16.

40. Lee Gomes, "Keeping the Earth Asteroid-Free Takes Science, Soft Touch," *Wall Street Journal,* Mar. 19, 2008, http://online.wsj.com/article/SB120588225372846789.html.

41. Schweickart et al., "Asteroid Threats," 20.

42. Schweickart et al., "Asteroid Threats," 20.

43. Star and Greisemer, "Boundary Objects," 508.

44. Ingold, "Globes and Spheres," 41.

45. George Myerson and Yvonne Rydin, *The Language of Environment: A New Rhetoric* (London: UCL Press, 1996).

## EPILOGUE. PRESERVATION IN THE AGE OF ENTANGLEMENT: STS AND THE HISTORY OF FUTURE URBAN NATURE

1. Michel Callon, "Some Elements of a Sociology of Translation: Domestication of the Scallops and the Fishermen of St Brieuc Bay," in *Power, Action and Belief: A New Sociology of Knowledge,* ed. John Law (London: Routledge and Kegan Paul, 1986).

2. In general on ANT, see Bruno Latour, *Science in Action: How to Follow Scientists and Engineers in Society* (Cambridge: Harvard University Press, 1987); for his later work see primarily Bruno Latour, *Pandora's Hope: Essays on the Reality of Science Studies* (Cambridge: Harvard University Press, 1999). An introduction to the theory is in Bruno Latour, *Reassembling the Social: An Introduction to Actor-Network Theory* (Oxford: Oxford University Press, 2005).

3. Bruno Latour, *Politics of Nature: How to Bring the Sciences into Democracy* (1999), trans. Catherine Porter (Cambridge: Harvard University Press, 2004).

4. Charles Darwin wrote in the concluding chapter of *The Origin of Species* (1859) about "an entangled bank" where biological diversity was evolving according to natural laws from "so simple a beginning" to "endless forms." See also, for its use as a metaphor of systems ecology, Joel B. Hagen, *An Entangled Bank: The Origin of Ecosystem Ecology* (New Brunswick: Rutgers University Press, 1992).

5. Samuel P. Hays, *Conservation and the Gospel of Efficiency: The Progressive Conservation Movement, 1890-1920* (Cambridge: Harvard University Press, 1959).

6. Alfred Runte, "The National Park Idea," *Journal of Forest History* 21 (1977): 2; Roderick Nash, *Wilderness and the American Mind*, 3rd ed. (1967; New Haven: Yale University Press, 1982); Roderick Nash, *American Environmentalism: Readings in Conservation History*, 3rd ed. (1976; New York: McGraw-Hill, 1990); Stephen R. Fox, *The American Conservation Movement: John Muir and His Legacy* (1981), new ed. (Madison: University of Wisconsin Press, 1985); Douglas H. Strong, *Dreamers and Defenders: American Conservationists* (1971), new ed. (Lincoln and London: University of Nebraska Press, 1988). Early classics are John Muir, *Our National Parks* (1901), new ed. (San Francisco: Sierra Club Books, 1991); and Aldo Leopold, *A Sand County Almanac: With Essays on Conservation* (1948), new ed., ed. and intro. Kenneth Brower (New York: Oxford University Press, 2001). For Scandinavia, where Sweden was the pioneering country, see Sten Selander, "Lule lappmarks nationalpark," *Bygd och Natur* (1940); Sten Selander, *Det levande landskapet i Sverige* (Stockholm: Bonniers, 1955); Désirée Haraldsson, *Rädda vår natur!. Svenska naturskyddsföreningens framväxt och tidiga utveckling*, Bibliotheca Historica Lundensis 63 (Lund, 1987); Sverker Sörlin, *Framtidslandet. Debatten om Norrland och naturresurserna under det industriella genombrottet* (Stockholm: Carlsson Bokförlag, 1988); Sverker Sörlin, "Stadens natur—det nya naturskyddet och sammanlevnadens politik," in *Naturvård bortom 2009: Reflektioner med anledning av ett jubileum*, ed. Lars J. Lundgren (Brottby: Kassandra, 2009); and especially the broad overviews in Lars J. Lundgren, *Staten och naturen: Naturskyddspolitik i Sverige 1869-1935, Part 1: 1869-1919* (Brottby: Kassandra, 2009), and *Part 2: 1919-1935* (Brottby: Kassandra, 2010).

7. See, e.g. Jane Carruthers, *The Kruger National Park: A Social and Political History* (Pietermaritzburg: University of Natal Press, 1995); A. Hall-Martin and Jane Carruthers, *South African National Parks: A Celebration* (Johannesburg: Horst Klemm, 2003); Libby Robin, "New Science for Sustainability in an Ancient Land," in *Nature's End: History and the Environment*, ed. Sverker Sörlin and Paul Warde (London: Palgrave MacMillan, 2009).

8. For Germany, see Franz-Josef Brüggemeier, Mark Cioc, and Thomas Zeller, eds., *How Green Were the Nazis? Nature, Environment, and Nation in the Third Reich* (Athens: Ohio University Press, 2005); Thomas M. Lekan and Thomas Zeller, eds., *Germany's Nature: Cultural Landscapes and Environmental History* (New Brunswick: Rutgers University Press, 2005). For Norway, see Arne Christian Stryken, *Er fjellene våre . . . : Om naturens egenverdi i historisk, politisk og etisk perspektiv* (Oslo: Andresen and Butenschön, 1994); and Ragnhild Sundby, "Natur og ressurser," in *Norges kulturhistorie*, vol. 8, *Underveis—mot nye tider* (Oslo: Aschehoug, 1981). For Finland, see *National Landscapes* (Helsinki: Department of the Environment, 1996).

9. A cornerstone text is Maurice Halbwachs, *Les cadres sociaux de la mémoire* (1925; Paris: Presses Universitaires de France, 1952), published as *On Collective Memory*, trans. and ed. Lewis A. Coser (Chicago: University of Chicago Press, 1992). See also Anssi Paasi, "The Institutionalization of Regions: A Theoretical Framework for Understanding the Emergence of Regions and the Constitution of Regional Identity," *Fennia* 164 (1986): 105-46; Paul Connerton, *How Societies Remember* (Cambridge: Cambridge University Press, 1989); Simon Schama, *Landscape and Memory* (London: Harper Collins, 1995); Sverker Sörlin, "Monument and Memory: Landscape Imagery and the Articulation of Territory," *Worldviews: Environment, Culture, Religion* 2 (1998): 269-79; Sverker Sörlin, "The Articulation of Territory: Landscape and the Constitution of Regional and National Identity," *Norsk Geografisk Tidsskrift* 53 (1999): 103-12.

10. Stephen Pyne, *How the Canyon Became Grand: A Short History* (New York: Viking Penguin, 1998); Libby Robin, *How a Continent Created a Nation* (Sydney: University of New South Wales Press, 2007); Robin, "New Science for Sustainability."

11. John McKenzie, *The Empire of Nature: Hunting, Conservation, and British Imperialism* (Manchester: Manchester University Press, 1998); Carruthers, *Kruger National Park;* William M. Adams, "Nature and the Colonial Mind," in *Against Extinction: The Story of Conservation* (London and Sterling, VA: Earthscan, 2004); William M. Adams, "Separation, Proprietorship and Community in the History of Conservation," in *Nature's End: History and the Environment,* ed. Sverker Sörlin and Paul Warde (London: Palgrave MacMillan, 2009); William M. Adams and Martin Mulligan, eds., *Decolonizing Nature: Strategies for Conservation in a Post-Colonial Era* (London and Sterling, VA: Earthscan, 2003). Roderick P. Neumann, "Dukes, Earls and Ersatz Edens: Aristocratic Nature Preservationists in Colonial Africa," *Environment and Planning D: Society and Space* 14 (1996): 79–98. See also several contributions in *Civilizing Nature: National Parks in Global Historical Perspective,* ed. Bernhard Gissibl, Sabine Höhler, and Patrick Kupper (New York: Berghahn Books, 2012), a volume that also speaks to the silences and absences addressed here through its international comparative focus and its embrace of theory.

12. William Cronon, "The Trouble with Wilderness; or, Getting Back to the Wrong Nature," in *Uncommon Ground: Rethinking the Human Place in Nature,* ed. William Cronon (New York: W. W. Norton, 1995), 69–90; William M. Adams, *Future Nature: A Vision for Conservation* (London: Earthscan, 1996); Adams and Mulligan, *Decolonizing Nature.*

13. Samuel P. Hays, *Beauty, Health, and Permanence: American Environmental Politics 1955–1985* (Cambridge: Cambridge University Press, 1987).

14. Henrika Kuklick and Robert E. Kohler, eds., *Science in the Field,* special issue, *Osiris* 11, 2nd series (Chicago: University of Chicago Press, 1996); Michael T. Bravo and Sverker Sörlin, "Narrative and Practice—An Introduction," in *Narrating the Arctic: A Cultural History of Nordic Scientific Practices,* ed. Michael T. Bravo and Sverker Sörlin (Canton, MA: Science History Publications, 2002).

15. For the United States see, e.g., Robert Gottlieb, *Forcing the Spring: The Transformation of the American Environmental Movement* (1993, 2005; Washington, DC: Island Press, 2009).

16. Richard A. Walker, *The Country in the City: The Greening of the San Francisco Bay Area* (Seattle: University of Washington Press, 2007).

17. Walker quotes, e.g., Michael Davis, *City of Quartz: Excavating the Future in Los Angeles* (London: Verso, 1990); Michael Davis, *Ecology of Fear: Los Angeles and the Apocalyptic Imagination* (New York: Metropolitan, 1998). But see also Martin Melosi, "The Place of the City in Environmental History," *Environmental History Review* (Spring 1993): 1–23; Martin Melosi, *The Sanitary City: Environmental Service in America from Colonial Times to the Present* (Pittsburgh: University of Pittsburgh Press, 2008); Christine Meisner Rosen and Joel Arthur Tarr, eds., *The Environment and the City: A Special Issue of the Journal of Urban History* 20 (May 1994); Joel Tarr, "The City and the Natural Environment," 1997, http://www.gdrc.org/uem/doc-tarr .html; Joel Tarr, "Urban History and Environmental History in the United States: Complementary and Overlapping Fields," in *Environmental Problems in European Cities of the 19th and 20th Century,* ed. Christoph Bernhardt (New York, Munich, and Berlin: Waxmann, 2001); Jens Lachmund, "Exploring the City of Rubble: Botanical Fieldwork in Bombed Cities in Germany after World War II," *Osiris* 18 (2003): 234–54.

18. See, e.g., Peter Hall, *Cities of Tomorrow: An Intellectual History of Urban Planning and Design in the Twentieth Century* (Oxford: Blackwell, 1988); Peter Hall, *Cities in Civilization: Culture, Technology and Urban Order* (London: Weidenfeld and Nicolson, 1998); Manuel Castells and Peter Hall, *Technopoles of the World: The Making of 21st Century Industrial Complexes* (London: Routledge, 1989); Joel Garreau, *Edge City: Life on the New Frontier* (New York: Anchor Books, 1991); Doreen Massey et al., *Hi-Tech Fantasies: Science Parks in Society, Science, and*

Space (London: Routledge, 2002); Annalee Saxenian, Regional Advantage: Culture and Competition in Silicon Valley and Route 128 (Cambridge: Harvard University Press, 1994); Saskia Sassen, ed., Global Networks, Linked Cities (New York: Routledge, 2002); Ann Markusen, "Sticky Places in Slippery Space: A Typology of Industrial Districts," Economic Geography 72 (1996): 3.

19. See Henrik Ernstson, Sander E. van der Leeuw, Charles L. Redman, Douglas J. Meffert, George Davis, Christine Alfsen, and Thomas Elmqvist, "Urban Transitions: On Urban Resilience and Human-Dominated Ecosystems," Ambio 39, no. 8 (2010): 531–45.

20. Cathy Wilkinson, "Historical Insights on the Adaptiveness of Metropolitan-Scale Spatial Planning to Global Scale Environmental Governance Initiatives," Earth System Governance Conference, Amsterdam, Dec. 2009; Cathy Wilkinson, "Social-Ecological Resilience and Planning: An Interdisciplinary Exploration" (PhD diss., Stockholm University, 2012); Cathy Wilkinson et al., "Strategic Spatial Planning and the Ecosystem Services Concept: An Historical Exploration," Ecology and Society 18, no. 1 (2013): 37 [online].

21. Sörlin, "Monument and Memory"; Sörlin, "Articulation of Territory."

22. On ANT see Latour, Science in Action; Latour, Reassembling the Social. For an application in an urban context, see J. Forsemalm, "Bodies, Bricks and Black Boxes: Power Practices in City Conversion" (PhD diss., Department of Ethnology, Gothenburg University 2007); Noel Castree and T. MacMillan, "Dissolving Dualisms: Actor-Networks and the Reimagination of Nature," in Social Nature: Theory, Practice, and Politics, ed. Noel Castree and Bruce Braun (Oxford: Blackwell, 2001).

23. Callon, "Some Elements of a Sociology of Translation"; Latour, Science in Action.

24. Etienne Wenger, Communities of Practice: Learning, Meaning, and Identity (Cambridge: Cambridge University Press, 1998).

25. G. Barker, ed., Ecological Recombination in Urban Areas (Peterborough: Urban Forum/ English Nature, 2000); Steve Hinchliffe and Sarah Whatmore, "Living Cities: Towards a Politics of Conviviality," Science as Culture 15 (2006): 123–38; Bruce Braun, "Environmental Issues: Writing a More-Than-Human Urban Geography," Progress in Human Geography 29 (2005): 635–50.

26. Donatella della Porta and Mario Diani, Social Movements: An Introduction, 2nd ed. (Malden, MA: Blackwell, 2006), 8–11; Manuel Castells, The City and The Grassroots (London: Arnold, 1983).

27. Examples from the literature are "the grassroots groups" of the South Park Campaign "claiming the creation of a protected green belt in the Southern Milanese periphery" (Mario Diani, "Networks and Social Movements: A Research Programme," in Social Movements and Networks: Relational Approaches to Collective Action, ed. Mario Diani and Doug McAdam [Oxford: Oxford University Press, 2003], 110); the diverse San Francisco Bay Area movement with professional organizations and "local groups working to preserve small neighbourhood natural areas" (C. K. Ansell, "Community Embeddedness and Collaborative Governance in the San Francisco Bay Area Environmental Movement," in Diani and McAdam, Social Movements and Networks, 125); the public-space-framed fight over People's Park in Berkeley (D. Mitchell, "The End of Public Space?: People's Park, Definitions of the Public, and Democracy," Annals of the Association of American Geographers 85 [1995]: 108–33); and the protection of biodiversity-rich areas in poor Cape Town suburbs (Henrik Ernstson, "The Social Production of Ecosystem Services: Environmental Justice and Ecological Complexity in Urbanized Landscapes," Landscape and Urban Planning 109 [2013]: 7–17). See also Henrik Ernstson and Sverker Sörlin, "Ecosystem Services as Technology of Globalization: On Articulating Values in Urban Nature," Ecological Economics (forthcoming 2013). doi: 10.1016/j.ecolecon.2012.1009.1012.

28. Michel Foucault, "Des Espace Autres" (1967), reprinted in Architecture, Mouvement, Continuité, Oct. 5, 1984.

29. The following present pertinent examples: Christine Alfsen-Norodom, B. D. Lane,

and M. Corry, eds., "Urban Biosphere and the Society: Partnerships of Cities," *Annals of the New York Academy of Sciences* 1023 (2004): 1–334; Richard Walker, *Nature in the City*, chap. 9, "Toxic Landscapes: Beyond Open Space," and chap. 10, "Green Justice: Reclaiming the Inner City."

30. Edward O. Wilson, *The Diversity of Life* (Cambridge: Harvard University Press, 1992).

31. Martin Bulmer, *The Chicago School of Sociology: Institutionalization, Diversity, and the Rise of Sociological Research* (Chicago: University of Chicago Press, 1984).

32. Eugene Cittadino, "The Failed Promise of Human Ecology," in *Science and Nature: Essays in the History of the Environmental Sciences*, ed. Michael Shortland (London: British History of Science Society, 1993).

33. Sharon E. Kingsland, *The Evolution of American Ecology, 1890–2000* (Baltimore: Johns Hopkins University Press, 2005), chap. 9.

34. B. Drayton and R. B. Primack, "Plant Species Lost in an Isolated Conservation Area in Metropolitan Boston from 1894–1993," *Conservation Biology* 10 (1996): 30–39.

35. See, e.g., Per Bolund and Sven Hunhammar, "Ecosystem Services in Urban Areas," *Ecological Economics* 29 (1999): 293–301; M. Alberti and J. M. Marzluff, "Ecological Resilience in Urban Ecosystems: Linking Urban Patterns to Human and Ecological Functions," *Urban Ecosystems* 7 (2004): 241–65; Thomas Elmqvist et al., "The Dynamics of Social-Ecological Systems in Urban Landscapes: Stockholm and the National Urban Park, Sweden," *Annals of the New York Academy of Sciences* 1023 (2004): 308–22; Alexander Ståhle and Hanna Erixon, *Regionens täthet och grönstrukturens potential: Det suburbana landskapets utvecklingsmöjligheter i en växande storstadsregion*, School of Architecture, KTH Royal Institute of Technology (Stockholm, 2008).

36. Gretchen C. Daily, ed., *Nature's Services: Societal Dependence on Natural Ecosystems* (Washington, DC: Island Press, 1997); R. Costanza, R. d'Arge, R. de Groot, S. Farberk, M. Grasso, B. Hannon, K. Limburg, S. Naeem, R. V. O'Neill, J. Paruelo, R. G. Raskin, P. Sutton, and M. van den Belt, "The Value of the World's Ecosystem Services and Natural Capital," *Nature* 387 (1997): 253–60.

37. Stephan Barthel, Johan Colding, Thomas Elmqvist, and Carl Folke, "History and Local Management of a Biodiversity-Rich, Urban, Cultural Landscape," *Ecology and Society* 10 (2005): 2.

38. Henrik Ernston, Sverker Sörlin, and Thomas Elmqvist, "Social Movements and Ecosystem Services: The Role of Social Network Structure in Protecting and Managing Urban Green Areas in Stockholm," *Ecology and Society* 13, no. 2 (2008): 39, http://www.ecologyand society.org/vol13/iss2/art39/.

39. Gunnar Brusewitz and Henrik Ekman, *Ekoparken: Djurgården—Haga—Ulriksdal* (Stockholm: Wahlström and Widstrand, 1995); Ernston et al. "Social Movements."

40. Henrik Ernston, "The Drama of Urban Greens and Regimes: Social Movements and Ecosystem Services in Stockholm National Urban Park" (Licentiate in philosophy thesis, Department of Systems Ecology, Stockholm University, 2007).

41. The notion of articulation of values was developed in Sörlin, "Articulation of Territory." Its application on urban preservation is in Henrik Ernston and Sverker Sörlin, "Weaving Protective Stories: Connective Practices to Articulate Holistic Values in Stockholm National Urban Park," *Environment and Planning A* 41 (2009): 6.

42. Ernston and Sörlin, "Weaving Protective Stories."

43. Gerard De Geer, *Om Skandinaviens geografiska utveckling efter istiden* (Stockholm: Norstedt, 1896), 150; Christer Nordlund, "'On Going Up in the World': Nation, Region and Land Elevation Debate in Sweden," *Annals of Science* 58 (2001): 17–50.

44. Karen Wonders, *Habitat Dioramas: Illusions of Wilderness in Museums of Natural History* (Uppsala: Acta Universitatis Upsaliensis, 1993); Lundgren, *Staten och nature*, vols 1 and 2.

45. Thomas Söderqvist, *The Ecologists: From Merry Naturalists to Saviours of the Nation* (Stockholm: Almqvist and Wiksell International, 1986).

46. See, e.g., Diani and McAdam, *Social Movements and Networks;* della Porta and Diani, *Social Movements.* See also Erik Andersson, Stephan Barthel, and Karin Ahrné, "Measuring Social-Ecological Dynamics behind the Generation of Ecosystem Services," *Ecological Applications* 17 (2007): 1267–78; and Erik Andersson, "Urban Landscapes and Sustainable Cities," *Ecology and Society* 11 (2006): 1.

47. Alexander Ståhle and Lars Marcus, "Compact Sprawl Experiments: Four Strategic Densification Scenarios for Two Modernist Suburbs in Stockholm," in *Proceedings of the 7th International Space Syntax Symposium,* ed. Daniel Koch, Lars Marcus, and Jesper Steen (Stockholm: KTH, 2009), http://www.sss7.org/Proceedings/05%20Spatial%20Morphology%20and%20Urban%20Growth/109_Stahle_Marcus.pdf.

48. Hanna Erixon, "Odla staden, bygg naturen," *Arkitekten* (Aug. 2006); W. M. Adams, *Future Nature.*

49. Alexander Ståhle, *Mer park i tätare stad: Teoretiska och empiriska undersökningar av stadsplaneringens mått på friytetillgång* (Stockholm: KTH School of Architecture, 2005); Ståhle and Marcus, "Compact Sprawl."

50. Braun, "Environmental Issues."

51. Steve Hinchliffe, Matthew B. Kearnes, Monica Degen, and Sarah Whatmore, "Urban Wild Things: A Cosmopolitical Experiment," *Environment and Planning D: Society and Space* 23 (2005): 643–58; Bruno Latour, "To Modernise or Ecologise? That Is the Question," in *Remaking Reality: Nature at the Millennium,* ed. Bruce Braun and Noel Castree (London: Routledge, 1998), 221–42.

52. Latour, *Pandora's Hope.*

53. Hinchliffe and Whatmore, "Living Cities."

54. Natalie Jeremijenko, project webpage: http://www.nyu.edu/projects/xdesign/ (accessed July 2008); Donna Haraway, *When Species Meet* (Minneapolis and London: University of Minnesota Press, 2008), esp. part 1, "We Have Never Been Human," 3–157. Sverker Sörlin, "The New Boundaries of Mankind," *Eurozine* (Aug. 2011): http://www.eurozine.com/articles/2011-07-15-sorlin-en.html, explores the current transformations in the views of humans as individuals and as global phenomenon impacting on the planet.

55. Dwight E. Baldwin, Judith de Luce, and Carl Pletsch, eds., *Beyond Preservation: Restoring and Inventing Historical Landscapes* (Minneapolis: University of Minnesota Press, 1994).

56. Sverker Sörlin and Paul Warde, "Making the Environment Historical—An Introduction," in Sörlin and Warde, *Nature's End,* esp. 12–15.

57. Stephan Barthel, Sverker Sörlin, and John Ljungkvist, "Innovative Memory and Resilient Cities: Echoes from Ancient Constantinople," in *The Urban Mind: Cultural and Environmental Dynamics,* ed. Paul J. J. Sinclair, Gullög Nordquist, Frands Herschend, and Christian Isendahl (Uppsala University: Department of Archaeology and Ancient History, 2010), 391–405.

CONTRIBUTORS

KEVIN C. ARMITAGE is associate professor at Miami University of Ohio, where he teaches interdisciplinary studies and environmental history. He is the author of *The Nature Study Movement: The Forgotten Popularizer of America's Conservation Ethic* (University Press of Kansas, 2009). When not teaching or writing, he can often be found exploring the wild lands of Colorado or Kentucky with his wife and daughter.

EUNICE BLAVASCUNAS is a Switzer Fellow and research associate at the College of the Atlantic in Bar Harbor, Maine. Since earning her PhD in cultural anthropology from the University of California, Santa Cruz, in 2008, she has held appointments at the University of Washington, Miami University, the University of Maine, and the Schoodic Education and Research Institute in Winter Harbor, Maine. She writes about forest and conservation conflicts in Poland and Maine.

STEPHEN BOCKING is professor of environmental history and policy and chair of the Environmental and Resource Science/Studies Program, Trent University. His research interests include the environmental history of science in a variety of contexts, including the Canadian Arctic, salmon aquaculture in the Pacific and Atlantic oceans, and biodiversity conservation in southern Ontario. His books include *Ecologists and Environmental Politics: A History of Contemporary Ecology* (Yale University Press, 1997) and *Nature's Experts: Science, Politics, and the Environment* (Rutgers University Press, 2004).

MICHAEL EGAN is associate professor of history at McMaster University. His work is in the interstices of the histories of science, technology, the environment, and the future. He is the author of *Barry Commoner and the Science of Survival: The Remaking of American Environmentalism* (MIT Press, 2007) and coeditor, with Jeff Crane, of *Natural Protest: Essays on the History of American Environmentalism* (Routledge, 2008).

**THOMAS D. FINGER** is a PhD candidate at the University of Virginia. His forthcoming dissertation, titled "Harvesting Power: American Agriculture, British Industry, and the Labor of Human Bodies, 1776–1918," uses the nineteenth-century Anglo-American grain trade to explore how historical power structures are produced through the organization of nature, technology, and capital. He has published articles in *Global Environmental Change, Journal of the Southwest,* and *Technology and Culture.*

**DOLLY JØRGENSEN** is a researcher in the Department of Ecology and Environmental Science at Umeå University in Sweden. She earned her PhD in history in 2008 from the University of Virginia after working for several years in the environmental engineering industry. She has written on a broad array of environmental history topics, including forestry management in medieval England, late medieval urban sanitation, the conversion of offshore oil structures into artificial reefs, and animal reintroduction in Europe.

**FINN ARNE JØRGENSEN** is associate senior lecturer in the history of technology and environment at Umeå University in Sweden. He holds a PhD in science and technology studies from the Norwegian University of Science and Technology and is the author of *Making a Green Machine: The Infrastructure of Beverage Container Recycling* (Rutgers University Press, 2011). His current research includes a historical study of Scandinavian leisure cabin culture.

**VALERIE A. OLSON** is assistant professor of anthropology at the University of California, Irvine. She researches how large-scale ecosystems are constituted as environmental, scientific, technical, and sociopolitical formations. She has completed an ethnography of US human spaceflight and its role in remaking the solar system as an ecological space and is conducting fieldwork on contingent transnational efforts to consolidate Gulf of Mexico ecosystem science and politics after the 2010 BP oil spill.

**SARA B. PRITCHARD** is assistant professor in the Department of Science and Technology Studies at Cornell University. She received her PhD in history from Stanford University with specialties in environmental history and the history of technology. Her first book, *Confluence: The Nature of Technology and the Remaking of the Rhône,* was published by Harvard University Press in 2011. She has also written about industrialization, conservation, the triple disaster at Fukushima, and the movement of hydraulic knowledge between France and French North Africa.

**TIAGO SARAIVA** is assistant professor in the Department of History and Politics at Drexel University. Previously he was a research fellow at the Institute of Social Sciences at the University of Lisbon and held several appointments as a visiting professor at the University of California, Los Angeles, and the University of California, Berkeley. His research explores the connections between geneticists' practices and food policies of fascist regimes.

**SVERKER SÖRLIN** is professor in the Division of History of Science, Technology and Environment at the KTH Royal Institute of Technology in Stockholm, Sweden, where he also initiated the KTH Environmental Humanities Laboratory. His current work is focused on the science politics of climate change and on the genealogy and applications of the concept of "environment" in social and scientific discourse. *The Future of Nature: Documents of Global Change*, coedited with Libby Robin and Paul Warde, is forthcoming from Yale University Press in 2013.

**DAVID TOMBLIN** currently teaches science and technology studies courses to engineering and science majors at the University of Maryland, College Park. He also works with the Consortium for Science Policy Outcomes at Arizona State University and Experts and Citizen Assessment of Science and Technology on a number of citizen engagement projects concerning emerging technologies.

**FRANK UEKOTTER** is the author of *The Age of Smoke: Environmental Policy in Germany and the United States, 1880–1970* (University of Pittsburgh Press, 2009) and the editor of *The Turning Points of Environmental History* (University of Pittsburgh Press, 2010). He was deputy director of the Rachel Carson Center for Environment and Society in Munich, Germany, and has been a reader at the University of Birmingham (United Kingdom) since 2013.

**ANYA ZILBERSTEIN**, who earned her PhD from MIT in history, anthropology, and science and technology studies, is assistant professor of history at Concordia University. She is completing work on her book *A Temperate Empire: Climate Change and Settler Colonialism in Early America*, a transatlantic history of greater New England in the seventeenth through the early nineteenth centuries that examines Enlightenment ideas about temperate and extreme (especially cold) weather and the failed improvement schemes that aimed to temper the regional climate and the people inhabiting it.

# INDEX

Note: Page numbers in italic type indicate illustrations

Ackefors, Hans, 110–11, 250n2
actants, 213, 225n1
actor networks, 158–59, 161
actor network theory (ANT), 151–52, 212–13;
    in Anglo-American grain trade, 153, *154*,
    154–55, 157–58; in environmental history,
    162, 227n4; in preservation efforts, 217,
    220; technological systems and, 162–63
Adams, William, 214, 222
aesthetics, and offshore oil structures, 62
agency, nature's *vs.* human, 4, 6
agnotology, 38–39, 48–50. *See also* ignorance
agriculture, 110, 137; conservation and, 88,
    90; corn production in Germany, 41–43;
    effects of hybrid seeds in, 44; environ-
    mental history and, 40, 157, 159–60;
    fungicides in, 105–9; karakul sheep and,
    139–40, 144–47, 150; knowledge in, 41–43;
    rabbit raising in Italy, 142–43; research in,
    44–45; science applied to, 88–89, 94–95;
    US, 155–60, 263n41; wheat, 153, 155–56
Agrosan fungicide, 106
Akrich, Madeleine, 73
Alexander, Czar (Russia), 122
Alicja and Marek, 121
alkylmercury, 108, 251n28
Alpert, Dede, 57–58, 61, 63
aluminum, in beverage containers, 69–71, 79

Alvarez, Luis, 202–3
Alvarez, Walter, 202–3
*American Game*, 94
American Indians: control over natural re-
    sources, 179; historical periods of, 181–92;
    knowledge production and, 181, 183. *See
    also* White Mountain Apaches
American Society for Environmental History,
    228n18
American Society of Agronomy, 87
Animal Breeding Institute, at the University of
    Halle, 137–41, 144
Antarctica, 196
Apache-Sitgreaves National Forest, 191–92
Apache trout, 184–85, 192–93
Arctic, 170; distinctiveness of, 164, 174–75;
    research in, 169, 175; situated and mobile
    science in, 168–70, 175–78
arctic ecology, 165, 174–75; importance of place
    in, 176–77; northern/circumpolar, 171–78
Arctic Land Use Research Program, 173
Arenberg, Jos, 250n2
Arizona Game and Fish Department (AGFD),
    186
Arizona Water Resources Committee, 189
artifacts, 217, 220; offshore oil rigs as, 59,
    62–63
artificial insemination, of karakul sheep,
    146–48
Aspling, Sven, 108
assemblage, of knowledge, 194